MODERN ASPECTS OF ELECTROCHEMISTRY
No. 8

LIST OF CONTRIBUTORS

G. BLYHOLDER
Department of Chemistry
University of Arkansas
Fayetteville, Arkansas, USA

J. WOJTOWICZ
Technical University
Warsaw, Poland
Present address:
Atlantic Industrial Research Institute
Halifax, Nova Scotia, Canada

A. A. HUMFFRAY
Chemistry Department
University of Melbourne
Melbourne, Victoria, Australia

LAZARO J. MANDEL
Department of Physiology
Yale University
New Haven, Connecticut, USA

A. T. KUHN
Chemistry Department
The University of Salford
Salford, Lancashire, UK

MODERN ASPECTS OF ELECTROCHEMISTRY

No. 8

Edited by

J. O'M. BOCKRIS
School of Physical Sciences
The Flinders University
Adelaide
South Australia

and

B. E. CONWAY
Department of Chemistry
University of Ottawa
Ottawa, Ontario

PLENUM PRESS · NEW YORK · 1972

Library of Congress Catalog Card Number 54-12732
ISBN 0-306-37648-2

© 1972 Plenum Press, New York
A Division of Plenum Publishing Corporation
227 West 17th Street, New York, N.Y. 10011

All rights reserved

No part of this publication may be reproduced in any
form without written permission from the publisher

Printed in the United States of America

Preface

This volume continues the development of the Modern Aspects series in the electrochemical field. The series is now 18 years old, and it is relevant to note the degree of evolution that electrochemistry has undergone during this time, for it affects the character of the articles chosen. The trend is towards development of interdisciplinary areas of electrochemical science, with full stress upon the many directions of applications of knowledge of electrode processes. The degree of import which should be attached to electrochemical science arises from the changes in technology which must be made during the next few decades. These clearly involve a massive electrification and the gradual elimination of the present fossil fuel economy, for both ecological and economic reasons. Research on the fundamental aspects of the field—slow in development to a modern standard—must be promulgated, but its justification is the provision of a basis for the needed future electrochemical technology.

One vast area of potential application of electrochemical concepts is omitted by the present attitude. It is, of course, the electrobiological aspect, perhaps, finally, the largest area of all for fruitful applications.

These concepts are reflected in the editors' choice of chapters. Quantum mechanical descriptions of surfaces must be bravely faced. Oscillatory aspects of electrochemical systems are often met in nature and demand attention at a fundamental level. Organic electrochemistry is in an ascending phase. With the electrobiological article, we hope to stimulate a beginning of electrodic applications in this area.

Finally, the article on some electrochemical aspects of pollution is a reminder that electrochemical science is not only at the center of future questions concerning energy shortage, and possibly an important part of energy conversion, but will play an increasing part in aspects of future industrial processes concerned with recycling and purification.

Adelaide J. O'M. Bockris
Ottawa B. E. Conway
May 1972

Contents

Chapter 1

QUANTUM CHEMICAL TREATMENT OF ADSORBED SPECIES
G. Blyholder

I. Introduction..	1
1. Relation of Adsorbed Species to Electrode Reactions	1
2. Theory of Electron Transfer.....................	4
3. Quantum Effects on Electrodes..................	7
II. Surface States of Crystals..........................	8
1. Band Structure of Crystals.....................	8
2. Surface States of Finite Crystals.................	13
III. Adsorbed Species................................	20
1. Extension of Crystal LCAO Calculations to Adsorbed Species.....................................	21
2. Less Approximate Formalisms...................	27
3. Surface Compound Approximation..............	30
4. Perturbation Approaches.......................	40
References...	44

Chapter 2

OSCILLATORY BEHAVIOR IN ELECTROCHEMICAL SYSTEMS
J. Wojtowicz

I. Introduction	47
II. Experimental Basis of Electrochemical Oscillations and Their Qualitative Interpretation	49
1. Anodic Processes Involving Metals	49
2. Anodic Oxidation of Nonmetallic Compounds	59
3. Oscillations in Cathodic Processes	63
4. Periodic Behavior in Nonaqueous Solutions	64
5. Electroosmotic Oscillations	65
III. Models of Electrochemical Oscillators	65
1. Mathematical Fundamentals	65
2. Kinetic or "Chemical" Models	77
3. "Electrical" Models	96
IV. Concluding Remarks	115
Appendix	116
References	117

Chapter 3

METHODS AND MECHANISMS IN ELECTROORGANIC CHEMISTRY
A. A. Humffray

I. Introduction	121
1. Synthetic Applications of Electroorganic Chemistry	122
2. Mechanistic Applications of Electroorganic Chemistry	126
3. Mass Transport Considerations	134
II. Methods of Investigating Organic Electrode Processes	135
1. Electrochemical Methods	136
2. Nonelectrochemical Methods	160
III. Intermediates in Electroorganic Chemistry	163
1. Anion Radicals and Anions	168

Contents ix

    2. Cation Radicals and Cations..................... 172
    3. Adsorption and Organic Electrode Processes....... 184
IV. Types of Reaction................................. 187
    1. Substitution Reactions......................... 187
    2. Addition Reactions............................ 193
    3. Elimination Reactions.......................... 198
    4. Cyclization Reactions.......................... 200
    5. Ring Opening, Expansion or Contraction......... 206
V. Stereospecificity in Electrode Processes.............. 208
    1. Substitution Reactions......................... 209
    2. Addition Reactions............................ 214
    3. Elimination Reactions.......................... 224
    4. Cyclization................................... 225
    5. Ring Opening................................. 226
VI. Applications for Commercial Synthesis.............. 227
References... 228

Chapter 4

ELECTROCHEMICAL PROCESSES AT BIOLOGICAL INTERFACES

Lazaro J. Mandel

I. Biological Oxidations............................. 239
II. Are the Elements for Electrochemical Reactions Present in Biological Cells?............................ 240
    1. Electronic Conduction in Biological Materials..... 241
    2. Electronic Conduction in Thin Lipid Films........ 248
    3. Concluding Remarks.......................... 253
III. Electrochemical Models of Biological Energy Conversion.. 254
    1. Introduction.................................. 254
    2. Brief Historical Survey........................ 255
    3. The Chemiosmotic Hypothesis.................. 255
    4. Electrodic Phosphorylation..................... 261
    5. Summary..................................... 267
IV. Electrochemical Reactions As Biological Regulators... 267
References... 270

Chapter 5

THE ROLE OF ELECTROCHEMISTRY IN ENVIRONMENTAL CONTROL

A. T. Kuhn

I.	Introduction.....................................	273
	Basic Problems in the Use of Electrochemical Methods	276
II.	Survey of Individual Applications.................	278
	1. Cathodic Processes............................	279
	2. Anodic Processes.............................	301
	3. Other Applications of Anodic Oxidation.........	313
	4. The Treatment of Gaseous Effluents.............	315
	5. Electroflotation...............................	321
	6. The Role of Electrochemistry in Corrosion Protection	325
	Conclusions.....................................	326
	Appendix.......................................	328
	1. Factors Influencing the Rate of Electrochemical Effluent Processes........................	328
	2. Conditions of Mass Transport Control.........	329
	3. Mass Transport under Diffusion Control.......	330
	4. Mass Transport under Convective Control.....	331
	5. Natural Convection..........................	333
	6. Conclusion..................................	333
References..		334
Index..		341

1

Quantum Chemical Treatment of Adsorbed Species

G. Blyholder

*Department of Chemistry, University of Arkansas
Fayetteville, Arkansas*

I. INTRODUCTION

The purpose of this chapter is to point out the current status of the quantum chemical treatment of surfaces and surface species so that its application to electrode processes may be more readily made. There has been no attempt made to exhaustively cover the literature, particularly that before 1960, but, rather, the more important developments are sketched. A number of review articles provide access to all but the more recent literature for both surface states of clean crystals[1-5] and adsorbed species.[5-8]

1. Relation of Adsorbed Species to Electrode Reactions

Adsorption on the electrode has long been recognized as having an effect on electrode processes.[9-11] To emphasize the importance of surface processes and to bring together workers in the fields of heterogeneous catalysis and electrochemistry, particularly those concerned with electrode reactions, a symposium on electrocatalysis was held in November 1968 and the proceedings are reported in Ref. 12. In some cases, after an electrolysis, an insulating film has been found coating the electrode. Sometimes, a chemical treatment easily removes the film, but at other times, heating to a high temperature is required to destroy an impermeable film. When a substance is being deposited on an electrode by electrolysis, the adsorption of components from the solution may result in the incorporation of extraneous components in the electrode matrix. Both physical and chemical adsorption processes have been observed to affect electrode processes. Delahay[9] has discussed the relationship of adsorption to electrode processes under the following major categories:

(i) Processes without adsorption of reacting species.
(ii) Processes with chemisorption of reacting species.
(iii) Processes with potential–dependent adsorption of reacting species.

Under each of these major categories, the following cases were considered:

(a) No specific adsorption of supporting electrolyte.
(b) Specific adsorption of supporting electrolyte or an additive: (1) ionic specific adsorption, (2) physical adsorption of an uncharged substance.
(c) Chemisorption of a species not participating directly in the electrode process.

The approach of a reactant and the departure of a product from the electrode surface are controlled by the nature of the electrical double layer at the surface. A detailed model of the double layer is shown in Fig. 1 as presented by Bockris et al.[13] While in some aspects this is a detailed model, it uses the simplest possible representation of the metal surface, i.e., an unstructured plane. From a quantum chemical treatment of the surface and adsorbed species, we would hope to obtain the electronic charge distribution at the surface which, together with the applied potential and nature of the solution, determines the structure of the double layer.

One specific case that has received attention is the effect of adsorbed layers of hydrogen or oxygen on polarograms using platinum electrodes.[9–11] Feldberg et al.[14] have proposed the following steps to rationalize data for oxygen:

Surface oxidation:

slow: $Pt + xH_2O \rightarrow Pt(OH)_x + xH^+ + xe$
fast: $Pt(OH)_x \rightarrow Pt(O)_x + xH^+ + xe$

I. Introduction

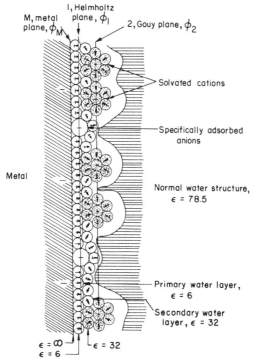

Figure 1. Model of the double layer (from Ref. 13, reproduced with permission).

Cathodic reduction:

fast: $Pt(O)_x + xH^+ + xe \rightarrow Pt(OH)_x$

slow: $Pt(OH)_x + xH^+ + xe \rightarrow Pt + xH_2O$

There is a complex relationship given by the current–overvoltage data for hydrogen evolution which depends on the proposed mechanism and type of hydrogen adsorption isotherm. Explanations have been sought in terms of the three potentially rate-determining steps[11]:

1. Reduction of H_3O^+ to give a surface-adsorbed hydrogen ion
   $H_3O^+ + M + e^- \rightarrow M-H + H_2O$
2. Reactive reduction of H_3O^+ to give H_2
   $H_3O^+ + M-H + e^- \rightarrow M + H_2 + H_2O$

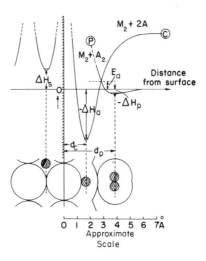

Figure 2. Potential energy curves for possible interactions at a metal surface (from Ref. 15, reproduced with permission).

3. Recombination of adsorbed hydrogen atoms

$$M–H + M–H \rightarrow 2M + H_2$$

A view[15] of hydrogen adsorption on a metal surface in terms of the potential energy of the interaction is shown in Fig. 2. In physical adsorption, the potential well ΔH_p is relatively shallow, about the order of magnitude of the heat of condensation, and is due to long-range forces, e.g., van der Waals forces, so the equilibrium distance from the surface is relatively large. Chemisorption often involves dissociation of adsorbed molecules and is characterized by somewhat greater heats of adsorption, usually 20–100 kcal/mole. The occurrence of an activation energy E_a for the transition from physical adsorption to chemisorption is dependent upon the system involved. We will describe later a number of quantum chemical calculations for adsorbed hydrogen. In the best of these for hydrogen adsorbed on platinum, there is little or no activation energy for adsorption into the atomic chemisorbed state.

I. Introduction

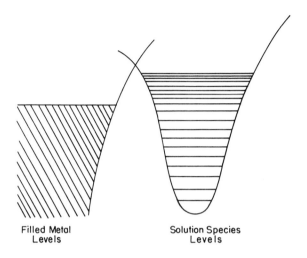

Filled Metal Levels   Solution Species Levels

Figure 3. Potential energy curve for interaction at an electrode surface.

2. Theory of Electron Transfer

The theory of the rate of electron transfer to and from an electrode may be approached from several points of view. Details will not be delved into here as these processes have been reviewed from a theoretical basis recently.[16,17] A graphical representation of the potential energy may be as indicated in Fig. 3 for an electron transfer process

$$A + e \rightarrow B \tag{1}$$

The rate of electron transfer is given by an equation of the form

$$\text{rate} = kC_i(e^{-\alpha(\eta F/RT)} - e^{(1-\alpha)(\eta F/RT)}) \tag{2}$$

where k is a rate constant, C_i is the reactant concentration in solution, α is a constant, F is the charge in a mole of electrons, and η is the overpotential. The rate constant, k, is proportional to the following terms:

$$\sum_{\varepsilon} n(\varepsilon) W_\varepsilon e^{-\Delta G_\varepsilon^{\neq}/RT} \tag{3}$$

where $n(\varepsilon)$ gives the occupation of metal energy levels denoted by ε, W_ε is the probability of electron transfer, and $\Delta G_\varepsilon^{\neq}$ is the free energy difference between the ground state of the reactant in solution and a state of the reactant with energy ε.

The probability of electron transfer W from state n to m is given by

$$W_{nm} = (2\pi/k) \sum_m \left| \int \psi_m^* V \psi_n \, dV \right|^2 \delta(E_n - E_m) \quad (4)$$

where V is the energy of perturbation causing the transition. The application of these equations in an exact manner is difficult in the absence of detailed knowledge about electrode surfaces, so the tendency has been to introduce empirical parameters to replace whole terms such as W_{nm} by a single parameter. The electron-transfer process involves mostly those metal electrode levels clustered about the Fermi level. The Fermi distribution law gives the electron occupation of these levels as

$$n(\varepsilon) = 1/[e^{-(\varepsilon - \varepsilon_F)/kT} - 1] \quad (5)$$

where ε_F denotes the Fermi level. Thus the highest occupied and lowest unoccupied levels, which are the ones most likely to be involved in electron-transfer processes, are mostly located within an energy range equal to kT on either side of the Fermi level.

Equation (4) indicates that a transition will occur with a reasonable probability only when the metal energy level and reactant energy level are matched. Adjustment of the metal energy level can occur through the action of the overpotential to change the Fermi level. On the reactant side the free energy of activation governs the probability of the reactant having an appropriate energy level available. The reaction coordinate is very complex in that it involves a number of small changes in bond distances in the inner coordination shell of each reactant, reorganization of solvent molecules outside the shell, changes in distance between the reactant and electrode, and changes in position and bond lengths in the solvent molecules outside the reactants. The major problem is to calculate the probability of finding the system in an appropriate configuration and calculating the probability of electron transfer in that configuration. The multiplication of these two probabilities averaged over all configurations produces the expression for rate constant. The determination of the potential energy surface for reaction is indeed a difficult problem in quantum mechanics. In this chapter that part of this problem concerned with calculating energy levels for adsorbed species is examined.

3. Quantum Effects on Electrodes

The quantum theory of clean crystal surfaces indicates the formation of surface states with energies located between the normal bands. Since electron transfer is usually to or from states near $\varepsilon_F \pm kT$ and ε_F is in the middle of a band for a normal metal, these surface states are not likely to be involved in electron transfer. However, these surface states can be important in determining which adsorbed species are stable on the surface and these affect the surface dipole and the structure of the electrical double layer. In the case of semiconductors, where ε_F is usually located between the filled valence band and the empty conduction band, the surface states, as well as playing the role indicated for metals, may also be electron-acceptor or -donor states for electrode reactions.

A number of quantum chemical treatments to be discussed treat the adsorbed species as a surface compound. These approaches have their greatest use in determining what species are adsorbed in a stable configuration on the surface and what the charge distribution is at the surface. This charge distribution affects the surface dipole, the double-layer structure, and how readily electrons can be transported through the surface. The stability of different species on the surface will influence the concentration of the reactant in contact with the surface.

Other treatments consider explicitly the band structure of the metal. In the model proposed by Grimley, the discrete levels of an adsorbed molecule or atom are assumed to be split into a series of virtual levels by the interaction with the metal band levels as shown in Fig. 4. The filling of these virtual levels is determined by the Fermi level of the metal. If there are a sufficient number of adsorbed species, a reactant R in Fig. 4 may have its orbitals come into contact with the adsorbed species. In this case, the density of levels which would enter into the calculation of the electron-transfer probability in equation (4) would be those of the virtual levels and not those of the metal. If the location of the Fermi level is such that the virtual levels are filled, then electron transfer to the metal becomes difficult. As the applied potential at the electrode is changed,

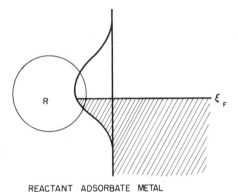

REACTANT ADSORBATE METAL

Figure 4. Energy levels representation for adsorbate-reactant interactions.

the Fermi level changes, which in turn changes the filling of the virtual levels. If the Fermi level is lowered, the virtual levels shared with the metal and thus responsible for the stability of the adsorbate may be emptied to the point that desorption occurs. Similarly, raising the Fermi level by an applied potential may fill antibonding orbitals to the point that there is decomposition within the adsorbate or desorption occurs. This, then, is one possible model for the action of overvoltage causing or being needed for desorption of an adsorbed reactant.

II. SURFACE STATES OF CRYSTALS

A large part of what is presently understood about the surfaces of crystals is derived from an understanding of the bulk crystal. We will proceed by quickly sketching the behavior of electrons in metallic or semiconductor crystals and then examine the consequences of introducing a surface. For our purposes, the behavior of an electron is satisfactorily described by its wave function, which specifies its average spatial distribution and its energy level. We will not be concerned here with time-dependent phenomena.

1. Band Structure of Crystals

The most outstanding feature of electrons in metals and semiconductors is their collection into energy bands. For a free electron,

II. Surface States of Crystals

the wave function ψ has the form

$$\psi = A \exp(i\mathbf{k} \cdot \mathbf{r}) \qquad (6)$$

and represents running waves with momentum $p = h\mathbf{k}/2\pi$. The wave vector $\mathbf{k}$ is a constant which can take on any value, while $\mathbf{r}$ is the position vector. In traveling through a crystal, an electron experiences a periodic potential. It was first demonstrated by Block[18] that the solutions to the Schrödinger equation with a periodic potential are of the form

$$\psi = \mu_k(\mathbf{r}) \exp(i\mathbf{k} \cdot \mathbf{r}) \qquad (7)$$

where the function μ_k has the periodicity of the lattice and depends on $\mathbf{k}$. Thus the plane wave function $\exp(i\mathbf{k} \cdot \mathbf{r})$ is modified to have the lattice periodicity. For a simple lattice, the positions of the atoms may be given by

$$R(l_1 l_2 l_3) = l_1 \mathbf{a}_1 + l_2 \mathbf{a}_2 + l_3 \mathbf{a}_3 \qquad (8)$$

where l_1, l_2 and l_3 are integers and $\mathbf{a}_1$, $\mathbf{a}_2$, and $\mathbf{a}_3$ are orthogonal unit vectors. A periodic potential is then defined as one that has the property

$$V(\mathbf{r} + l_1 \mathbf{a}_1 + l_2 \mathbf{a}_2 + l_3 \mathbf{a}_3) = V(\mathbf{r}) \qquad (9)$$

The wave equation for the problem is

$$\nabla^2 \psi + (8\pi^2 m/h^2)(E - V)\psi = 0 \qquad (10)$$

There are several approaches which may be taken to solving this equation. One approach is to assume a model potential and seek solutions of the form of equation (7). A second approach is to assume a wave function that is a linear combination of functions which would each be appropriate for an electron localized at a particular lattice site. Both of these approaches will be discussed since they both lead to surface states, although of different types.

In the first approach, a potential like that of the Kronig–Penney model[19] is used. In one dimension, this may take the form of a square-wall potential as shown in Fig. 5. The functional forms of μ_k which will make ψ in equation (7) solutions to equation (10)

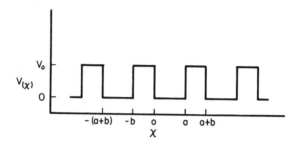

Figure 5. Potential energy for the Kronig–Penney model of a metal.

with various potentials are well known. The constants in the general solution are evaluated from the equations that result from requiring that the wave function and its first derivative be continuous at the potential boundaries, e.g., at $x = 0$ and $X = a$. Periodicity also requires that the value of the wave function be the same at $X = 0$ and $X = a + b$. In order that these conditions be met, it turns out that only certain values of the wave vector **k** and hence only certain energy ranges or bands are allowed.

The approach to solving the Schrödinger equation which starts with the wave function assumed to be made up of local atomic functions is usually referred to as the tight-binding approximation in physics texts and as the LCAO-MO (linear combination of atomic orbitals—molecular orbital) method in chemistry texts. The wave function is written as

$$\psi = \sum_l c_l \phi_l \qquad (11)$$

where l stands for the three indicies l_1, l_2, l_3 of equation (3), c is a constant, and ϕ is an atomic wave function localized at the lattice point $l(l_1 l_2 l_3)$. The atomic wave function ϕ_l satisfies the wave equation

$$[(-h^2/8\pi^2 m)\nabla^2 + v_l]\phi_l = \varepsilon \phi_l \qquad (12)$$

where v_l is the potential due to the atomic core at lattice point l. The total potential is assumed to be the sum of the local potentials v_l

II. Surface States of Crystals

plus the electron repulsions of the valence electrons. For the sake of mathematical expediency, it is often assumed that

$$\int \phi_l^* \phi_{l'} \, d\mathbf{r} = \delta_{ll'} \quad (13)$$

i.e., that overlap integrals between atomic orbitals may be neglected. Conceptually, this is a difficult approximation to justify since bonding between atoms depends upon this overlap. However, the qualitative results obtained using this type of approximation are in accord with experiments, as has been amply demonstrated with Hückel molecular orbital calculations. One way around this difficulty is to construct orthogonalized functions which are linear combinations of the atomic orbitals and while not completely localized about one lattice point, are largely so. The method of doing this was worked out by Wannier[20] for crystals and so these orthogonalized functions are often referred to as Wannier functions.

Substituting ψ from equation (13) into equation (10) and multiplying both sides of the equation by ϕ^* and then integrating over the electron coordinates gives the secular equation (known as the difference equation in physics texts):

$$(E - \varepsilon - \alpha)c_l = \beta[c(l_1 + 1, l_2, l_3) + c(l_1 - 1, l_2, l_3)$$
$$+ c(l_1, l_2 + 1, l_3) + c(l_1(l_2 - 1), l_3)$$
$$+ c(l_1, l_2, l_3 + 1) + c(l_1, l_2, l_3 - 1) \quad (14)$$

where

$$\alpha = \int \phi_l^*(V \quad v_l)\phi_l \, d\mathbf{r} \quad (15)$$

and

$$\beta = \int \phi_l^*[(-h^2/8\pi^2 m)\nabla^2 + V]\phi_{l'} \, d\mathbf{r} \quad (16)$$

Usually, only exchange integrals β are retained as nonzero when l and l' are adjacent to each other. Equation (14) is usually solved with the Born–von Karman boundary conditions, in which the infinite crystal is divided into cubes with N atoms on a side. The coefficients c are constrained to have translational symmetry from one cube to the next so that

$$c_l = c(l_1 \pm N, l_2, l_3) = c(l_1, l_2 \pm N, l_3) = c(l_1, l_2, l_3 \pm N) \quad (17)$$

With these boundary conditions, the solution to equation (14) takes the form

$$c_l = e^{i(k_1 l_1 + k_2 l_2 + k_3 l_3)a} \quad (18)$$

where

$$k_i = 2\pi n_i / Na \quad (19)$$

with a being the lattice spacing and n_i an integer between zero and N. Thus the wave function becomes

$$\psi = \sum_l (e^{i(k_1 l_1 + k_2 l_2 + k_3 l_3)a}) \phi_l \quad (20)$$

and the energy levels from equation (14) become

$$E = \varepsilon + \alpha + 2\beta(\cos k_1 a + \cos k_2 a + \cos k_3 a) \quad (21)$$

Thus the energy levels lie in a band defined by $\varepsilon + \alpha \pm 6\beta$. When the lattice parameter a is large, the values of α and β are small, so the energy levels are all those of the isolated atoms. As the atoms are compacted into the crystal, the increasing value of β splits the energy levels into a band. This is shown diagrammatically in Fig. 6, where the energy bands arising from two different atomic levels are

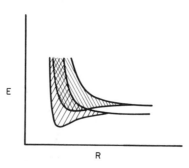

Figure 6. Development of energy bands as a function of internuclear distance in a metallic crystal.

II. Surface States of Crystals

shown. If the interatomic distance becomes short enough, the two bands will cross, a situation which can result in the formation of surface states, as we will see shortly.

2. Surface States of Finite Crystals

As we have seen, whether the nearly-free-electron approach or the LCAO approach is taken, the energy levels for electrons in an infinite periodic crystal are arranged into bands. The introduction of a surface in the crystal changes the boundary conditions in such a way that some new solutions to the Schrödinger equation arise that correspond to surface states. Surface states have finite, nonnegligible amplitudes only near the surface. Based upon their origin, these states may be roughly divided into intrinsic and extrinsic states. Intrinsic surface states arise solely from the interruption of the lattice. Their physical properties depend somewhat on how the boundary conditions which replace the usual Born–von Karman cyclic boundary conditions are applied, e.g., whether the lattice potential is interrupted at a maximum or a minimum. Extrinsic surface states are those whose existence is fostered by the occurrence of an extra potential at the surface. The presence of a foreign adsorbed atom which perturbs the potential sufficiently to cause a surface state would be an example of an extrinsic surface state. When the lattice is interrupted by a surface, the asymmetric potential may be expected to shift surface atoms from those locations which would be an extension of a perfect lattice. This shift in atom position results in a variation of the lattice potential which can be instrumental in the formation of surface states. Some authors regard these as intrinsic surface states because no foreign atoms have been introduced, while others regard them as extrinsic because the perfection of the lattice has been disturbed.

(i) *Nearly-Free-Electron Approach*

The possible occurrence of surface states was first demonstrated by Tamm[21] in 1932. He used the nearly-free-electron approach with a Kronig–Penney-type potential in which the periodic square wells are separated by δ-function potential spikes, i.e., infinitely high and infinitely narrow potential barriers between wells. Tamm considered a semiinfinite lattice, i.e., only one terminating surface, and terminated the last well with a finite potential rather than the

δ-function potential. Solution of the equations for this model with the boundary conditions that the wave functions and its first derivative smoothly join across the boundary inside and outside the crystal, gives an eigenstate between the allowed bands. The eigenfunction for this state has its maximum value at the surface and decays with distance from the surface into as well as out of the crystal. Amplification of these results have been provided by Maue[22] and Fowler.[23] In further developments, Goodwin[24] expanded the lattice potential $V(\mathbf{r})$ in a Fourier series so that in one dimension

$$V(x) \sum_{n=-\infty}^{+\infty} V_n e^{n2\pi/a} \qquad (22)$$

where a is the lattice constant and $V_n = V_{-n}{}^*$, the asterisk indicating a complex conjugate. Goodwin demonstrated that two surface states appear in each band gap, one from each adjacent band, and that the separation of these two states decreases rapidly as the distance between the two surfaces bounding the crystal increases. In general, the surface state is filled if the band from which it originates is filled and there is one surface state for each surface lattice site in a three-dimensional crystal. In one way or another, the surface states so far discussed have arisen because of a sufficiently large variation in the potential at the surface. Such surface states are referred to as Tamm states even though it may be noted that this is not a precise definition.

Shockley[25] has examined with a general treatment the problem of a one-dimensional lattice with a periodic potential. Again, solutions to the equations resulting from matching wave functions across the potential discontinuity are obtained. He showed that surface states appeared if the bulk bands crossed (see Fig. 6) and if there was a sufficiently small surface perturbation. Such states have come to be known as Shockley states. This situation is in distinct contrast to the previous one, where a sufficiently large variation in the potential at the surface was required to produce Tamm surface states. The condition to produce Shockley states has been associated with covalent bonding and thus with the elemental semiconductors Ge and Si. In some sense, these surface states may be associated with dangling, unsaturated covalent bonds at the surface of an undistorted crystal. However, there is some question about the relevance of these calculations since

spontaneous gross rearrangement of surface lattice atoms have been observed.[26]
More realistic models for potential functions than square wells and δ-functions have been investigated to some extent. A sinusoidal potential for a monatomic crystal terminated by a step potential of arbitrary height has been looked at by Statz[27] and Levine.[28] Their results are qualitatively in agreement with those using more approximate potential models as far as the types of surface states found is concerned.

(ii) LCAO Approach

The nearly-free-electron approach has the advantage that while it is conceptually close to the classical picture of a metal, we can see how the occurrence of surface states arises naturally out of the introduction of a free surface. We turn now to the LCAO approach, which we have already seen also leads to the electronic band structure of metals. The LCAO approach not only leads to the same types of surface states as previously found, but in addition it is able to handle in a natural manner adsorbed atoms and thus chemisorption. Furthermore, it lends itself to calculations which handle surface species without having to consider the entire crystal, a step which reduced calculations about real surfaces from the impractical to the practical realm.

The early application of LCAO methods to the study of surface states was developed by Goodwin.[29,30] In equations (11)–(21), the LCAO or tight-binding approximation is applied to the problem of an infinite crystal. A finite crystal in the $\mathbf{a}_3$ direction is created by introducing surfaces at $l_3 = 0$ and $l_3 = N$. Thus the crystal is bounded by (001) surfaces but extends to infinity in the $\mathbf{a}_1$ and $\mathbf{a}_2$ directions. The secular equation derived just like equation (14) for $l_3 - 0$ now becomes

$$(E - \varepsilon - \alpha')c(l_1, l_2, 0) = [c(l_1 + 1, l_2, 0 + c(l_1 - 1, l_2, 0)$$
$$+ c(l_1, l_2 + 1, 0) + c(l_1, l_2 - 1, 0)$$
$$+ c(l_1, l_2, 1)] \tag{23}$$

A similar equation is found for $l_3 = N$. In deriving equation (23), the integral α' is different from α, which expresses the interaction of an electron on a particular atom with the potential due to all other atomic cores and electrons when that particular atom is a surface atom. Somewhat arbitrarily, to keep the mathematics simple, the interaction integrals β have been kept the same for surface and interior atoms. For the two directions parallel to the surface, the boundary conditions are

$$c(l_1, l_2, l_3) = c(l_1 \pm N, l_2, l_3) = c(l_1, l_2 \pm N, l_3) \qquad (24)$$

For surface states localized at the $l_3 = 0$ surface, i.e., a state that will decay rapidly going into the crystal, the boundary condition at $l_3 = N$ may be chosen as

$$c(l_1, l_2, N) = 0 \qquad (25)$$

Now, the solutions to equation (14) are subject to the conditions imposed by equations (20) and (23)–(25) and yield

$$c = e^{i(l_1 k_1 + l_2 k_2)a} \sin(N - l_3) k_3 a \qquad (26)$$

where k_1 and k_2 are given by equation (19) as before and k_3 may be obtained from equation

$$\cos k_3 a + \sin k_3 a \cot N k_3 a = (\alpha' - \alpha)/\beta \qquad (27)$$

There are N values of k_3 which may be obtained from equation (27). If $|(\alpha' - \alpha)/\beta| < 1$, there are no surface states because equation (27) yields N real roots for k_3. These may be combined with the N^2 values for k_1 and k_2 from equation (26) to give N^3 states whose energies lie in the same energy bands as the infinite crystal.

However, for $|(\alpha' - \alpha)/\beta| < 1$, the solution to equation (27) gives $N - 1$ real roots and one imaginary root. The real roots give the usual volume band solutions, while the imaginary root gives the surface states. Since there are N^2 values of k_1 and k_2 to be combined with the values of k_3, the result is $N^2(N - 1)$ energy states in the volume bands and N^2 surface states with energies outside the volume bands. The energies for the surface states are given by

$$E = \varepsilon + \alpha + 2\beta(\cos k_1 a + \cos k_2 a) \pm 2\beta \cosh \xi a \qquad (28)$$

where $\xi a = \log|\alpha' - \alpha/\beta|$ and $k_3 a = i\xi a$ for the plus sign or $k_3 a = \pi + i\xi a$ for the minus sign. These states are referred to as Tamm states since they are the result of a surface potential term being sufficiently large. In the case described here, the perturbation was in the α integral with β unchanged. Surface states are also produced if α remains the same and β is different for surface atoms. Extensions of these investigations into the occurrence and properties of Tamm and Shockley states have been carried out by workers such as Artmann,[31] Hoffmann,[32] and Lippmann.[33] More sophisticated general theories of localized surface states have come from Koutecky[5,34] and Koutecky and Tomasek.[35] Extensions have included work on AB-type crystals and inclusion of next-nearest-neighbor interactions. While the bulk of the work has dealt with one-dimensional models, some calculations have been made for three-dimensional crystals. Work has also been done on the surface states of ionic crystals.[2] In most calculations concerning surface states, the integrals are treated as parameters and very few attempts at numerical evaluation from first principles have been attempted. This being the case, one may wonder just what connection the theory has to real crystals. The existence of surface states, particularly for semiconductors, is a well-established experimental fact. However, Davison and Levine[1] have noted that for Si and Ge, there is little correlation at the present time between theory and experiment. This lack of exactness on the part of current theories is one of the reasons that we have taken the space to show that both the nearly-free-electron and the LCAO approaches lead to similar types of surface states. Thus the particular mathematical approximations used do not seem to qualitatively change the results.

(iii) State Density and Charge at Surfaces

One of the fundamental quantities that occurs in the equation for the probability of an electron transition to or from an ion and an electrode is the density of energy levels on the electrode. If a transition were to be to or from a surface state, the density of surface states would be required. The calculations for three-dimensional crystals do provide rough estimates for the density of surface states. In Fig. 7 (from Davison and Levine[1]) some of the possible situations that can arise are shown. This figure shows a plot of the energy versus the wave vector k_{11} parallel to the surface. The volume bands, shown shaded, have an additional wave vector normal to the surface so that they are quasicontinuous with a central band gap. For a surface band distribution represented by curve A, the surface bands are concentrated in a very narrow energy range, perhaps

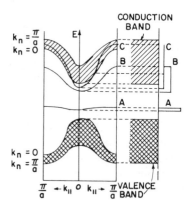

Figure 7. Energy vs wave vector and density of states for some possible surface state models (from Ref. 1, reproduced with permission).

about 0.1 eV wide. Thus with one surface state per surface atom giving 10^{14} states per cm^2, the density of states is the order of magnitude of 10^{15} per cm^2 per eV.

For curve B, the surface states are only slightly perturbed from the volume band edges, so that the surface states are more spread out and the density drops to around 10^{14} per cm^2 per eV. Just looking at the right-hand side of Fig. 7, it might appear that the surface states overlap the volume band and should merge into it, but so long as the band edges do not cross in k-space, the surface states remain separate. In curve C, the band edges cross and not all of the surface states remain localized.

Comparison of equations (21) and (28) indicates that the creation of a surface shifts the energy levels in the bulk of the crystal, thus causing a surface energy for the crystal. In addition, the distribution of the valence electrons is shifted so as to effect the spatial charge at the surface. There is an outward spread of charge which gives rise to a positive charge inside the surface and a negative charge outside as shown in Fig. 8 (from Horiuti and Toya[36]). This negative double layer which is formed increases the work function of the metal. Smoluchowski[37] has shown that the electron density $\rho(z)$

II. Surface States of Crystals

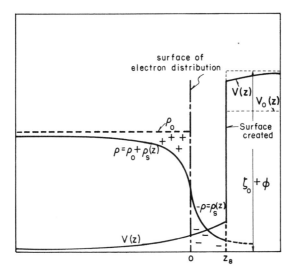

Figure 8. Electron density and potential normal to the surface for a metal (from Ref. 36, reproduced with permission).

is approximately represented by

$$\rho(z) = \rho_0 - \tfrac{1}{2}\rho_0 e^{Cz} \quad \text{for} \quad z < 0 \tag{29}$$

$$\rho(z) = \tfrac{1}{2}\rho_0 e^{-Cz} \quad \text{for} \quad z > 0 \tag{30}$$

where C is a constant and z is the distance outward from the surface. As can be seen in Fig. 8, the electron density at $z = 0$ is $\tfrac{1}{2}\rho_0$. For the idealized situation before the spread, the potential of an electron is given by

$$V_0(z) = 0 \quad \text{for} \quad z < z_D \tag{31}$$

and

$$V_0(z) + V_0' = \xi_0 + \phi - 4\pi e^2 \rho_0/C^2 \tag{32}$$

where ξ_0 is the kinetic energy of the electron at the Fermi level, ϕ is the work function, and z_B is the location of the potential wall where the surface is created. Sugiyama[38] has calculated z_B as

$$z_B k_F = \frac{3}{4}\left[\frac{\pi}{2} + \left(\frac{\phi}{\xi_0} - 1\right)\sin^{-1}\left(\frac{\xi_0}{\xi_0 + \phi}\right)^{1/2} - \left(\frac{\phi}{\xi_0}\right)^{1/2}\right] \tag{33}$$

where k_F is the wave number of an electron at the Fermi level. For platinum, it was estimated that $z_B = 0.35$ Å. For $z > z_B$, the potential $V(z)$ is the sum of $V_0(z)$ and $V_{\rho_s}(z)$, where ρ_s is the excess electron density caused by the spread. The potential $V_{\rho_s}(z)$ is obtained by integrating Poisson's equation

$$d^2 V_{\rho_s}/dz^2 = -4\pi e^2/\rho_s(z) \tag{34}$$

with the boundary conditions

$$V_{\rho_s}(z) = 0 \quad \text{at} \quad z = -\infty$$

$$dV_{\rho_s}(z)/dz = 0 \quad \text{at} \quad z = \pm\infty$$

and $V_{\rho_s}(z)$ is continuous at $z = 0$. The result is

$$V_{\rho_s}(z) = (2\pi e^2/C^2)\rho_0 e^{\beta z} \quad \text{for} \quad z < 0 \tag{35}$$

$$V_{\rho_s}(z) = -(2\pi e^2/C^2)\rho_0 e^{-\beta z} + (4\pi e^2 \rho_0/C^2) \quad \text{for} \quad z > 0 \tag{36}$$

The charge distribution and potential at the metal surface then appear as in Fig. 8.

III. ADSORBED SPECIES

Full quantum mechanical treatments of metallic crystal surfaces are particularly difficult because the presence of metallic character strongly suggests that all atoms in the crystal should be included in any calculation. The crystal is a many-particle system, including nuclei and electrons in continual motion, so that the solutions to the complete many-particle Schrödinger equation should be used to describe the states of the system. Assuming the Born–Oppenheimer approximation, the nuclear and electronic motions are treated separately, with the electronic energy being solved for under the assumption that the nuclei are stationary. Even with the nuclei stationary, the Schrödinger equation to be solved is a many-electron problem. This problem is still much too difficult, so each electron is assumed to move in the field of the static nuclei and the average field of all other electrons. It is in this already drastically simplified one-electron orbital formalism that we have been discussing surface states and will for the most part discuss adsorption. Even here, the calculations have usually been for simplified models involving only one orbital and electron per crystal atom in the

III. Adsorbed Species

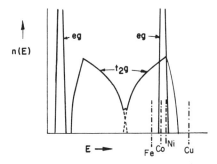

Figure 9. Density of states for a metal d band (from Ref. 89, reproduced with permission).

LCAO approximation. Thus most calculations do not deal with a particular adsorbent and adsorbate and so even the integrals appearing in these equations are not evaluated but merely used as parameters such as α and β in equation (14). For real transition metals, there are nine orbitals in the valence shell to be considered and these have particular spatial relationships to each other. The band structure for transition metals is complex; an example of just the d bands which are superimposed on the s and p bands is shown in Fig. 9.

1. Extension of Crystal LCAO Calculations to Adsorbed Species

In considering chemisorption, it is evident that the LCAO technique applied to calculating surface states can be directly applied by regarding the surface atom to be an adsorbed species. This procedure has received considerable attention from a number of workers, especially Koutecky[5] and Grimley.[7] All that has been previously said about surface states is now applied to chemisorption. If the adsorbed atom has potential-energy terms sufficiently different from those of the crystal lattice, Tamm localized states may be produced. If the adsorbed atom does not produce a sufficient perturbation to give Tamm states, Shockley localized states may still be produced. However, it should not be thought that localized states are necessary to have chemisorption, since the adsorption bond may involve only delocalized states, as seems likely in many cases of hydrogen adsorption.

(i) Localized States

For those cases where a localized state is necessary for adsorption, several observations can be made. In general, for a three-dimensional crystal, there are N^2 surface states, where N^2 is also the number of surface atoms. This situation would allow complete coverage of the surface with an adsorbate. Surface bands occurring as in curves A and B of Fig. 7 correspond to this situation. The spread in energy for curve A is quite small so all adsorption energies would be similar, thus resulting in the heat of adsorption not changing much as the fractional surface coverage changed. In contrast to this, in curve B, the energy of the surface state changes considerably. In this case, the heat of adsorption would be expected to fall off as the fractional coverage increases, a common occurrence for the adsorption of many adsorbates.[39] This is not the only possible explanation for variation in heats of adsorption with coverage. More often, this variation is attributed to surface heterogeneity, but we see here that even a homogeneous surface could exhibit a variation in heat of adsorption with coverage. It has been found that some gases will only adsorb on a surface to an extent which corresponds to considerably less than full coverage, as for example N_2 on some transition metals. A possible explanation for this would be the occurrence of surface states as in curve C of Fig. 7. Because the surface state band crosses the boundary of the volume band, many of the surface states are delocalized into the volume band. Thus if a localized state is necessary for a stable adsorption bond, partial surface coverage would completely saturate the available localized states.

The interaction of chemisorbed atoms with each other via the crystal on which they are adsorbed and not through direct interaction with each other has been discussed by Koutecky[40] and Grimley.[7] When two atoms are adsorbed on a one-dimensional lattice, for some values of the parameters, as the atoms are brought closer together, but still kept far enough apart to neglect direct interactions, the two localized states produced become separated by a wider energy gap. If the localized state must be occupied by an electron pair to produce a stable adsorption bond, then as the atoms draw closer together, the upper localized state becomes higher in energy and may eventually cross into a volume band. Thus a falling heat of adsorption with increasing coverage may be accounted for

III. Adsorbed Species

since adatoms would be forced to be closer together. Also, in the extreme of the loss of a surface state, the surface would be saturated with less than one adatom per surface atom. It should be emphasized that these effects are through the crystal and not due to direct interactions of the adsorbed species.

If the adsorbed atoms are held by one-electron bonds, then for the energy-level behavior just described, as the adatoms approach each other, the lower state drops in energy so that the atoms are more tightly held. This would appear as an attraction between the two adsorbed atoms.

Depending on the values of the parameters that go into the LCAO treatments, a wide variety of types of bonding of a foreign atom to a surface may occur. One may get homopolar bonds in which the electron charge involved in the bond is equally shared between the crystal and the adatom, or the adatom may have any fractional charge either positive or negative. The adsorption bond may involve localized or delocalized electrons. Just because localized states exist in a crystal does not mean the adsorption will necessarily involve them. It is possible to have localized crystal surface states and separately adsorbed atoms whose adsorption bonds do not involve these states.

(ii) Adsorbed Hydrogen

While much of the quantum chemistry of adsorbed species has been general in nature, almost to the point of lacking usefulness, the adsorption of hydrogen on metals has been treated in several different approximations including the general LCAO approach by Toya[36,41] and co-workers as follows. The wave function ψ_r is formulated for N metal electrons plus one electron from a hydrogen adatom as

$$\psi_r = a_0 \phi(\mathbf{k}_1, \ldots, \mathbf{k}_N, 1s) + \sum_{\alpha,i} a_{\alpha s, si} \phi(\mathbf{k}_1, \ldots, \mathbf{k}_{\alpha i}, \ldots, \mathbf{k}_N, 1s)$$
$$+ \sum_\alpha b_{\alpha s} \phi(\mathbf{k}_1, \ldots, \mathbf{k}_N, \mathbf{k}_\alpha) + \sum_i c_{si} \phi(\mathbf{k}_1, \ldots, 1s, \ldots, \mathbf{k}_N, 1s) \quad (37)$$

where the $a_0, a_{\alpha s, si}, b_{\alpha s}$ and c_{si} are constants that satisfy the normalization condition

$$a_0^2 + \sum |a_{\alpha s, si}|^2 + \sum |b_{\alpha s}|^2 + \sum |c_{si}|^2 = 1 \quad (38)$$

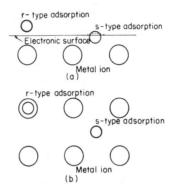

Figure 10 Model for r- and s-type adatom on a metal (from Ref. 36, reproduced with permission).

The functions $\phi(\mathbf{k}_1,\ldots,\mathbf{k}_N,1s)$ represent Slater determinants of N Block wave functions designated by the wavenumber vectors $\mathbf{k}_1,\ldots,\mathbf{k}_n$ and the $1s$ wave function of hydrogen atoms for the ground state of the $N+1$ electron system. The second term corresponds to the excitation of an electron from the $\mathbf{k}_i$ state below the Fermi level to the excited state $\mathbf{k}_\alpha$. The third and fourth terms correspond to (M^-H^+) and (M^+H^-), respectively. This wave function is used to describe an r-type adatom as shown in Fig. 10. Calculations indicate the adatom is located about 2.5 Å from the surface metal atom and has a negative polarization of about 0.02 units of charge. Hence r-type adsorption increases the work function of a metal and results in a pronounced repulsion between adatoms. Comparison of parameters for various metals indicates the energy of the adsorbed state is lower on Ni than on Pt.

Also shown in Fig. 10 is s-type adsorption, in which the hydrogen atom may be thought of as dissolved in the metal because it is treated as a proton and an electron in the conduction band. The wave function is written as

$$\psi_s = b_0\phi(\mathbf{k}_1,\ldots,\mathbf{k}_N,\mathbf{k}_{N+1}) + \sum_{i\alpha} b_{\alpha i}\phi(\tilde{\mathbf{k}}_1,\ldots,\tilde{\mathbf{k}}_{\alpha i},\ldots,\mathbf{k}_{N+1}) \quad (39)$$

where the second term represents an excited state with an electron in the α level which is above the Fermi level. The extra electron density around the proton $\delta\rho$ is evaluated as

$$\delta\rho = (\lambda^3/8\pi)e^{-\lambda r} \quad (40)$$

III. Adsorbed Species

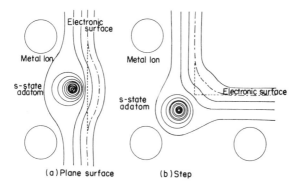

Figure 11. Electron density contours for s-type adatoms (from Ref. 36, reproduced with permission).

and the heat of dissolution of the hydrogen atom is given by

$$-I + \phi + \lambda e^2/4 \qquad (41)$$

where $1/\lambda$ is a constant determined to be 0.30 Å for Ni and 0.34 Å for Pt, I is the ionization energy for a hydrogen atom, and ϕ is the metal work function. Contours of the electron density for the adsorbed hydrogen atom on a plane surface and at a step are shown in Fig. 11. A positive dipole moment of 0.06 D is calculated for an s-type adatom, so that s-type adsorption decreases the work function of a metal.

(iii) Ionic Adsorption

The electronic interaction is considered to have relatively little effect on the initial surface states of the adsorbent when ionic adsorption occurs, in the treatment by Mark[2] of ionic adsorption on semiconductors, which are themselves treated in the ionic approximation. Attention will be focused on the electronic energy level of the adsorbate when an electron has been transferred to or from the band structure of the adsorbent. The potential energy level of an ion in free space is given by the ionization potential $-I$ or the electron affinity $-A$, depending on whether an electron has been gained or lost. Near an ionic lattice, the potential is shifted by the electrostatic interaction with the lattice. The lattice interaction is obtained from the Madelung constant C_a for the ion occupying a lattice site one lattice distance a_0 above the surface. The value C_a may be obtained from the expression

$$C_a = C - C_s \qquad (42)$$

where the Madelung constant C is defined by

$$C = \sum_{i,j,k} (q_{i,j,k}/R_{i,j,k}) = \sum_{i,j,k} Q \qquad (43)$$

and the Madelung constant for an ion in the surface C_s is defined by

$$C_s = \sum_{i \geq 0, j, k} Q \qquad (44)$$

In the above, $q_{i,j,k}$ is the charge on an ion at index position (i, j, k) and $R_{i,j,k}$ is the distance from the origin. The free-space potential is then corrected by ΔV_a, where

$$\pm \Delta V_a = Z_a C_a e/a_0 \qquad (45)$$

with the plus sign for a position above a negative lattice site and the minus sign for a positive lattice site. In the case of some covalent bonding in the lattice, the charge Z_a for the lattice ion may be fractional. If the adsorbate is located between lattice sites but still at distance a_0, the potential correction could have any value between $\pm \Delta V_a$.

A possible potential energy diagram is shown in Fig. 12. On the left, the valence bands for the adsorbent are shown crosshatched, while the conduction band is shaded. The Fermi level and intrinsic surface states occur between the valence and conduction bands and the zero of potential is placed at the vacuum level. The electrophobic adsorbate on the right is characterized by a small ionization potential and a positive electron affinity. Since its corrected potential on the surface $-I_D(s) = -I_D \pm \Delta V_a$ is above the Fermi level, this adsorbate will donate electrons to the surface. The electrophilic adsorbate is characterized by a large ionization potential and a negative electron affinity. Thus its corrected ionization potential and corrected electron affinity on the surface $-A_A(s) = -A_A \pm \Delta V_a$ lie below the Fermi level to produce acceptor states. Such electrophobic adsorbates as hydrogen and alkali metals are most likely to be donors, while electrophilic adsorbates like oxygen and halogens are expected to be acceptors.

III. Adsorbed Species

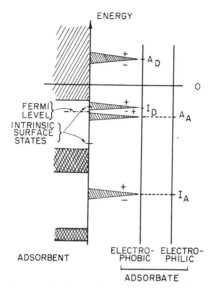

Figure 12. Potential energy diagram for chemisorption on an ionic surface (from Ref. 2, reproduced with permission).

As the surface index of a crystal plane increases, equation (42) indicates that C_a and hence $|\Delta V_a|$ will increase. This results in greater stability for chemisorbed species on high-index surface planes. During crystal growth, the highest growth rate will be for these high-index planes, so that the low-index planes will have the greatest extent, as is indeed observed in the crystal habit of most ionic solids. There can also be effects on electrode processes. If a component is present in solution which will inhibit electron transfer when adsorbed, the greatest concentration of the adsorbed species would be on the high-index surface planes so that most of the electron transfer would have to occur on the low-index planes. On the other hand, if adsorption of a species promoted the electron-transfer process, the reaction might proceed most readily on the high-index planes, where the chemisorption process occurred most readily.

2. Less Approximate Formalisms

As noted in the introduction to Section III, the work so far discussed has been carried out with drastic approximations. Recently, some work has been done to examine more refined treatments than the LCAO approach in the Hückel approximation. A series of papers[42-46] based on the method originally developed to deal with dilute alloys[47] has appeared. However, for the most part, these have dealt with the development of a formalism for handling the adsorption problem rather than calculations for particular systems. Using the Anderson[47] type of Hamiltonian, Grimley[42] concluded that when an adsorbate had energy levels near the Fermi level, the adsorbate–adsorbent interaction could shift energy levels so that the extent to which they were filled would be changed by adsorption. He expected important effects of this kind for alkali metals atoms and hydrogen but not for CO adsorbed on tungsten and nickel. He also concluded[43] that the electronic disturbance in the metal caused by the adsorbent would be long-range, oscillatory, and in many cases nonisotropic. It was noted that the interaction of a second adatom with the electronic disturbance of another adatom could be interpreted[44] in terms of a force between the adatoms and would affect the heat of adsorption as a function of coverage.[45] Starting from an Anderson[47] type of Hamiltonian, Newns[46] treated hydrogen chemisorption considering only d-band electrons. However, Politzer and Kasten[48] have shown that s and p electrons from the metal cannot be neglected in considering the tungsten–hydrogen interaction. A valence bond formalism capable of handling electron correlation has been suggested,[49] but no calculations were made.

Because of the approximate nature of the calculations indicating the formation of surface states, the question arises as to whether a more refined calculational model would produce qualitatively similar surface states. The nature of the states obtained when a linear chain of six hydrogenlike atoms was examined[50] with both the Hückel approximation and a SCF-LCAO treatment using Roothaan's equations.[51] The results of the Hückel treatment for this quite small chain showed that a localized state developed when the electron-attracting power of the end atom was sufficiently increased, i.e., the absolute value of α, the Coulomb integral, was increased. This state appears with about the same variation in α and corresponds to the Tamm surface states found earlier. Thus a quite small chain exhibits the same surface, or in this case end state, formation as a semiinfinite chain.

III. Adsorbed Species

Having shown that a six-atom chain reproduces the main results of the semiinfinite chain, the validity of the simple Hückel method itself was examined by comparison to the SCF-LCAO-MO calculation. The conclusion was reached that semiquantitatively the same localized states and bonding properties were produced. In the SCF procedure, the total wave function was represented as a single Slater determinant constructed from spin orbitals whose space parts ϕ_i were of the LCAO form

$$\phi_i = \sum_u c_{iu} X_u \qquad (46)$$

where the c_{iu} are coefficients and the X_u are $1s$ atomic orbitals. The c_{iu} were determined from the secular equation

$$\sum_v (F_{uv} - \varepsilon_i S_{uv})c_{iv} = 0 \qquad (47)$$

in which $S_{uv} = \int X_u^* X_v \, d\mathbf{r}$ and the ε_i which are the molecular orbital energies were determined from the determinational equation

$$|F_{uv} - \varepsilon S_{uv}| = 0 \qquad (48)$$

In these equations

$$F_{uv} = \int X_u^* \left[-\tfrac{1}{2}\nabla^2 - \sum_\alpha V_\alpha \right] X_v \, d\mathbf{r} + \sum_i [2(uv|ii) - (ui|iv)] \qquad (49)$$

where V_α is the potential due to nucleus α and the electron repulsion integrals in the second term are given by

$$(uv|ij) = \int X_u^*(1) X_v^*(1) \cdot (1/r_{12}) \cdot \phi_i(2)\phi_j(2) \, d\mathbf{r}_1 \, d\mathbf{r}_2 \qquad (50)$$

All one- and two-center integrals were obtained from standard tables.[52] Three- and four-center electron repulsion integrals were obtained by using the Mulliken approximation.[53] Three-center nuclear attraction integrals not involving atom number 1 were taken from Hirschfelder and Weygandt[54]; all other three-center nuclear attraction integrals were computed using an approximation given by Barker and Eyring.[55] While this calculation suffers from all the faults of a one-electron orbital method, it is reassuring to find surface states produced when many integrals are evaluated from first principles and electron repulsion terms are included.

3. Surface Compound Approximation

Having noted that all of the atoms in a metallic crystal should be included in any calculation for adsorbed species but that these equations are only soluble under the most drastic approximations, the question arises as to whether it would be better to consider less of the crystal with the possibility remaining open to calculate more terms in the Hamiltonian exactly. The work of Blyholder and Coulson[50] indicates that small, finite models do indeed give the same surface states when treated in the Hückel approximation as calculations for semiinfinite crystals. Further, they showed[50] that the Hückel results qualitatively mimicked the results of SCF calculations that included electron repulsion terms explicitly. With this indication that approximate calculations for small, finite models are not unreasonable, a number of treatments of surface compounds which do not include much of the crystal will be considered.

(i) Diatomic Treatment

In the extreme case, an atom adsorbed on a surface may be treated as if it and the surface atom to which it is bonded were a diatomic molecule. In this spirit, Ely[56] has applied the Pauling[57] approximation for the bond strength of a diatomic molecule. In this approximation, the bond energy is put equal to the arithmetic mean of the bond energies of the two homonuclear diatomic molecules from which the new bond is formed with a correction for the electronegativity differences of the two atoms. For hydrogen adsorption on tungsten, the formula for the W–H binding energy, $D(W-H)$, is

$$D(W-H) = \tfrac{1}{2}[D(W-W) + D(H-H)] + 23.06(X_W - X_H)^2 \quad (51)$$

where X is the electronegativity. Considering only nearest-neighbor interactions for a face-centered cubic metal atom with 12 nearest neighbors, an equal division of the binding energy between each two bound atoms leads to

$$D(W-W) = E_s/6 \quad (52)$$

where E_s is the metal sublimation energy. Equation (52) has also been used for body-centered cubic lattices where there are eight nearest and six next-nearest neighbors. Ely estimated the difference in

III. Adsorbed Species

electronegativity by the approximation[58]

$$\mu = X_W - X_H \tag{53}$$

where μ is the surface dipole moment. The surface dipole moment may be obtained from its relationship to the work-function change $\Delta\phi$ on adsorption,

$$\Delta\phi = 4\pi n\mu \tag{54}$$

where n is the number of adsorbed atoms per unit surface area. This procedure gives heats of adsorption in reasonable agreement with experimental values.[8] The use of Mulliken's electronegativity values has been introduced by Stevenson.[59] The Mulliken electronegativity is defined by

$$X = \tfrac{1}{2}(eI + eA) \tag{55}$$

where e is the electron charge, I is the ionization potential, and A is the electron affinity. Even though Stevenson's method gave slightly better results, the crudeness of this type of calculation has been criticized.[8] One of the difficulties with this approach is that it assumes simple metal hydrogen bonds, which Toya's work,[36] presented earlier, shows is not always the case.

The diatomic molecule model of adsorption has been treated by an approximate quantum mechanical calculation by Higuchi et al.[60] The Schrödinger equation is solved in a variational procedure using the wave function as a linear combination of covalent, ψ_c, and ionic functions, ψ_i, given by

$$\psi = c_c\psi_c + c_i\psi_i \tag{56}$$

to yield

$$1/c_i = 1 + [(E - H_{ii})/(E - H_{cc})] \tag{57}$$

where E is the bonding energy of the diatomic molecule, $H_{cc} = \int \psi_c^* H \psi_c \, d\tau$ is the covalent energy, and $H_{ii} = \int \psi_i^* H \psi_i \, d\tau$ is the ionic energy term. Rather than evaluate integrals from first principles, they were approximated as

$$H_{ii} = eA - eI + (\tfrac{8}{9})(e^2/R_{MA}) \tag{58}$$

and

$$H_{cc} = \tfrac{1}{2}[D(M-M) + D(A-A)] \tag{59}$$

where R_{MA} is the metal–adsorbed atom distance and $D(M-M)$ and $D(A-A)$ are bond energies for metal–metal and adsorbed-atom–adsorbed-atom covalent bonds, respectively. The distance R_{MA} was taken as the sum of the covalent radii for the metal atoms and adsorbed atoms. The ionic fraction c_i^2 was calculated from the observed surface dipole moment μ from the relation

$$\mu = c_i^2 e R_{MA} \tag{60}$$

The agreement of adsorption energies calculated from equation (57) with experimental values was reasonable over a wide variety of systems. For alkali and alkaline earth metals adsorbed on transition metals, the value of c_i^2 was close to one, while for hydrogen on transition metals, it varied from 0.02 to 0.09.

(ii) Charge-Transfer Complex

Another semiempirical approach is based on Mulliken's charge-transfer-complex theory,[61] which uses a wave function that is a combination of a no-bond structure and a dative bond (charge-transfer) structure designated M^+A^- or M^-A^+ depending on the direction of charge transfer. The difference in energy between the dative and no-bond states is given by

$$E_d - E_{nb} = eI - eA - e^2/4R \tag{61}$$

The application to adsorption systems has been considered by Mignolet,[62] Matsen et al.,[63] and Brodd.[64] The bond energy for adsorption is given by

$$D = \tfrac{1}{2}\{E_{nb} - E_d + [(E_{nb} - E_d)^2 + 4\beta^2]^{1/2}\} \tag{62}$$

where the interaction integral $\beta = \int \psi_{nb}^* H \psi \, d\tau$ is related to the surface dipole moment and overlap integrals are neglected. Brodd assumed that R, the distance of the adsorbed atom from the surface, had an unrealistically small value equal to the atomic radius of the adatom. This theory may also be objected to on the grounds that covalent bonding is neglected.

(iii) Semiempirical Valence Bond Approach

A semiempirical method based on the valence bond method has been used by Sherman and Eyring[65] to calculate a potential energy surface for H_2 adsorption and dissociation on carbon and by

III. Adsorbed Species

Sherman et al.[66] for adsorption of benzene and hydrogen on nickel. In the valence bond method, the wave function is a linear combination of wave functions representing the various ways the electrons can be paired into bonds. The total wave function is made antisymmetric with respect to electron permutations by use of the appropriate determinantal form. The coefficients which determine the weighting of the different electron pair structures in the final wave functions and the total energy are obtained by applying the variational principle and solving the resulting secular equation. When all overlap integrals and all exchange integrals corresponding to permutations of more than one electron pair are neglected, the formula for the total interaction energy given by London[67] for four atoms is

$$E = A_1 + A_2 + B_1 + B_2 + C_1 + C_2 + \{\tfrac{1}{2}[(\alpha_1 + \alpha_2 - \beta_1 - \beta_2)^2$$
$$+ (\alpha_1 + \alpha_2 - \gamma_1 - \gamma_2)^2 + (\beta_1 + \beta_2 - \gamma_1 - \gamma_2)^2]\}^{1/2} \quad (63)$$

The quantities A_1 through C_2 represent Coulomb energies and the Greek letter quantities α through γ represent exchange energies for bonding between pairs of atoms as illustrated in Fig. 13. The coulomb energy A_1, for example, would contain integrals such as $\int \psi_W(1)\psi_X(2) H \psi_W(1)\psi_X(2)\, d\tau_1\, d\tau_2$, where ψ_W and ψ_X are atomic orbitals centered on atoms W and X, respectively. The exchange energy α_1 would contain integrals of the type

$$\int \psi_W(1)\psi_X(2) H \psi_W(2)\psi_X(1)\, d\tau_1\, d\tau_2.$$

The explicit expressions for the Coulomb and exchange energies need not concern us here since the values are evaluated by a semiempirical procedure rather than through evaluation of the integrals.

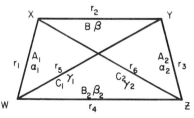

Figure 13. Distance and energy terms for 4 atom interactions.

The empirical procedure of Eyring and Polanyi[68] to evaluate these quantities from Morse[69] potentials for diatomic molecules was used. The Morse potential has the form

$$E = D'(2e^{-a(r-r_0)} - e^{-2a(r-r_0)}) \qquad (64)$$

where D' is the heat of dissociation plus the zero-point vibrational energy ($\frac{1}{2}hw_0$), r_0 is the equilibrium bond distance, and $a = 0.1227w_0(M/D')^{1/2}$, where w_0 is the vibrational frequency and M is the reduced mass, $M_1M_2/(M_1 + M_2)$. Thus from an experimentally determined dissociation energy, vibrational frequency, and equilibrium bond distance, the Morse potential can be plotted as a function of internuclear separation. For one bond, for example, a W–X bond in Fig. 13, the potential energy is given by $E = A_1 + \alpha_1$. Eyring and Polanyi[68] assumed that the Coulomb energy A_1 would always be a constant fraction of the total energy. Thus if the energy E is known from the Morse potential, the Coulomb and exchange energies can be determined for any internuclear distance and then the total potential energy in equation (63) can be determined for any arbitrary distances in a configuration such as occurs in Fig. 13. If X and Y are hydrogen atoms and W and X are nickel atoms, the potential energy surface for the reaction

$$\begin{array}{cc} \text{H–H} & \text{H H} \\ & | | \\ \text{Ni–Ni} & \text{Ni–Ni} \end{array} \rightarrow \qquad (65)$$

can be plotted. Using the fixed percentage of Coulomb energy as 14% for H–H, 20% for Ni–H, and 30% for Ni–Ni, it was calculated[66] that the heat of adsorption for H_2 on Ni was 4.6 kcal/mole and the activation energy for dissociation was 24 kcal/mole when the Ni–Ni distance was fixed at 2.38 Å. Using the somewhat different values of 11% for the Coulomb energy of H–H, 24% for Ni–H, and 37% for Ni–Ni, Okamoto et al.[70] calculated activation energies of 75 and 57 kcal/mole for Ni–Ni distances of 2.49 and 3.52 Å, respectively.

(iv) Hückel LCAO Calculations

Hückel-type molecular orbital calculations have been applied to a variety of surface complexes that include 1–10 surface metal atoms in addition to the adsorbed molecule. For CO adsorbed on

III. Adsorbed Species

Ni, Blyholder[71] has shown that a simple MO model can qualitatively account for the shifts in infrared frequencies observed upon adsorption of CO, adsorption on different surface sites, and shifts resulting from coadsorbing other gases and alloying the Ni with Cu. This model for CO adsorbed on a cluster of nine metal atoms also accounts[72] for the shifts in position of the two principal CO stretching bands as the metal is varied from V to Ni in the first-row transition metals. This latter case uses a model of the π-electron system which includes only one p orbital from each of oxygen and carbon and one orbital from each metal atom. An extended Hückel-type calculation was done in which the molecular orbitals are assumed to be linear combinations of atomic orbitals, and a secular determinantal equation, $|H_{ij} - ES_{ij}| = 0$, was solved for the orbital energies and coefficients. The diagonal effective Hamiltonian elements H_{ii} are taken as valence-state ionization potentials. The off-diagonal elements H_{ij} were determined from the equation

$$H_{ij} = 0.80 S_{ij}(H_{ii} + H_{jj}) \qquad (66)$$

In this qualitative model, no attempt to reproduce actual metal geometry and orbital orientation was made so the overlap integrals S_{ij} were arbitrarily chosen to match normal covalent compounds.

Laforgue et al.[73] considered a Hückel π-electron model for ethylene interacting with either groups of one to four metal atoms or a chain or rectangular array of up to 25 atoms. From variations in the energies of interaction with the number of metal atoms, they concluded that particle size, particularly for particles less than 100 Å in diameter, could affect adsorption energy. From their consideration of various geometries of interaction, they suggest the most favored orientation would be one where the metal atom was in the same plane as the four hydrogen atoms of ethylene. This is not in accord with current considerations of ethylene as a ligand in transition metal complexes. It appears that the simple Hückel method that considers only π electrons is not suitable to determining orientation effects.

The adsorption of CO, HCN, and C_2H_4 was examined by Dunken, Dunken, and Opitz[74-79] in the Hückel π-electron approximation with one orbital per atom. Parameters for noncarbon atoms were adjusted relative to carbon with coefficients suggested by

Streitwieser.[80] Calculations were made for the structures

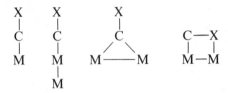

where X is O, N, or C. The π-electron energies gave the stabilities of various structures in the order

cyclic > bridge > linear

CO > HCN > C_2H_4

In general with increasing electronegativity of the metal, the adsorption energy decreased. Correlations were also made between bond orders and vibrational frequencies. Extended Hückel[81] calculations for adsorbed OH groups using all valence electrons for H and O and up to four electrons for the metal were reported by Dunken and Opitz[79] for different first-row transition metals, but the trends in stability varied as the number of metal electrons in the calculation was changed. Using the simple Hückel method, these authors also reported the results of adsorption on the corner, edge, or center of a 3 × 3 array of atoms and concluded that adsorption energy was in the order:

center > corner > edge

The effect of adsorption site has also been considered in a Hückel calculation for a 6 × 6 array consisting of identical atoms each with a single s-type valence electron.[50] The bond order for an extra atom added to the array at a corner position was shown to be greater than for one at the middle of an edge. This was correlated with the free valence of the adsorbent atom to verify correspondence with the intuitive idea that adsorption should be strongest where the free valence is greatest. When the electronegativity of the adatom was increased, a localized surface state was produced. The contribution to the bond order for the adsorption bond from the localized state was greater for adsorption in the middle of the edge than at the corner, so that if a judgment were made on the basis of localized states only, wrong relative bond orders for adsorption on the two

III. Adsorbed Species

different sites would be predicted. Previous workers[5,7] have tended to concentrate on the production of localized states, but this work demonstrates that even in cases where the localized state is primarily responsible for adsorption, the contributions from the nonlocalized states can be critical in determining the most favorable adsorption site.

(v) All-Valence-Electron Extended Hückel Calculations

The extended Hückel methods differ from simple Hückel methods, in which nonnumerical parameters are used for Coulomb and resonance integrals and overlap is neglected, in that diagonal matrix elements are estimated from atomic properties such as ionization potentials, nondiagonal elements are taken as proportional to overlap integrals, and overlap integrals for nonadjacent atoms are kept in the calculation. Extended Hückel calculations have had some success in organic chemistry in situations where charge distributions are fairly uniform, but in its simple form, its deficiencies are well recognized. Due to its neglect of electron repulsion terms, the total molecular energy is *not* equal to the sum of the occupied orbital energies, which is the only total energy term available in the extended Hückel calculation. Therefore, it is most reliable in comparing a series of similar structures, but even here, if electron repulsion terms vary in the series, the comparative orbital sum energies will be off.

Robertson and Wilmsen[82] have done extended Hückel calculations for various organic fragments adsorbed on Pb surfaces. They found that electron-rich groups like methyl mercaptan were not stable on the surface, while electron-poor groups like acetoxy radical and ethyl radical were stable. Adsorption of an ethyl group on a Pb atom next to an already covered Pb atom was 45 or more kcal/mole less favorable than adsorption on an isolated site. The calculated mobility for ethyl groups was low because an energy barrier to migration greater than 35 kcal/mole was found.

A modified extended Hückel treatment was used by Bennett et al.[83] to investigate hydrogen atom interaction with graphite. The most stable position for a hydrogen atom was found to be directly over a carbon atom, but there was little activation energy to travel along a line to the position above the next carbon atom. The position above the center of a hexagon was unfavorable. In considering the formation of CH_4 at an edge position, it was found that the heat of addition of a second hydrogen atom to a carbon atom was only one-fourth as great as the first. Nonetheless, CH_4 formation was an energetically favorable process.

The adsorption of hydrogen atoms on first-row transition metals has been considered by Politzer and Kasten.[48] The adsorbed hydrogen atoms were found to have a negative charge, in agreement with the negative surface potentials that have been observed for chemisorbed hydrogen on metals. One of the most interesting results of this study was the finding that metal s orbitals played a major role in the bonding. Workers in chemisorption have tended to concentrate on metal d orbitals and electrons to the exclusion of s and p orbitals. The possible impropriety of this should be borne in mind in considering the next section, where geometric arrangements at the surface are discussed entirely in terms of d-orbital orientation. It might also be noted that there is a sizable literature on molecular orbital treatments of transition metal complexes that indicates s and p metal orbitals are as important as d orbitals in bonding to transition metals.

(vi) *Surface Orbital Orientation*

Orientation effects at a metal surface have been considered by Bond,[84] Dowden,[85] and Tamm and Schmidt[86] in the framework of the general theory of directed valence[87] where orientation is determined by maximum orbital overlap. Bond used the results of describing the formation of electron bonds in terms of the overlap of molecular orbitals which are directional because of the existence of the crystal field.[88,89] The orbital orientations of Troost[88] were used for the face-centered cubic structure. Figure 14 shows the directions of emergence of orbitals at the (100) face. The $1s$ orbital of a hydrogen atom could overlap with an e_g orbital projecting vertically from a surface atom to form an adsorbed state, or the hydrogen could go into the octahedral hole where it would overlap five e_g orbitals to give a strongly bound state. A possible mode of adsorption of ethylene on the (100) face is shown in Fig. 15. The bonding is a result of a σ bond formed by electron donation from the filled bonding orbital of the olefin to the partially vacant metal e_g orbital and a π bond formed by back-donation from a partially filled t_{2g} orbital of the metal into the vacant π antibonding orbital

III. Adsorbed Species

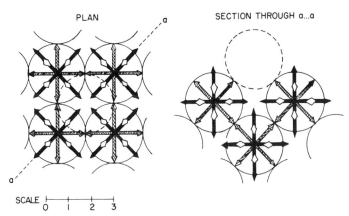

Figure 14. Diagrammatic representation of the emergence of orbitals at the (100) face of a face-centered cubic metal. Filled arrows: e_g orbitals in plane of paper; hatched arrows: t_{2g} orbitals in plane of paper; open arrows: t_{2g} orbitals emerging at 45° to plane of paper. The broken circle shows the position of an atom in the next layer above the surface layer. In both the plan and section, an e_g orbital emerges normal to the plane of the paper from each atom. The scale applies to nickel (from Ref. 84, reproduced with permission).

of ethylene. Other planes and adsorbents were considered and some planes such as the (111) face were found to be poorly suited to ethylene adsorption. Although only a qualitative picture, this approach gives a good feel for possible surface geometries and orbital utilization which should guide more detailed calculations.

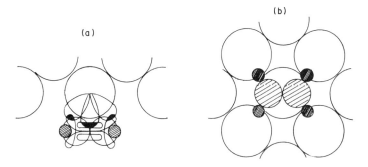

Figure 15. Representation of π-adsorbed ethylene on the (100) face of a nickel: (a) plan; (b) section showing orbital overlap (from Ref. 84, reproduced with permission).

4. Perturbation Approaches

Perturbation theory in various forms has been applied to adsorption problems. Jansen has developed[90] and applied[91] an exchange perturbation method to H_2 adsorption. Unfortunately, in this form, the method was unable to explain the dissociation of hydrogen on a platinum surface.

A further development of Jansen's perturbation theory has been made by Avoird[92,93] and applied[94,95] to adsorption systems. Calculations were made for the two models shown in Fig. 16. Model A represents the interaction of a rare-gas atom c with two metal atoms a and b, while model B represents hydrogen atoms c and d adsorbing on two metal atoms. The interaction energy between the adsorbate and the metal atoms is calculated by a perturbation method in which the first-order energy is essentially the same as a valence-bond calculation which contains the exchange interactions as the major contribution to the chemical bond. The second-order energy calculation produces the van der Waals attraction and the second-order exchange energy, so that physical adsorption as well as chemical adsorption should be covered by the model. Each metal atom is represented by an unpaired d electron approximated as occupying a spherical orbital with the two metal atoms at the nearest- or next-nearest-neighbor distance of a face-centered cubic lattice for Ni, Pd, or Pt. The interaction energy is developed as a cluster expansion:

$$E = \sum_{i<j} E_{ij} + \sum_{i<j<k} E_{ijk} + \cdots \qquad (67)$$

The term E_{ij} represents the interaction energy as if only atoms i and j were present. The wave function is developed as the appropriate antisymmetric function formed from ordered products of molecular

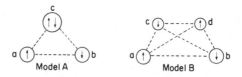

Figure 16. Models used by Avoird in perturbation calculation.

III. Adsorbed Species

functions. For model B, the product function is

$$\phi_0 = \phi_a(1)\phi_b(2)\phi_c(3)\phi_d(4) \tag{68}$$

and the interaction potential is

$$V = \sum_{\alpha < \beta} (1/R_{\alpha\beta}) - \sum_{\alpha,i} (1/r_{\alpha i}) + \sum_{i<j} (1/r_{ij}) \tag{69}$$

such that

$$H = H_0 + V \tag{70}$$

where H_0 and H are the unperturbed and final Hamiltonians for the system. H_0 is defined with the adsorbate an infinite distance from the adsorbent. The first-order interaction energy $E^{(1)}$ is given by

$$E^{(1)} = N^{-1} \int (A\phi_0)(AV\phi_0) \, d\tau \tag{71}$$

where A is a projection operator to select that part of the total wave function appropriate to the molecular symmetry and N is a normalization constant. The second-order energy expression is

$$E^{(2)} = N^{-1} \sum_{k \neq 0} \left\{ \int (A[V - E^{(1)}]\phi_0)(A\phi_k) d\tau \right. \\ \left. \times \int (A\phi_k)(A[V - E^{(1)}]\phi_0) \, d\tau / (E_0 - E_k) \right\} \tag{72}$$

which is simplified by use of the Unsold[95] approximation to

$$E^{(2)} = (-1/\Delta E_{av}) \left[N^{-1} \int (AV\phi_0)(AV\phi_0) \, d\tau - (E^{(1)})^2 \right] \tag{73}$$

The average energy for model B is given by

$$\Delta E_{av} = \Delta E_\infty (1 + CS^2) \tag{74}$$

where ΔE_∞ is taken as twice the estimated ionization energy of a d electron plus twice the ionization energy of a hydrogen atom. The total overlap function S^2 is given by

$$S^2 = S_{ab}^2 + S_{ac}^2 + S_{ad}^2 + S_{bc}^2 + S_{bd}^2 + S_{cd}^2 \tag{75}$$

The results for hydrogen show the formation of a physically bonded state which passes smoothly over into a chemically bound state with dissociation of the hydrogen into atoms. In agreement with experiment, little or no activation energy for dissociation of H_2 is indicated. This calculational model is among the most successful in obtaining results for adsorbed systems but is also one of the longest and most complex from a calculational viewpoint.

Grimley[96,97] has looked qualitatively at the application of perturbation theory using semiempirical parameters for adsorbed species. His approach is to examine the new situation resulting from the interaction of an adsorbed molecule and metal energy levels within the framework of one-electron orbital theory. In the first paper, two approaches to chemisorbed carbon monoxide are presented. In one approach, the interaction energy between the CO molecular orbitals and the metal band states is given by the perturbation equation

$$E - H_0 = \sum_k [|V_{0k} - H_0 S_{0k}|^2/(H_0 - H_k)] \quad (76)$$

which is simplified to

$$E - H_0 = 4 \int_{\varepsilon_v}^{\varepsilon_F} [\omega_\pi(\varepsilon)/(\varepsilon + A)] \, d\varepsilon - 2 \int_{\varepsilon_F}^{\varepsilon_{max}} [\omega_\sigma(\varepsilon)/(\varepsilon + I)] \, d\varepsilon \quad (77)$$

where A and I are, respectively, the electron affinity and ionization potential of CO. The first term on the right is for electron transfer from metal levels ε below the Fermi level ε_F to the vacant 2π level of CO. The right-hand term is for electron transfer from the filled CO σ level to the empty metal levels above the Fermi level. The energy interaction terms are defined by

$$\omega_\sigma(\varepsilon) = \sum_k |V_{0k} - H_0 S_{0k}|^2 \, \delta(\varepsilon - \varepsilon_k), \qquad k > k_F \quad (78)$$

and

$$\omega_\pi(\varepsilon) = \sum_k |V_{0k} - H_0 S_{0k}|^2 \, \delta(\varepsilon - \varepsilon_k), \qquad k < k_F \quad (79)$$

where V_{0k} is the interaction term between the ground state and level k of the metal. In evaluating the interaction energy, the terms $\omega_\sigma(\varepsilon)$ and $\omega_\pi(\varepsilon)$ were treated as semiempirical parameters and given "reasonable" values. The difference in heat of adsorption of CO on Fe and W is presumed to arise in this treatment because of differences in the density of states in the metal bands. Since this density of states also enters into the expressions for the probability

III. Adsorbed Species

of electron transfer from an electrode, there should be a correlation between adsorption energy and electron-transfer rates. In the other model, in this paper[96] a surface compound is considered to be formed so the interaction is written between a CO molecule and one surface metal atom. Again, the interaction energy terms are treated as parameters to be given "reasonable" values. Qualitatively, the results of these two models were similar. It was recognized that the levels from the surface compound should be considered to further interact with the band structure of the metal.

Another formulation within which to examine the interaction of an adsorbed atom or molecule with the metal band structure has recently been presented.[97] The coupling of an atom or molecule to a metal is treated as converting the originally discrete levels of the molecule into a series of levels, referred to as virtual levels, as indicated in Fig. 4. The virtual levels may have fractional occupancies determined by the fraction of levels falling below the Fermi level ε_F of the metal. This occupancy n_A is given by

$$n_A = \int_{-\infty}^{\varepsilon_F} \rho_A(\varepsilon) \, d\varepsilon \tag{80}$$

where $\rho_A(\varepsilon)$ is the density of the virtual levels. The level density of the combined system is given by

$$\rho = \rho_M^{(0)} + \rho_A + \Delta\rho_M \tag{81}$$

where $\rho_M^{(0)}$ is the density for the noninteracting metal and $\Delta\rho_M$ is the change in metal density caused by the interaction. The model Hamiltonian is one developed for dilute alloys[47] and contains terms for the separated molecule and metal–electron repulsion J and interaction terms from perturbation theory. Ignoring the provision the original theory makes for having different numbers of electrons with different spins, the binding energy D for an atom is

$$D = E_A - \int_{-\infty}^{(0)} (\rho_A + \Delta\rho_M)\varepsilon \, d\varepsilon + J(n_A)^2 \tag{82}$$

where E_A is the level of the noninteracting atom. The expression for ρ_A in this approach is

$$\rho_A = -\pi^{-1} Im(\xi - \varepsilon_A - q)^{-1} \tag{83}$$

where

$$q = \sum_k [|V_{Ak}|^2/(\xi - \varepsilon_k)] = \alpha - i\Gamma \tag{84}$$

and
$$\varepsilon_A = E_A + Jn_A \tag{85}$$
The level density change for the metal is given by
$$\Delta\rho_M = -\pi^{-1} Im\{(\xi - \varepsilon_A - q)^{-1} \sum_k [|V_{Ak}|^2/(\xi - \varepsilon_k)^2]\} \tag{86}$$
The application of this theory depends upon being able to evaluate $\Gamma(\varepsilon)$, the simplest formula being
$$\Gamma(\varepsilon) = \pi \sum_k |V_{Ak}|^2 \delta(\varepsilon - \varepsilon_k) \tag{87}$$
where V_{Ak} is the interaction term between the adsorbate and metal band level k. In considering the application of this formalism to an adatom of Na or S on the Ni(100) face, it was unfortunately found that the equations were too difficult to handle in any but the most approximate manner. In order to evaluate $\Gamma(\varepsilon)$ from equation (87), the interaction energy integral V_{Ak} was reduced to an empirical parameter and the sum over energy levels for the metal was reduced to a single level, chosen to be at the d-band center. Within this model, values for the parameters could be found to give a reasonable value for the heat of dissociation of a Na atom from the surface. To be useful, this approach requires a lot more work to determine energy levels for the surface complex. However, this approach has the advantage of providing the framework for discussing the interaction of these levels, once determined, with the metal band structure.

REFERENCES

[1] S. G. Davison and J. D. Levine, *Solid State Physics* **25** (1970) 1.
[2] P. Mark, *Catalysis Reviews* **1** (1968), 165.
[3] S. G. Davison and M. Steslicka, *Int. J. Quantum Chem.* **4** (1971) 455.
[4] M. Tomasek and J. Koutecky, *Int. J. Quantum Chem.* **3** (1969) 249.
[5] J. Koutecky, *Adv. Chem. Phys.* **9** (1965) 85.
[6] A. D. Crowell, *The Solid–Gas Interface, Vol. 1*, Ed. by E. A. Flood, Marcel Dekker, New York, 1967.
[7] T. B. Grimley, *Adv. Catalysis* **12** (1960) 1.
[8] P. M. Gundry and F. C. Tompkins, *Quart. Rev. (London)* **14** (1960) 257.
[9] P. Delahay, *Double Layer and Electrode Kinetics*, Interscience, New York, 1965.
[10] R. N. Adams, *Electrochemistry at Solid Electrodes*, Marcel Dekker, New York, 1969.
[11] A. W. Adamson, *Physical Chemistry at Surfaces*, 2nd Ed., Interscience, New York, 1967.
[12] Proceedings of the Symposium on Electrocatalysis, November 1968, *Surface Science* **18** (1969).

References

[13] J. O'M. Bockris, M. A. V. Devanathan, and K. Muller, *Proc. Roy. Soc. (London)* **274** (1963) 55.
[14] S. W. Feldberg, C. G. Enke, and C. E. Bricker, *J. Electrochem. Soc.* **1963**, 110, 826.
[15] G. C. Bond, *Catalysis by Metals*, Academic Press, New York, 1962.
[16] D. B. Matthews and J. O'M. Bockris, *Modern Aspects of Electrochemistry, Vol. 6*, Ed. by J. O'M. Bockris and B. E. Conway, Plenum Press, New York, 1971; J. O'M. Bockris and A. K. N. Reddy, *Modern Electrochemistry*, Plenum Press, New York, 1970; J. O'M. Bockris, *J. Chem. Ed.* **48** (1971) 352.
[17] T. N. Anderson, H. Eyring (Ch. 3), and V. G. Levich (Ch. 12), in *Physical Chemistry: An Advanced Treatise, Vols. 9A and 9B*, Ed. by H. Eyring, Academic Press, New York, 1970.
[18] F. Block, *Z. Physik* **52** (1928) 555.
[19] R. de L. Kronig and W. G. Penney, *Proc. Roy. Soc. (London)* **A130** (1930) 499.
[20] G. Wannier, *Phys. Rev.* **52** (1937) 191.
[21] I. Tamm, *Z. Physik* **76** (1932) 849.
[22] A. W. Maue, *Z. Physik* **94** (1935) 717.
[23] R. H. Fowler, *Proc. Roy. Soc. (London)* **A141** (1933) 56.
[24] E. T. Goodwin, *Proc. Cambridge Phil. Soc.* **35** (1939) 205.
[25] W. Shockley, *Phys. Rev.* **56** (1939) 317.
[26] J. W. May, *Ind. Eng. Chem.* **57** (1965) 19.
[27] H. Statz, *Z. Naturforsch.* **5a** (1950) 534.
[28] J. D. Levine, *Phys. Rev.* **171** (1968) 701.
[29] E. T. Goodwin, *Proc. Cambridge Phil. Soc.* **35** (1939) 221.
[30] E. T. Goodwin, *Proc. Cambridge Phil. Soc.* **35** (1939) 232.
[31] K. Artmann, *Z. Physik* **131** (1952) 244.
[32] T. A. Hoffmann, *Acta Phys. Acad. Sci. Hung.* **2** (1952) 195.
[33] B. A. Lippmann, *Ann. Phys.* **2** (1957) 16.
[34] J. Koutecky, *Phys. Rev.* **108** (1957) 13.
[35] J. Koutecky and M. Tomasek, *Phys. Rev.* **120** (1960) 1212.
[36] J. Horiuti and T. Toya, in *Solid State Surface Science, Vol. 1*, Ed. by M. Green, Marcel Dekker, New York, 1969.
[37] R. Smoluchowski, *Phys. Rev.* **60** (1941) 661.
[38] A. Sugiyama, *J. Phys. Soc. Japan* **15** (1960) 965; **16** (1961) 1327.
[39] D. O. Hayward and B. M. W. Trapnell, *Chemisorption*, 2nd Ed., Butterworths, Washington, D.C., 1964.
[40] J. Koutecky, *Trans. Faraday Soc.* **54** (1958) 1038.
[41] T. Toya, *J. Res. Inst. Catalysis, Hokkaido U.*, **8** (1960) 209.
[42] T. B. Grimley, *Proc. Phys. Soc. (London)* **90** (1967) 751.
[43] T. B. Grimley, *Proc. Phys. Soc. (London)* **92** (1967) 776.
[44] T. B. Grimley, *J. Am. Chem. Soc.* **90** (1968) 3016.
[45] T. B. Grimley and S. M. Walker, *Surface Science* **14** (1969) 395.
[46] D. M. Newns, *Phys. Rev.* **178** (1969) 1123.
[47] P. W. Anderson, *Phys. Rev.* **124** (1961) 41.
[48] P. Politzer and S. Kasten, Abstracts, 2nd North American Meeting of the Catalysis Society, February 1971, Houston.
[49] J. R. Schrieffer and R. Gomer, *Surface Sci.* **25** (1971) 315.
[50] G. Blyholder and C. A. Coulson, *Trans. Faraday Soc.* **63** (1967) 1782.
[51] C. C. J. Roothaan, *Rev. Mod. Phys.* **23** (1951) 69.
[52] R. C. Sahni and J. W. Cooley, NASATN D-146, Vol. 1, 1959, and Vol. 2, 1960.
[53] R. S. Mulliken, *J. Chem. Phys.* **46** (1949) 479.
[54] J. O. Herschfelder and C. N. Waygandt, *J. Chem. Phys.* **6** (1938) 806.

[55] R. S. Barker and H. Eyring, *J. Chem. Phys.* **22** (1954) 1182.
[56] D. D. Ely, *Disc. Faraday Soc.* **8** (1950) 34.
[57] L. Pauling, *The Nature of the Chemical Bond*, Cornell Univ. Press, Ithaca, 1939.
[58] Malone, *J. Chem. Phys.* **1** (1933) 197.
[59] D. P. Stevenson, *J. Chem. Phys.* **23** (1955) 303.
[60] I. Higucki, T. Ree, and H. Eyring, *J. Am. Chem. Soc.* **79** (1957) 1330.
[61] R. S. Mulliken, *J. Am. Chem. Soc.* **74** (1952) 811.
[62] J. C. P. Mignolet, *J. Chem. Phys.* **21** (1953) 1298.
[63] F. A. Matsen, A. C. Makrides, and N. Hackerman, *J. Chem. Phys.* **22** (1954) 1800.
[64] R. J. Brodd, *J. Phys. Chem.* **62** (1958) 54.
[65] A. Sherman and H. Eyring, *J. Am. Chem. Soc.* **54** (1932) 2661.
[66] A. Sherman, C. E. Sun, and H. Eyring, *J. Chem. Phys.* **3** (1934) 49.
[67] F. London, *Z. Elektrochem.* **35** (1929) 552.
[68] H. Eyring and M. Polanyi, *Z. Physik. Chem. (Leipzig)* **B12** (1931) 279.
[69] R. M. Morse, *Phys. Rev.* **34** (1924) 57.
[70] G. Okamoto, J. Horiuti, and K. Hirota, *Sci. Papers Inst. Phys. Chem. Res. (Tokyo)* **29** (1936) 223.
[71] G. Blyholder, *J. Phys. Chem.* **68** (1964) 2772.
[72] G. Blyholder and M. Allen, *J. Am. Chem. Soc.* **91** (1969) 3158.
[73] A. Laforgue, J. Rousseau, and B. Imelick, *Adv. Chem. Phys.* **8** (1965) 141.
[74] H. Dunken and H. H. Dunken, *Z. Chemie* **6** (1966) 234.
[75] H. H. Dunken and C. Opitz, *Z. Chemie* **6** (1966) 390.
[76] H. H. Dunken and C. Opitz, *Z. Phys. Chemie* **60** (1968) 25.
[77] H. Dunken and H. H. Dunken, *Z. Phys. Chem. (Leipzig)* **239** (1968) 161.
[78] H. Dunken, *Z. Chemie* **10** (1970) 158.
[79] H. H. Dunken and C. Opitz, in *Reprints of Fourth Internat. Congress on Catalysis Moscow, 1968*, Vol. 1, p. 20, Rice Univ. Press, Houston, 1969.
[80] A. Streitwieser, Jr., *Molecular Orbital Theory for Organic Chemists*, Wiley, New York, 1961.
[81] R. Hoffmann, *J. Chem. Phys.* **39** (1963) 1397.
[82] J. Robertson and C. W. Wilmsen, *J. Vac. Sci Tech.* **8** (1971) 53.
[83] A. J. Bennett, B. McCarroll, and R. P. Messmer, *Surface Sci* **24** (1971) 191.
[84] G. C. Bond, *Disc. Faraday Soc.* **41** (1966) 200.
[85] D. A. Dowden, *Quimica Fisica de Procesos en Superficies Solidas*, Liberia Cientifica Medinaceli, Madrid, 1965, p. 177.
[86] P. W. Tamm and L. S. Schmidt, *J. Chem. Phys.* **54** (1971) 4775.
[87] C. A. Coulson, *Valence*, University Press, Oxford, 1953.
[88] W. R. Trost, *Can. J. Chem.* **37** (1959) 460.
[89] J. B. Goodenough, *Magnetism and the Chemical Bond*, Interscience, New York, 1963.
[90] L. Jansen, *Phys. Rev.* **162** (1967) 63.
[91] L. Jansen, in *Molecular Processes on Solid Surfaces*, Ed. by E. Drauglis, R. D. Gretz, and R. I. Jaffee, p. 49, McGraw-Hill, New York, 1969.
[92] A. van der Avoird, *J. Chem. Phys.* **47** (1967) 3649.
[93] A. van der Avoird, *Chem. Phys. Letters* **1** (1967) 411.
[94] A. van der Avoird, Thesis, Technical University Eindhoven, 1968.
[95] A. van der Avoird, *Surface Sci.* **18** (1969) 159.
[96] T. B. Grimley, in *Molecular Processes on Solid Surfaces*, Ed. by Drauglis, Gretz, and Jaffee, McGraw-Hill, New York, 1969.
[97] T. B. Grimley, *J. Vac. Sci. Tech.* **8** (1971) 31.

2

Oscillatory Behavior in Electrochemical Systems

J. Wojtowicz*

Technical University, Warsaw, Poland

I. INTRODUCTION

A closed chemical system which is not in thermodynamic equilibrium will tend to relax asymptotically toward such a state as soon as the existing constraints are removed (e.g., when contact between the components is established, or, in an electrochemical system, when a galvanic cell circuit is closed). The resulting transient can be either purely monotonic, which is the most common behavior, or the system in its relaxation to equilibrium can pass through an infinite or a finite number of maxima and minima. However, in the latter case, the amplitude diminishes steadily with time, and it has been proved by the methods of irreversible thermodynamics that, in a closed system, sustained oscillations can occur neither around the equilibrium,[1] nor for appreciable displacements from it, i.e. "in the large."[2]

There exist, however, no similar restrictions concerning the behavior of systems open to fluxes of matter, and indeed a great number of cases of sustained oscillations are known among various classes of reacting systems. Considerable experimental material concerning oscillations has been collected in the field of electrochemical processes, and several propositions for their theoretical interpretation have been published. It is of interest that some of the first observations of periodic chemical phenomena, apart from the

---

*Present address: Atlantic Industrial Research Institute, Halifax, Nova Scotia.

classical case of Liesegang's rings, which are periodic in a different sense, arose in electrochemical systems.

Oscillations in biological systems, e.g., with regard to population variation, have been treated by Volterra.[3a] In recent years, intensive study of oscillations has been concentrated especially in the field of kinetics of complex enzymatic reactions and other phenomena in biology (see, e.g., Ref. 3b), with obvious reference to their relation to the periodicity in biochemical functions in living organisms.

Oscillations in chemical (or electrochemical) systems deserve study not only because of their direct cognitive interest, but because analysis of the causes of periodic behavior can provide useful information on the possible mechanism of the process, the individual steps involved, and the nature of the coupling between them. It may often happen that an otherwise apparently reasonable and well-founded scheme will have to be discarded because oscillations cannot be explained in terms of the mechanism proposed.

Practical aspects of studying oscillations can also be envisaged. Although it has been stated[4] that it is normally assumed that steady-state operation is always the most desirable, it has been shown[5] that, "The time-average conversion obtained from an oscillating reactor is sometimes superior to the steady-state output. This result implies that under certain circumstances, an oscillator will have a performance better than that expected from the optimum steady-state design."[4] While this quotation concerns nonisothermal reactors, there does not seem to be any fundamental reason why these expectations should not apply equally well to isothermal oscillating systems.

The first case of periodicity in an electrochemical system was described by Fechner as early as 1828.[6] It concerned periodic deposition and dissolution of silver on iron in an acidified solution of silver nitrate. A comprehensive review of papers published during the next 100 years has been given in a book.[7a] In the preface to the book, the authors state, however, that "... most of the cases have been regarded as curiosities or anomalies and many more have been forgotten entirely." This attitude has since changed radically, and the reality of electrochemical oscillations, together with their significance as the typical rather than the exceptional behavior of some systems, is now generally recognized.

Except in cases of trivial relaxation oscillations (see p. 73), periodic phenomena arise because of the cross-couplings and positive feedbacks that exist between various elementary stages of a process. The picture is usually complicated and it is seldom possible to account for kinetic oscillations adequately without recourse to a mathematical model and use of analog methods for obtaining solutions.

In electrochemical systems, the situation is further complicated by the fact that there always exists a closed loop with respect to one of the reagents, namely electrons, so that the elements of the external circuit must, of necessity, usually influence the course of events at the electrode under consideration.

The present chapter is primarily concerned with models of oscillating electrochemical systems and the various ways of approach to the rational interpretation of observed periodic phenomena. Presentation of the models will be preceded by a review of more typical or more important experimental material and its qualitative interpretation as offered by different authors. This review of the experimental aspects of the phenomena will by no means be complete, and it is intended that it should serve only as a background to the theories subsequently discussed.

II. EXPERIMENTAL BASIS OF ELECTROCHEMICAL OSCILLATIONS AND THEIR QUALITATIVE INTERPRETATION

1. Anodic Processes Involving Metals

A large proportion of the cases of periodicity in electrochemical systems described in the literature concerns oscillations encountered during anodic polarization of various metals. They are directly connected with the characteristic instability of passivating films, porous or otherwise, under certain conditions. Review of a selection of the considerable number of papers which have treated this problem should suffice to give insight into the kinds of qualitative interpretation which usually accompany the purely phenomenological description of the effects observed. The forms of some typical oscillations of current and potential in various systems are shown in Fig. 1.

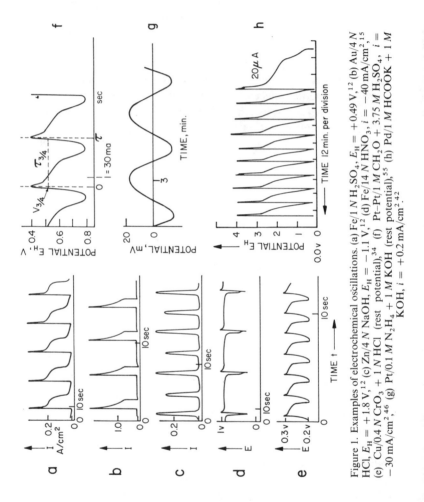

Figure 1. Examples of electrochemical oscillations. (a) Fe/1 N H$_2$SO$_4$, $E_H = +0.49$ V,[12] (b) Au/4 N HCl, $E_H = +1.8$ V,[12] (c) Zn/4 N NaOH, $E_H = -1.1$ V,[12] (d) Fe/14 N HNO$_3$, $i = -40$ mA/cm^2,[15] (e) Cu/0.4 N CrO$_3$ + 1 N HCl (rest potential),[34] (f) Pt–Pt/1 M CH$_2$O + 3.75 M H$_2$SO$_4$, $i = -30$ mA/cm^2,[46] (g) Pt/0.1 M N$_2$H$_4$ + 1 M KOH (rest potential),[55] (h) Pd/1 M HCOOK + 1 M KOH, $i = +0.2$ mA/cm^2.[42]

II. Electrochemical Oscillations and their Qualitative Interpretation

Oscillations observed at electrodes undergoing passivation are related to the characteristic shape of their polarization curves, which show a very sharp change of current at a certain value of potential (Flade potential[7b]). This means that the electrode–solution interface can exist as a stable system only in one of two extreme states—the active, or the passive—and that intermediate configurations are unstable.[8a] If two iron electrodes, one completely passivated, the other active, are brought into contact, either the active one becomes passivated, or the passive one turns active. The result depends on the position of the mixed potential[8b] of the two electrodes in relation to the Flade potential.

In order that oscillations between the two states can occur, some factor must be operative which causes the Flade potential to become more positive when the electrode is in the passive state while it becomes less positive when the electrode is active. When passivation is caused by anodic polarization, oscillations result because of coupling with an essentially nonelectrochemical process which tends to reactivate the electrode. If the metal is passivated by chemical action, cathodic polarization would act as reactivating factor.[9]

Iron polarized anodically in H_2SO_4 is an example of the first case.[10-12] Local currents of high density (of the order of 10 A cm^{-2}) which flow when the electrode is in the active state cause the local concentration of H^+ ions to drop rapidly. This affects the Flade potential, which is shifted by 58 mV negatively per unit increase of pH. When the electrode potential becomes more positive than the Flade potential, passivation occurs and the current density falls to a very small value (a few μA cm^{-2}). However, owing to an increased diffusion rate, the concentration of H^+ ions is soon rebuilt, and the electrode potential becomes again less positive than the actual Flade potential; the electrode is then activated, high currents flow, and the cycle starts anew. This qualitative description forms the basis of detailed models[13] which are discussed below (p. 81).

In the opposite case of iron passivated "chemically" in HNO_3, with superimposed cathodic polarization, the coupling term is attributed to the formation of HNO_2 and its effect on the electrochemical reduction of the passive film.[14-16]

Oscillations at anodically polarized copper were investigated by several authors. A semiquantitative model of Bonhoeffer and Gerischer[17] is described on p. 80.

Cooper and Bartlett[18] investigated the behavior of copper anodes in HCl solutions with constant voltage applied across the cell. The observed oscillations of the current and anode potential were interpreted in terms of the formation and characteristics of a porous layer of CuCl. The chloride deposition starts at random nuclei and spreads over the whole surface. The current decreases, causing the anolyte concentration at the base of the pores to increase. This, in turn, lowers the potential there to a value at which the formation of CuCl no longer occurs. The existing layer then dissolves until the surface is bare enough for the cycle to begin again.[18] This interpretation, contrary to that of Bonhoeffer and Gerischer[17] (see p. 80), does not employ the concept of two solid phases, CuCl and Cu_2O, since under the conditions of the experiment, "potentials were lower than necessary for oxide formation."

Heterogeneity of the electrode surface as a basis for the origin of oscillations was investigated by Meunier.[19] He found[20] that copper under electropolishing conditions (H_3PO_4) exhibited oscillations of its potential only when its surface had a certain degree of roughness (no oscillations were observed on previously polished samples). When unevenly distributed current was passed at a perfectly polished electrode, the potential began to oscillate as soon as a visible oxide layer had been formed at zones of lower current density.[21]

Although these findings seem to substantiate the view that heterogeneity is closely related to oscillatory behavior, in this case, it is difficult to accept as generally applicable the author's statement that, "It does not seem possible to obtain oscillations if the current density has exactly the same value at all points on an electrode."[19]

Oscillations of potential at a nickel anode in Ni_2SO_4 solution containing chlorides were described by Förster and Krüger,[22] and also by Hoar and Mowat[24] under electropolishing conditions. In the former work, oscillations appeared when at a given current density, the chloride concentration was below a certain limit. The oscillatory effects were accompanied by the periodic formation and disappearance of a dark brown surface layer. The following mechanism was proposed. The charge-transfer reaction $2Cl^- \rightarrow Cl_2 + 2e^-$ (I) was regarded as being followed by $Cl_2 + Ni \rightarrow Ni^{2+} + 2Cl^-$ (II). Reaction (II) is considered to be fast and depolarizes reaction (I). As long as there is adequate supply of

II. Electrochemical Oscillations and their Qualitative Interpretation

chloride ions, the anode dissolves, i.e., it is in the "active state." However, when transport of Cl^- is too slow at some points, the local potential rises and two other charge-transfer processes become operative: $SO_4^{-2} \rightarrow SO_4 + 2e^-$, $SO_4 + H_2O \rightarrow 2H^+ + SO_4^{2-} + O$ and $Ni^{2+} \rightarrow Ni^{3+} + e^-$. The Ni^{3+} ions undergo immediate hydrolysis and a film of $Ni(OH)_3$ (probably NiO.OH) is said to deposit on the electrode surface. Because of this blocking effect on the parts of the electrode initially free from deposit, the current density increases so that the inhibiting layer starts to form at these places as well. However, the simultaneously occurring reaction $Ni^{3+} + Cl^- \rightarrow Ni^{2+} + Cl$ provides a pathway for dissolution of the layer. The current density then decreases, the more so because the surface has become rough, and process (I) again becomes predominant. After a time, locally high current density or insufficient supply of Cl^- will induce passivation which will spread "autocatalytically" over most of the surface.

Electrochemical periodicity has often been observed during electropolishing of various metals. Usually the periodic formation and dissolution of surface films is again considered responsible for the phenomenon.

For example, Hoar and Mowat,[24] who investigated the behavior of a nickel anode in 50% H_2SO_4, reported that when a constant current density only just large enough to produce electropolishing is applied, violent oscillations of the potential difference across the bath (that is, of the anode potential, since the cathode "single" potential and the electrolyte ohmic drop are constant) are noticed just after the rapid potential rise. They offered the following qualitative interpretation. If the rate of growth of the surface film is greater than the rate of its dissolution, the film will actually be formed. This will cause the potential to rise, rendering an alternative charge-transfer reaction possible, i.e., oxygen evolution. As the current available for the film formation process decreases and the rate of its (chemical) dissolution (which is constant) becomes greater than the rate of its formation, the film gradually disappears. The potential now falls to a value characteristic of the active condition of the metal, oxygen ceases to be evolved, and formation of the film can again take place.

This interpretation, similar to several others, *assumes* (rather than *explains*) that a mixed electrode process cannot exist permanently under the conditions of the experiment. In other words, it is not clearly indicated *why* the rate of formation of the film cannot adjust itself to some steady value corresponding to the rate of its dissolution, the rest of the current producing free oxygen and the electrode remaining partially covered with the oxide film. This type of question was stressed by Wojtowicz et al.[42] in their discussion of periodic processes in anodic oxidation of organic substances at Pt electrodes. They pointed out that most oscillatory systems will arise when the rate equations for coupled pathways in the reaction have no (stable) steady-state solution.

According to Dmitriev and Rzhevskaya,[25] oscillations which occur during the anodic dissolution of copper in phosphoric acid electropolishing cannot be connected with formation of a phase film on the electrode surface. If this were so, passivation would be facilitated in more dilute acid and the time period during which the potential remained in the region corresponding to the passive state would be extended. In fact, the opposite is true. The limiting current density at which oscillations appear decreases with the increasing acid concentration and the oscillations lose their typical relaxation character (see p. 73). For the acid of density 1.72, the oscillations become almost harmonic.

The same case (copper electropolished in H_3PO_4) was interpreted by Pointu[26] in terms of charging and discharging of a condenser equivalent to Metal|CuO|viscous liquid layer.

Oscillations of current which arise during electropolishing of silver were also described by Francis and Colmer.[27] Formation and dissolution of films of either AgCN or Ag_2O (depending on current density) were again considered the basis of the effects.

Two alternative mechanisms which might possibly lead to the oscillations observed during anodic polarization of silver but in chloride solutions under potentiostatic or galvanostatic conditions were discussed by Lal et al.[23] The first is based on the observation that the rate of nucleation of AgCl rises sharply when supersaturation exceeds a certain critical value. It is assumed that nucleation does not occur at all when the AgCl concentration is below this value and that the overpotential, being a function of Ag^+ concentration, depends on the rate of nucleation, the "manner in which ions (Ag^+) disappear from the solution by deposition on the previously precipitated AgCl surface," and on the rate of growth of nuclei. Although it is claimed that the general shape of the periodic changes of the potential can be accounted for in this way, it was found that the overpotential was caused largely by ohmic resistance. Consequently, preference was given to the second mechanism, in which the mechanical properties of the AgCl film were regarded as responsible for the discontinuity of the behavior of the system. While the film grows, internal strains are built up and finally cause disruption of the film. Because of the increased facility of access of Cl^- ions to the electrode surface, the potential decreases (under galvanostatic conditions) or the current increases (under potentiostatic conditions). However, the buildup of the film soon starts again and the process will repeat itself. The observed linear relation between frequency of oscillations and current density was explained by assuming that the charge transferred during each cycle is equivalent to the amount of material in the film formed up to the moment of its rupture.

In the case of tin anodized galvanostatically in NaOH solution, Shams El Din and El Wahab[28] showed that a layer of $Sn(OH)_2$ is formed at comparatively low current density, while at a higher current density (the actual value depending on NaOH concentration), $Sn(OH)_4$ is produced causing total passivity. In both situations, the electrode acquires a well-defined equilibrium potential. At intermediate current densities, the potential oscillates between these two values. This was ascribed to "the coupling between the anodic formation and chemical dissolution of $Sn(OH)_4$."

Hackerman[29] has expressed the opinion that periodicity (at least in the case of passivated iron) is connected with adsorption processes rather than with the formation of a new phase. The observed frequencies are incompatible with the relative slowness of reactions with a solid phase.*

The case of pitting corrosion represents a clear example of heterogeneity effects (cf. Meunier[19,20]; see p. 52) where oscillations are often observed. For example, Piggott et al.[30] have described the behavior of titanium polarized anodically in formic acid. At the "breakdown potential" (when blister rupture and formation of pits occur) oscillations of potential appear, which increase in amplitude as the pits propagate into the metal. The periodic effects were

*The *initial* stages of passivation are to be regarded as an adsorption since only a monolayer or less of blocking O-species is involved.[28a,b]

ascribed to the changing ohmic drop in the pits, enhanced by the formation of bubbles of O_2 and CO_2, as well as by precipitation of titanium hydroxide. "Dispersal of these species into the bulk solution will result in a sudden decrease in the ohmic potential. It is evident that as the pits propagate into the metal, this effect will become more pronounced."[30]

Oscillations of potential have also been reported in the case of stainless steels.[31–33] On steel Fe16Cr undergoing corrosion in NaCl + H_2SO_4 under galvanostatic conditions, oscillations occurred only then when the concentration of SO_4^{2-} ions was considerably greater than that of Cl^- ions. Szklarska-Smialowska and Janik-Czachor[33] suggested an explanation based on the possible inhibiting action by SO_4^{2-} ions, which does not allow the potential to settle at a stable value (breakdown potential) to which the system tends normally at lower concentrations of SO_4^{2-}.

Oscillations are so inherently associated with pitting corrosion that they may even serve as a test of the susceptibility of a metal to pitting in a given solution. Thus, it has been suggested[31] that "the less an alloy is prone to pitting corrosion, the lower is the oscillation frequency and the more positive are the upper and lower limits of the potential variations. The absence of periodic variations of the potential shows that the alloy in the given medium is not susceptible to pitting corrosion. A method based on charging curves may consequently be employed to determine the tendency of alloys to suffer pitting corrosion and to evaluate the efficiency of inhibitors."

Interesting experiments on coupling of oscillating electrodes were performed by Franck and Meunier.[34] Two cobalt electrodes in HCl + CrO_3 solution would, in general, differ somewhat in the frequency of the oscillations of their potentials. When they are coupled, however (e.g., by establishing a connection through a low resistance), both frequencies become almost identical and nearly equal to the average of the two values exhibited previously. The difference (small, if the areas of the two electrodes are similar) is "equalized,"[34] i.e., after every few cycles an additional oscillation of the potential of one of them develops, bringing oscillations into phase. These effects are also observed when coupling is effected by means of a condenser, or simply by placing two electrodes sufficiently close to each other in the same solution. In the last case, however, the frequency after coupling is higher than either of the two original

II. Electrochemical Oscillations and their Qualitative Interpretation

frequencies. This is caused by a new path now available for the local currents. Thus, currents generated at, say, electrode A will now enter in part electrode B and flow through its body, where the resistance is lower than in the solution. As a result, the currents will increase, causing the state of different parts of electrodes to change more rapidly and the potential to oscillate with a higher frequency.

It was also demonstrated that if a completely passivated electrode is coupled with an active one (both independently stable in their respective states), either one of them, or both, begin to oscillate. An explanation of these phenomena was given[34] in terms of instability of mixed electrodes with a two-phase structure of the surface.

Synchronization of two different oscillations requires that a certain definite amount of charge is exchanged through the coupling element. If the period of oscillation is of insufficient duration, then during each cycle time, a certain deficiency of charge is built up. The deficiencies are additive until they reach a certain limiting value when "equalization" occurs by development of one or more additional oscillations.

Inhomogeneous distribution of the potential on oscillating electrodes of this type was demonstrated by direct measurements. Tips of two reference electrode capillaries were placed at some distance apart and the difference of potential between them $(E_2 - E_3)$ as well as the electrode potential (E_1) were measured. Results obtained in the case of Fe in H_2SO_4 + HCl are depicted in Fig. 2.

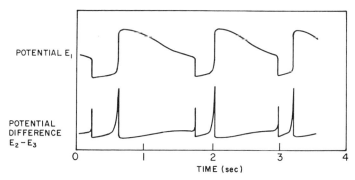

Figure 2. Inhomogeneous distribution of potential at an oscillating Fe electrode (in 3.16 N H_2SO_4 + 0.05 N HCl).[34]

The difference $E_2 - E_3$ was nonzero only then when the potential at point 2 was different from the potential at point 3, i.e., when the state of the electrode at these two points was different. The finite differences were observed only during these very short periods of time when the average (mixed) potential was passing from the one extreme value to the other.

Similarity to the well-known relaxation (see p. 73) oscillations in some electrical circuits is indicated; thus systems which exhibit relaxation oscillations operate, in principle, according to the so-called "all-or-nothing" rule so that, in general, they can oscillate only with one characteristic amplitude and do not possess properties which could lead to a change of amplitude in a continuous manner.

An entirely different explanation of periodicity was proposed by Indira et al.,[35] who associated oscillations of potential with changing structure of semiconducting films formed on the electrode surface. In the case of nickel covered with a nonporous, electronically nonconductive oxide film, an excess of cations is built up on the metal side of the film, while on the solution side (sulfuric acid), cations entering the solution leave negatively charged vacancies. The film is thus nonstoichiometric on both extreme planes, with a stoichiometric region (SR) in between. As the stoichiometric nickel oxide is a poor conductor, a large field exists in the SR:

> "When the field reaches a sufficiently high value, internal field emission occurs from the valence band of the nickel oxide on the solution side of the SR to the conduction band of nickel oxide on the metal side of the SR, across the SR. This causes a breakdown of the SR, forming highly conducting (because of high defect concentration) filaments. This brings down the potential drop across the SR and the overall potential drop. The SR then again begins to form. The repeated formation and breakdown of the SR is thus the main cause of oscillations."[35]

The potential of silver polarized anodically in 10 M HCl with 10 mA cm^{-2} exhibits irregular oscillations of high amplitude (over 5 V) and high frequency (100 kHz). It was again considered that the film has a three-zone structure, but as AgCl has a comparatively high electronic conductance, a good proportion of the associated electrons in the silver-ion-excess region is in the conduction band, making it an N-type conductor. Similarly, the associated holes in the silver-ion-deficient region make it P-type. In between, a badly conducting PN junction (PNJ) can be expected to be formed, resulting in a considerable potential drop across the PNJ domain.

When the field exceeds a certain limit, the PNJ fails by an avalanche type of breakdown and causes highly conducting filaments to form. This brings down the potential and the PNJ again builds up.

2. Anodic Oxidation of Nonmetallic Compounds

Oscillations have often been reported in connection with anodic oxidation of hydrogen or organic fuels.

Thalinger and Volmer[36] were the first to describe oscillations of current during oxidation of a nonmetallic substance, H_2, on a smooth Pt electrode. Oscillations occurred when the resistance in series with the cell was suitably adjusted. They were of constant amplitude, but their period decreased in time from several minutes to tenths of seconds. Under galvanostatic conditions, Armstrong and Butler[37] observed related oscillations of potential at bright Pt in dilute H_2SO_4. The potential oscillated between the equilibrium value and $+0.4$ to $+0.8$ V, depending on the activity of the particular electrode. It was stated that the potential remains near its equilibrium value as long as there is enough hydrogen available. When, however, hydrogen is being consumed "faster than it is replenished by diffusion," the potential rises to some higher value. At this stage, hydrogen previously "dissolved in the electrode" is oxidized. It was considered that this gave an opportunity for the molecular hydrogen to reach the electrode again by diffusion and so reestablish the reversible potential.

Similar behavior was described by Sawyer and Seo,[38] who obtained very regular oscillations in the case of Pt in 1 N HCl. Since they found that stirring had no effect on the oscillatory behavior, they proposed a different interpretation. At the higher potentials, a surface oxide is formed. Simultaneously occurring reduction by dissolved hydrogen brings the surface back to its active, oxide-free state. Now, oxidation of molecular hydrogen can readily occur at (or nearly at) reversible potential. However, as continued exposure to hydrogen takes place, the electrode becomes deactivated and the potential again becomes positive. The phenomenon continues to repeat itself. Undoubtedly, the formation and removal of surface oxide at Pt is an important process in the oscillation and similar behavior was found by Wojtowicz et al.,[42] who directly determined the changes in surface oxide coverage during oscillatory oxidation of HCOOH.

Many cases of oscillatory behavior in the anodic oxidation of organic substances are now known. The first observations of this kind seem to be those of Müller,[39–41] who investigated the oxidation of formic acid under galvanostatic and "quasipotentiostatic" conditions. His interpretations of the periodic effects were based on the assumption that the current could be maintained at a low potential as long as there was adsorbed hydrogen available on the electrode surface.

Oscillations of current and of the potential during anodic oxidation of sodium formate were described by Wojtowicz et al.[42] and a tentative explanation of the result, in terms of mathematical models, is discussed on p. 88. In this work, an attempt was made to examine the intrinsic reasons for oscillatory behavior in electrochemical reactions in terms of feedback effects in the reaction sequence and thus to answer the basic question why no steady state is attained. The importance of autocatalytic effects in periodicity, i.e., the role of chemical positive feedback,[42,80] was first pointed out by Bonhoeffer.[74]

The origin of oscillations of potential which develop at sufficiently high current densities during oxidation of methanol on platinum were treated by Pavela.[43] At the lowest value of potential (0.69 V), the quantity of methanol at the electrode is small and, as it is readily oxidized, the potential quickly rises to the value (0.87 V) at which the initial stage (PtOH) of surface oxidation of Pt commences. Reaction of this oxide with methanol yields formaldehyde and formic acid. It was stated that oxidation of these substances gives methanol time to diffuse to the electrode, and this leads to a periodic process. It was estimated that the charge transferred in one cycle constituted only $\sim 1\%$ of the charge corresponding to full occupation of the electrode surface by methanol. This result, it was claimed, showed that during one oscillation period, only a fraction of the electrode surface takes part in the reaction. This is in agreement with the above explanation, since the diffusion front of methanol cannot have time to enter so-called[43] pores of the electrode surface.

Buck and Griffith,[44] who investigated the oscillations accompanying the oxidation of methanol, formaldehyde, formic acid, and other organic fuels, followed Sawyer and Seo's theory.[38] Criticizing another tentative explanation based on the existence of two different paths in the process (direct oxidation at low potentials, and oxidation via platinum surface oxide at higher potentials), they directed attention to the problem examined by Wojtowicz et al.[42] that whether or not the chemical reaction of the fuel with Pt surface oxide is slower than diffusion, the processes ought to seek a steady state at some intermediate potential as in any other type of reaction. It is difficult to see why a similar criticism cannot be equally well applied to the mechanism of Sawyer and Seo itself.

Oscillations of potential during oxidation of formaldehyde on Pt were also observed by Shropshire,[45a] who suggested an interpretation based on his theory of carbonaceous fuel oxidation. According to this view, two kinds of sites exist on the electrode surface, differing in their ability to adsorb water and fuel molecules. Shropshire's argument can be reduced to the following: on sites of type A, adsorption of water and formation of electrosorbed oxygen occur at a lower potential than on sites of type B. Because the adsorption of fuel molecules on sites A proceeds at a lower rate than on sites B, a bifunctional oxidation is possible and is primarily effected through[45b] a reaction between adsorbed oxygen on A and fuel molecules on B, i.e., at a comparatively low potential (galvanostatic conditions). However, sites A become progressively blocked by the intermediates, which can desorb sufficiently fast only at higher potentials. To keep the current constant and render the deposition of oxygen on sites B possible, the potential must rise. Desorption of intermediates can now take place. Soon, the conditions are restored under which the process can again proceed at lower potentials.

Here again, no clear reason is proposed why some stable steady-state with the appropriate coverages on both kinds of sites is not attained.

Further important work was carried out in the anodic oxidation of formaldehyde by Hunger[46a] under conditions where potential oscillations arose in galvanostatic oxidation in 3.5 M H_2SO_4 at smooth and platinized Pt electrodes. No essential differences in the behavior of the two kinds of electrodes were observed, however, if the real area of the surfaces was taken into account. A typical shape of a potential wave is shown in Fig. 1(f). Two lines can be drawn[46] which border the oscillation region and are the loci of the peaks of the potential waves over a range of current densities.

Estimation of the charge corresponding to one cycle showed[46] that, independently of the current imposed, 3/4 of it is transferred in the lapse of time $T_{3/4}$ between the moments when the potential passes successively through the same value at which the oscillations began (marked as $V_{3/4}$ in Fig. 1(f)). This observation suggests[46] the following interpretation of the occurrence of the oscillations. The scheme shown below is assumed, where A and B represent schematic intermediates:

$$CH_2(OH)_2(\text{sol'n}) \rightarrow CH_2(OH)_2(\text{ads}) \rightarrow \begin{array}{c} A + H^+ + e^- \\ \downarrow \\ B + H^+ + e^- \\ \downarrow \\ HCOO + H^+ + e^- \\ \downarrow \\ CO_2 + H^+ + e^- \end{array}$$

Decomposition of the HCOO radical is assumed to be a comparatively slow reaction, so that for currents exceeding a certain limiting value (corresponding to $V_{3/4}$), the surface concentration of HCOO builds up to an extreme value, when a rapid catalytic decomposition is said to occur,* CO_2 is evolved, and H atoms left on the surface are easily oxidized. Thus, starting from $V_{3/4}$, the potential rises (inhibition by HCOO), and then drops to low values (oxidation of H), where adsorption of formaldehyde is enhanced. As soon as hydrogen has been oxidized, the surface concentration of HCOO begins to build up again, at first slowly, then at an increasing rate, so that the potential rises again.

Measurements of the open-circuit potential of electrodes with preadsorbed formaldehyde, immersed in H_2SO_4, have shown that oscillations do not arise if the only formaldehyde available is that adsorbed on the electrode surface. In fact, at least 428 $\mu C\ cm^{-2}$ is needed for oscillation to occur, while saturation of a smooth Pt surface with formaldehyde corresponds to 182 $\mu C\ cm^{-2}$. It was concluded that during the oscillation itself, adsorption of the reagent from the solution must occur (cf. Ref. 42).

The involvement of chemisorbed oxygen in the generation of oscillations is also recognized. Increasing oxygen coverage at higher potentials is considered responsible for the sharp increase of charge corresponding to one cycle, which is observed with larger currents.

*Cf. the delayed gas evolution reported by Conway and Dzieciuch[46b] for decomposition of HCOOH at Pd.

Current oscillations during *potentiostatic* oxidation of ethylene at a rotating-disc electrode were investigated by Wojtowicz et al.[42] and found to be influenced by mass transport conditions. Two attempts were made at their explanation in terms of different mechanistic models. These are discussed on pp. 89 and 90.

3. Oscillations in Cathodic Processes

Although the majority of the observed cases of electrochemical oscillations originate in anodic processes, several examples of cathodic oscillating processes have also been reported.

One case of oscillations arising through a combined action of cathodic polarization and chemical oxidation (Fe in HNO_3) has already been mentioned (p. 51) and oscillations of current observed at zinc in NaOH at -1.1 V were depicted in Fig. 1. A few early observations (including oscillations during the reduction of nitrobenzene[47]) are described in the useful book by Hedges and Myers.[7a]

Oscillations which arise in the polarographic reduction of anions $S_2O_8^{2-}$, $Fe(CN)_6^{3-}$, $Pt(Cl_4)^{2-}$, and CrO_4^{2-} were investigated by Frumkin et al.[48,49] They have a typical relaxation character and the necessary condition for their generation was a high resistance (several kΩ) in series with the cell. As De Levie emphasizes,[50] "In all these cases, the current–voltage curves exhibit a *region of negative resistance* due to the double-layer repulsion, and the oscillations occur precisely in that region." De Levie himself described oscillations encountered during the reduction of In^{3+} in NaSCN on mercury. This reaction served as an illustration to his theoretical treatment of electrochemical oscillations (see p. 107). The general significance of electrical negative-resistance characteristics in electrochemical oscillation due to passivation or self-inhibition[30a] of the reaction was pointed out by Wojtowicz and Conway.[80]

Potential oscillations of a cathode in an acid tin plating bath were observed by Clarke and Bernie,[51] who pointed to the similarity between their polarization curve and curves normally found in cases of anodic passivation which lead to oscillations. *Cathodic* passivity was tentatively explained as follows. The addition agent to the bath forms a soluble ion by proton addition to its amino-nitrogen group. The deposition potential of tin in the solution used is -0.192 V. At somewhat lower (more negative) potentials, the ions of the brightener addition agent adsorb strongly, transport of Sn^{2+} ions becomes difficult, and the current, after attaining a maximum value at -0.250 V, begins to fall when the potential is made still more cathodic. (The alternative process of H_2 evolution is also difficult because of its high overvoltage on tin.) Finally, the current becomes almost independent of potential as the film thickness increases so that field within it does not change. The electrode then becomes completely passivated. Cathodic passivity also occurs in some electroorganic reductions where an intermediate or product of the reaction becomes strongly adsorbed, e.g., in the reduction of benzophenone.

Irregular fluctuations of potential (of the order of a few millivolts) during cathodic polarization of a platinized Pt electrode in 1 N HCl have been observed when H_2 was bubbled through the solution.[52] The effect was connected with a rate-limiting diffusion of H_2 molecules from the interface. Stirring resulted in diminished surface concentration of H atoms and decreased polarization. In the case of a palladized Pd electrode, the effect was barely noticeable. This was explained by the ability of palladium to act as a "sink" for hydrogen atoms and thus exert a buffering influence on any changes of H concentration.

Oscillations are not an exclusive property of externally polarized electrodes. The rest potential of an electrode on which cathodic and anodic processes simultaneously occur (or at which a chemical reaction by a component of the solution can take place) may also exhibit oscillations. Naturally, these can never be truly sustained oscillations, although damping may be very weak and practically unobservable during a short experiment. The following systems may be noted in which oscillations of the rest potential have been observed: copper in $HNO_3 + HCl$,[7a] cobalt in $CrO_3 + HCl$,[34,53], iron in $CrO_3 + H_2SO_4$,[54] iron in $H_2SO_4 + HCl$,[34] and platinum in $N_2H_4 + KOH$.[55]

4. Periodic Behavior in Nonaqueous Solutions

Only a few cases of oscillations have been reported in nonaqueous electrolytes. Galvanostatic anodization of lead in 1 M $LiAlCl_4$–propylene carbonate solution produces[56] periodic oscillation in voltage, followed by passivation. Damped oscillations of potential were observed in the case of palladium in 1 M HCOOK in anhydrous formic acid.[57] Degn[58] mentions unpublished experiments on oxidation of H_2 on platinum in acetonitrile saturated with HCl, during which oscillations were observed. Passivation of iron in solutions of HCl in dimethylsulfoxide was investigated by Posadas et al.,[59] who found in transient experiments at a constant c.d. a periodic fluctuation of the electrode potential when the overvoltage exceeded 1.5 V. This effect was independent of both the HCl concentration and any gas saturation in the solution.

5. Electroosmotic Oscillations

Oscillations of potential difference which are not connected with any electrode process but rather with electroosmotic transport phenomena have been discovered recently. Teorell[60] observed that when a constant current was passed through two KCl solutions of different concentrations, separated by a porous glass membrane, periodic changes of the potential difference across the membrane were developed. Interaction between ionic and volume flows were suggested as the reason for the oscillations. Shashua[61] described oscillations of potential difference across a polyelectrolyte double membrane under constant-current conditions. Oscillations were of a rather high frequency (20 Hz), with sharply defined spikes of short (1 msec) duration. This case was treated theoretically by Katchalsky and Spangler[62] (see p. 94).

III. MODELS OF ELECTROCHEMICAL OSCILLATORS

1. Mathematical Fundamentals

As has been indicated above, only in exceptional cases can a mechanism be declared oscillatory through simple common sense or *a priori* reasoning. A purely descriptive "explanation" may be in essence correct, but the proof that it can, in fact, lead to oscillations can be obtained only by investigation of the properties of a corresponding mathematical model. The usual difficulty with any qualitative explanations is that they rarely give a satisfactory basis for understanding why a combination of various coupled processes should not give rise to a steady-state situation in the process or reaction.

The necessity for reference to mathematical models is directly related to the fact that differential equations (derived on the basis of some mechanistic assumptions) which may have *periodic solutions* under certain conditions, can lose this property if appropriate relations concerning the constants involved are not fulfilled.

Sustained oscillations in real physical systems cannot be described by linear differential equations. Harmonic oscillation can very seldom be accepted as an adequate approximation for the behavior of a real system. In general, mathematical models which represent real oscillators are *nonlinear*.

The theory of nonlinear oscillations constitutes an important and constantly expanding branch of applied mathematics. Any attempt to give in this text a review, however sketchy, of various qualitative and quantitative methods by which nonlinear problems are treated is, of course, neither appropriate nor possible. The reader is referred to the many special texts available, of which only a few are quoted here.[63-69] An excellent review of the literature of the subject is given in an appendix to one of them[63] by its editor.

However, for readers who are relatively unfamiliar with the subject, it may prove useful if certain concepts and fundamental results of the qualitative (topological) approach to problems of stability and oscillations are presented here, naturally without any claim to completeness or rigor. In this way, at least, certain terms used in the text will not remain unexplained (see also the appendix).

(i) Autonomous Systems: Phase-Plane Representation

The following set of first-order, ordinary differential equations which do not explicitly contain an independent variable is called "autonomous":

$$\dot{x}_i = f(x_1, x_2, \ldots, x_i) \tag{1}$$

where $\dot{x}_i = dx_i/dt$. Equations of this class are normally found in chemical kinetics if the reaction can be considered as independent of mass-transport conditions. Differential equations of the second order

$$f(x, \dot{x}, \ddot{x}) = 0 \tag{2}$$

which are common in problems connected with electrical circuits can be converted into an equivalent set of two autonomous equations by putting $\dot{x} = y$.

III. Models of Electrochemical Oscillators

The discussion will be restricted to the set of two autonomous equations

$$\dot{x} = P(x, y), \qquad \dot{y} = Q(x, y) \qquad (3)$$

Extension of the results to sets of more than two equations is possible, though difficult, but application of simple topological methods is then no longer feasible.

If functions P and Q are not linear, it is seldom possible to obtain solutions by analytical methods. Apart from numerical and analog methods, which are obviously unsuitable as a basis for any general discussion, various approximate methods of solving nonlinear equations have been developed which can be used with varying degrees of difficulty or success in different cases.

However, the general character of solutions of equations (3) can be ascertained by investigating the behavior of a point $R[x(t), y(t)]$, which on the plane xy, or "phase-plane," represents the actual time-dependent state of the system (3). The trajectories traced by R are defined by

$$dy/dx = P(x, y)/Q(x, y) \qquad (4)$$

The solution of this relation may again present difficulties, but a good "phase portrait" can be obtained using the method of isoclines (e.g., see Refs. 63, 64). A selected point (x, y) is marked on the plane, the value of the derivative dy/dx at this point is calculated from equation (4), and a short, straight sector of this slope is then drawn. When this is repeated for a sufficiently large number of points, trajectories which are tangential to the sectors can be traced quite accurately.

A closed trajectory of R represents a (sustained) periodic solution of equation (3). If it has a form of a spiral, oscillations are either damped or their amplitude increases indefinitely.

When the functions $P(x, y)$ and $Q(x, y)$ are linear, the shapes of the trajectories are determined by the type of the equilibrium point* (the so-called singular point or singularity) i.e., the point

---

*This term, which will be used frequently in the material to follow, corresponds to the steady-state condition in chemical or electrochemical kinetics. Its usage as a mathematical term (see the appendix) must not be confused with that in equilibrium chemical thermodynamics, where a true equilibrium rather than a special kinetic condition is involved.

where $\dot{x}$ and $\dot{y}$ vanish. In order to investigate the character of a singularity, the coordinates are translated so that the origin lies at the equilibrium point: $\bar{x} = x - x_0$, $\bar{y} = y - y_0$, where x_0 and y_0 are the solutions of $P(x, y) = 0$, $Q(x, y) = 0$ and $\bar{x}, \bar{y}$ denote variables in the new coordinate system.

Let the transformed equations be

$$\dot{\bar{x}} = A\bar{x} + B\bar{y}, \qquad \dot{\bar{y}} = C\bar{x} + D\bar{y} \tag{5}$$

where A, B, C, and D are constants determined by the coefficients of the original equation. The character of the equilibrium point is determined by the properties of the roots λ_1 and λ_2 of the charac-

Table 1

Property of roots	Character of equilibrium point	Typical shape of trajectories
Real $\lambda_1 \leq \lambda_2 \leq 0$	Stable node	
$0 < \lambda_1 \leq \lambda_2$	Unstable node	
$\lambda_1 < 0 < \lambda_2$	Saddle	
Complex $\mathrm{Re}(\lambda) < 0$	Stable focus	
$\mathrm{Re}(\lambda) > 0$	Unstable focus	
$\mathrm{Re}(\lambda) = 0$	Center	

III. Models of Electrochemical Oscillators

teristic equation

$$\begin{vmatrix} A - \lambda & C \\ B & D - \lambda \end{vmatrix} = 0 \quad (6)$$

By putting $p = -(A + D)$ and $q = AD - BC$, we have

$$\lambda^2 + p\lambda + q = 0 \quad (7)$$

The various different cases which can arise are listed in Table I and conditions for p and q, corresponding to each case, are shown in Fig. 3.

In the case of linear systems, the general character of the phase portrait, i.e., relations in the phase plane, as determined by the type of the equilibrium point, is maintained throughout the whole xy plane. This does not necessarily apply to nonlinear systems and, in general, their phase portraits are much more complicated.

Useful information about a *non*linear system can be obtained by considering a corresponding linear system. According to Lyapunov's theorem, an equilibrium point of a nonlinear system, and an equilibrium point of a system obtained by its linearization

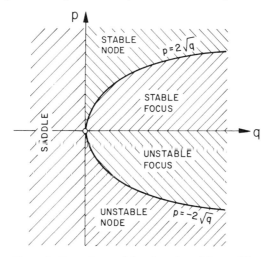

Figure 3. Dependence of the character of the equilibrium point of an autonomous system on parameters p and q.

around the point in question, are of the same type. Consequently, in the vicinity of equilibrium, the trajectories of the nonlinear system will closely resemble those of its corresponding linear form. One possible exception is a center; when a linearized system has a center, the equilibrium point of the original system can be either a center or a focus (see the appendix).

The above method does not form a sufficient basis for determining the shapes of trajectories outside the regions close to equilibrium points. The behavior of a representative point R "in the large," i.e., well away from equilibrium, is of fundamental importance in ascertaining whether a given system has periodic solutions or not. Contrary to the case of linear systems, which generate sustained oscillations only if their singularities are centers, nonlinear systems can develop configurations known as "limit cycles."

A limit cycle is a closed trajectory such that no trajectory in the close neighborhood of it is also closed. Every trajectory from the surrounding regions either winds itself upon the limit cycle as $t \to \infty$ (a stable limit cycle), or unwinds from it (an unstable limit cycle). Examples of limit cycles are shown in Fig. 4. It is clear that limit cycles represent sustained oscillations which are determined by the differential equations themselves, and are independent of the initial conditions (contrary to the oscillations around a center).

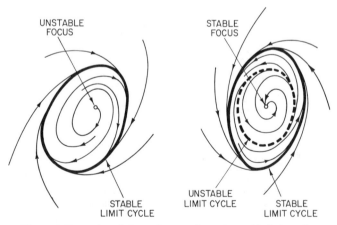

Figure 4. Examples of phase-plane configurations with limit cycles.

III. Models of Electrochemical Oscillators

A sufficient condition for the existence of a limit cycle is given by the Poincaré–Bendixson theorem, which may be summarized as follows.

(a) If a trajectory C remains in a closed domain D without approaching any singularity, then C is either a closed trajectory, or tends to a closed trajectory.

(b) In the case of a ring domain D bounded by two concentric circles C_1 and C_2 (Fig. 5), for a closed trajectory to exist in D, it suffices that: (i) through every point of C_1 and C_2 there enters a trajectory into D (or is leaving it): (ii) no singularities exist either in D or on C_1 or C_2.

Although the Poincaré–Bendixson theorem is almost self-evident when one refers to a hydrodynamic analogy of sinks or sources and flow lines, its practical application is limited by the usually considerable difficulties of establishing a D which would fulfill the requirements of the theorem. No generally applicable method exists and each particular case has to be treated individually.

Certain necessary conditions have also been formulated and their application is comparatively easy. In this way, a proof may be obtained that in a given nonlinear system (nonconservative, i.e., one which has no center), sustained oscillations are impossible.

The Poincaré–Bendixson theorem states that, for a limit cycle to exist, the condition

$$N + F + C - S = 1 \qquad (8)$$

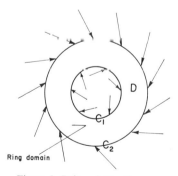

Figure 5. Poincaré–Bendixson theorem.

must be fulfilled, where N, F, C, and S denote, respectively, the number of nodes, foci, centers, and saddle points inside the limit cycle. This theorem does not specify any requirements as to the stability of the singular points. However, it is evident that if, in the region bounded by a limit cycle, there is only one singular point (which is the most common case), then if the cycle is to be stable, it must be either an unstable focus, or an unstable node, unless inside the boundary an unstable cycle exists.

The following two negative criteria are due to Bendixson and to Dulac:

(i) If the expression $\partial P/\partial x + \partial Q/\partial y$ does not change its sign in certain regions of the xy plane, then no closed trajectories can exist in this region.

(ii) [A generalization of (i)]: If a function $B(x, y)$, continuous and with continuous derivatives, exists such that in a certain region of the xy plane the expression $\partial(BP)/\partial x + \partial(BQ)/\partial y$ does not change its sign, then no closed trajectories can exist in this region.

It may sometimes be convenient to use two other negative criteria given by Higgins[70] as follows.

First Higgins' Criterion. If a closed trajectory lies in a region of uniform character in which one, and only one, singular point exists, then the following conditions must obtain:

$$\partial P/\partial y \times \partial Q/\partial x < \partial P/\partial x \times \partial Q/\partial y < 0 \qquad (9)$$

(A region of uniform character is defined as a domain where none of the partial derivatives, $\partial P/\partial x$, etc., changes sign.)

Second Higgins' Criterion. (a) In the region bounded by a closed trajectory, subregions must exist where

$$\partial P/\partial x \times \partial Q/\partial y < 0 \qquad (10)$$

(b) A closed trajectory must pass through regions where

$$\partial P/\partial y \times \partial Q/\partial x < 0 \qquad (11)$$

The above criteria have straightforward and important interpretations in terms of chemical kinetics. For instance, $\partial P/\partial x = \partial \dot{x}/\partial x > 0$ denotes that net production of x is *autocatalytic*, and $\partial Q/\partial x = \partial \dot{y}/\partial x < 0$ means that net production of y is *inhibited* by x.

III. Models of Electrochemical Oscillators

(ii) *Quasidiscontinuous, or "Relaxation," Oscillations*[63]

Excluding the exceptional case of a system with a center, sustained oscillations which can be fully described by analytical functions can occur only when a stable limit cycle is developed on the phase plane. However, many physical systems exhibit, under certain conditions, periodic behavior which cannot be described by continuous and differentiable functions. These so-called relaxation, or discontinuous, oscillations are the most commonly encountered kind of oscillations in electrochemical, as well as in other heterogeneous systems. In their phase-plane representation, they are characterized by very fast movements of the representative point on some sections of the cycle.

Frank-Kamenetskii[71] differentiates between kinetic oscillations and oscillations determined by mass conservation laws, which he calls "trivial." The following example of the trivial type of oscillations is given in regard to combustion processes: "If a lower inflammation limit with respect to concentration exists, the admission of a reactant mixture to a reaction vessel can lead to inflammation when the critical concentration is reached. If after this, the reaction goes to completion sufficiently rapidly, the further supply of initial reactants to the vessel may lead to a series of successive explosions." Another case of oscillations which, by the same token, could also be classified as "trivial" occurs when a diffusion-limited current fluctuates in an irregular manner because of turbulence effects. This kind of oscillation was described by Vetter[72] (for reduction of iodine) and by Hull and Lewis[73] (for the case of evolution of hydrogen). Typical processes of the "trivial" kind would be oscillations connected with the periodic growth and breakdown of a passivating layer, when the internal stresses which were progressively building up in the growth period exceed a certain sharply defined limit.[23]

However, several cases of relaxation oscillations are known which cannot be accounted for in a similarly simple way, and it is evident that the inherent properties of the kinetics of the system, as well as conditions of transport processes and electrical characteristics of the elements of the circuit, are involved in the generation of kinetic periodicity.

Different ways of approach to the problem of modeling relaxation oscillations have been proposed. They are treated extensively in two of the texts quoted.[63,64] The fundamental concepts of only one of them will be presented here.

The "discontinuous theory of relaxation oscillations" is a method best suited for practical application. Its basic assumption is that the velocity of the representative point on certain sections of its path is infinitely great. In other words, R jumps from one analytical section of its trajectory to another, instead of moving fast, but according to some law (i.e., a differential equation) which, being unknown, is replaced by a discontinuity. The discontinuous theory is closely related to the theory of degeneration of differential equations. The common point in both is the recognition of the fact that, in constructing a mathematical model of the real physical system, one normally neglects, either intentionally or not, some of its features which, it may reasonably be assumed, are not essential for the main purpose of the treatment. For example,[64] in the description of the behavior of an RC circuit with constant emf, one can neglect the very small parasitic inductance of the leads. Thus, instead of the full equation,

$$L\ddot{q} + R\dot{q} + (1/C)q = E \qquad (12)$$

where q is the charge and other symbols represent, as usual, inductance, resistance, and capacitance, it is possible to deal with the degenerate system

$$R\dot{q} + (1/C)q = E \qquad (13)$$

This procedure is admissible so long as we are not interested in what happens at the moment when the emf is switched on. The original (full) equation has, of course, two initial conditions. If the condenser is not charged before the source of the emf is connected, they are $q_0 = 0$ and $\dot{q}_0 = 0$. However, the degenerate equation has only one integration constant and one initial condition $q = 0$. Now, by differentiating its solution, which is $q = EC[1 - \exp(-t/RC)]$, it is found that for $t = 0$, there results $\dot{q}_0 = E/R$. This means simply that if the degenerate equation is to represent correctly the behavior of the circuit at all times, we must introduce "artificially" an additional condition, not contained in the equation itself, that at time $t = 0$, there occurs a discontinuous change of charge.

A classical example[63] of relaxation oscillations according to the formulation of Minorsky[64] may be quoted. A perfectly elastic

III. Models of Electrochemical Oscillators

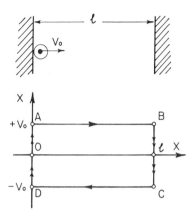

Figure 6. Elastic rebound as a case of relaxation oscillation.

ball, rolling on a frictionless horizontal plane between two equally elastic walls, is considered. It is assumed that the mass of the walls is infinite and the direction of motion is at right angles to the walls (Fig. 6a). The problem is considered from the standpoint of the classical theory of elastic collisions. The ball strikes the wall with velocity $+v_0$, rebounds from it with an equal and opposite velocity $-v_0$, then strikes again the other wall, and so on. There exists thus a periodic process represented, as usual, by a phase-plane diagram (Fig. 6b). Along the pathways AB and CD, the representative point moves according to the equation $\ddot{x} = 0$ with constant, finite velocity; however, on pathways BC and DA, its acceleration is infinite. In other words, what happens during the very short time when the ball is in contact with the wall is intentionally ignored, and the elastic shock is merely specified by a relation between the conditions that exist immediately *before* it and immediately *after* it. These conditions result from the principles of conservation of momentum and kinetic energy, but not from the differential equation of the motion itself.

Relaxation oscillations are thus described on the xy plane (or, in general, in the n-dimensional space) by cycles composed of analytical arcs and discontinuous jumps. Relaxation cycles do not, however, possess the properties of limit cycles, and the conditions under which limit cycles may be generated do not apply here. In particular, it is not necessary that in the region bounded by a relaxation cycle, singular points exist which fulfill Poincaré's theorem.

Instead, the discontinuous type of theory introduces the so-called Mandelstam condition. It states that only such jumps are allowed that do not demand a discontinuous change of the energy stored in the system. The violation of this condition would need an infinitely large source of power, which, of course, is not possible. For instance, in the case of the degenerate RC circuit discussed above, the potential must remain invariant during a jump because the energy stored is $(1/2)CV^2$. By the same principle, in an RL circuit, the current must remain constant during the jump.

The full portrait of a phase plane with a relaxation cycle is discussed in detail in Andronov's book.[63] It is presented here in a greatly abbreviated form in order to provide a basis for some cases in the subsequent discussion.

Suppose that an "accurate" model of a real system is represented by

$$\dot{x} = G(x, y), \qquad \mu \dot{y} = F(x, y) \qquad (14)$$

where μ denotes a small parameter which we want to disregard in the approximate treatment. The simplified, degenerate model will thus be represented by

$$\dot{x} = G(x, y), \qquad 0 = F(x, y) \qquad (15)$$

The model can be considered adequate if substitution of the motion of the representative point in a small neighborhood of the curve $F(x, y) = 0$ by its motion on the curve does not change the properties of the model to any significant extent.

Consider the phase portrait of the original system. Outside the curve $F(x, y) = 0$, but within its neighborhood to the order of μ^a, $0 < a < 1$, the condition $|F(x, y)| > \mu^a$ applies. It follows that $|\dot{y}| > \mu^{a-1}$, which means that because μ is very small, the velocity of the representative point will be very large, and in the limit,

$$|\dot{y}| \xrightarrow[\mu \to 0]{} \infty$$

By writing $dy/dx = F(x, y)/\mu G(x, y)$, it is seen that if $F(x, y)$ is not zero (i.e., for points outside the $F = 0$ curve),

$$dy/dx \xrightarrow[\mu \to 0]{} \infty \qquad (16)$$

III. Models of Electrochemical Oscillators

results. Trajectories of the representative point outside the close neighborhood of the curve $F(x, y) = 0$ will thus be nearly straight and almost vertical. They will be traversed very fast, at velocities which will be greater the closer is the point R situated with respect to $F = 0$ at a given moment. In the limit, these "fast trajectories" will appear as vertical straight lines on which R will be moving with "infinitely" large velocity.

The direction of these fast motions can be found as follows. The curve $F = 0$ can be considered as the loci of equilibrium points of the fast motions. If a point is stable, a corresponding trajectory will terminate on it, and vice versa, i.e., originate from it. Introducing a new scale of time, defined as $\bar{t} = t/\mu$, the approximate equations of the fast motions of R are

$$dy/d\bar{t} = F(x_0, y), \qquad x = x_0 \, (=\text{const}) \tag{17}$$

Linearizing the first equation around a selected point (x_0, y_0), we obtain

$$dy/d\bar{t} = (\partial F/\partial y)(y - y_0) \tag{18}$$

Point (x_0, y_0) is a stable equilibrium point of the motion on $x = x_0$ if, for $y - y_0 > 0$, there is $dy/d\bar{t} < 0$, and for $y - y_0 < 0$, there is $dy/d\bar{t} > 0$. Hence a condition for (x_0, y_0) to be a stable equilibrium point is

$$\partial F/\partial y < 0 \tag{19}$$

The phase portrait is thus a continuum of "fast" vertical trajectories and a curve $F(x, y) = 0$, on which R moves "slowly." Fast trajectories are directed toward those sections of the curve where $\partial F/\partial y < 0$, and depart from those sections for which $\partial F/\partial y > 0$.

An example is given in Fig. 7. *ABCD* is a relaxation cycle. It is composed of two analytical sections *AB* and *CD*, and two discontinuous jumps *BC* and *DA*. If, by the application of some external force (or impulse), point R is shifted from F to, say, M, it will jump back on the curve (M') as soon as the force is removed, and will continue its movement around the cycle.

2. Kinetic or "Chemical" Models

To investigate oscillations arising in an electrical circuit, one has, in general, to take into consideration the properties of all the

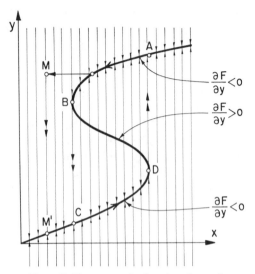

Figure 7. Phase portrait of a relaxation cycle.

elements of the circuit. Circuits involving electrochemical cells are, of course, no exception. It is therefore strictly incorrect, in principle, to speak of a single electrode uniquely as the site where oscillations are generated.

However, if the appropriate restrictions are imposed on the rest of the circuit, the problem can be reduced in the sense that only the characteristics of the cell, or even of one particular electrode, will in fact determine whether oscillations can be generated. This attitude is taken by several authors, although sometimes this is not stated explicitly.

Models of oscillating systems are often based on a set of two autonomous differential equations. Although it is difficult to visualize an electrochemical process the transient behavior of which could be exactly described by only two time-dependent variables, such a treatment is admissible so long as reasonably sufficient evidence exists that certain stages of the process can be considered as being always in equilibrium. In any case, sets of more than two nonlinear equations become unmanageable as far as any general discussion is concerned, and can be investigated only by assigning definite numerical values to their constants.

III. Models of Electrochemical Oscillators

The first attempt at elucidation of an electrochemical oscillating process in terms of a mathematical model is due to Bonhoeffer,[74a] and gives useful insight into the origin of oscillations in a simple case. This work, although concerned with a generalized chemical reaction A → B, with two intermediates X and Y, forms the basis for later papers by this author and his collaborators in which specific electrochemical processes are discussed.

It is assumed that concentrations of X and Y (denoted by x and y) are very small compared to concentrations of A and B (in other words, there is an inexhaustible supply of A, and B has no effect on the course of reaction). Hence, the kinetics are described by two equations

$$\dot{x} = f(x, y), \qquad \dot{y} = g(x, y) \tag{20}$$

Functions f and g are considered as composed each of two terms: one, P, which expresses the rate of the reaction steps responsible for production of X (or Y), and the other, R, which represents the rate of the steps in which a given intermediate is converted into the other one or into B. We thus have

$$\dot{x} = P_x(x, y) - R_x(x, y), \qquad \dot{y} = P_y(x, y) - R_y(x, y) \tag{21}$$

Let it be assumed[74a] that at a certain point (x_1, y_1), the intermediate X begins to be formed ($P_x > 0$) but does not undergo any decomposition, while y remains at a constant value. When x becomes increased to x_2, Y begins to form ($P_y > 0$), but its decomposition rate remains zero, and x does not change. After y has risen to y_2, x begins to decrease ($R_x > 0$), with $P_x = 0$ and $y = y_2$. Finally, at $x = x_1$, decomposition of Y takes over ($R_y > 0$), and the cycle is closed when y has dropped to y_1. The sequence of events is illustrated in Fig. 8(a). If the rates of all elementary processes are of the same order, functions $x = x(t)$ and $y = y(t)$ will have the shape shown in Fig. 8(b).

In spite of the simplified argument, this work constituted a step in the right direction in so far as the usefulness of the phase-plane representation and the importance of the concept of the limit cycle were recognized (although a Volterra[3a]–Lotka[74b] model was incorrectly given as an example of a system which develops a limit cycle). The author also pointed out that instability of the equilibrium

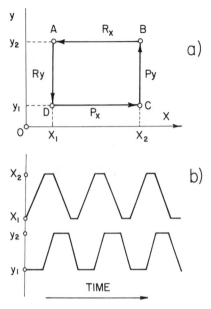

Figure 8. Bonhoeffer's treatment of a system of two reactions.

point, as a necessary condition for a limit cycle, may be realized in the presence of some *autocatalytic effect*.

The method of arriving at the approximate shape of the functions $f(x, y)$ and $g(x, y)$, by considering separately the reactions of type P and R (and treating each variable in turn as a parameter), was used to explain the oscillations observed during cathodic polarization of iron in HNO_3,[16] and during the anodic polarization of copper in HCl.[17] Feasibility of a cyclic kinetic configuration (limit cycle oscillation) in both systems was demonstrated. We shall confine ourselves to presenting the essential points of the argument used in the case of processes at copper.

Comprehensive experimental investigation of the electrochemical reactions in this case led the authors to the following picture. In the "active" state of the electrode, the reaction $Cu \rightarrow Cu^+ + e$ is dominant. As soon as the solubility product of CuCl has been locally reached, the chloride begins to precipitate, forming a porous layer on the electrode surface. The high local

III. Models of Electrochemical Oscillators

current density leads to increase of alkalinity of the electrolyte in the pores (high rate of H^+ migration) and creates conditions favorable to the precipitation of Cu_2O. This brings about a further increase of current density and precipitation of Cu_2O is accelerated. Finally, the potential in the pores attains the value at which Cu^{2+} ions may be formed. This is the ultimate situation in the "passive" state. In the meantime, the rate of formation of CuCl has substantially decreased, becoming lower than the rate of its dissolution (determined by the rate of diffusion of Cu^+ ions to the bulk of solution). As the chloride gradually disappears, conditions which induced the formation of Cu_2O cease to exist and it also begins to dissolve, at an accelerating rate. It is thus seen that both the formation and dissolution of the oxide can be considered as "autocatalytic." On the basis of this mechanism, the authors constructed curves showing the probable dependence of the rates of formation of Cu_2O ($\dot{x}$) and CuCl ($\dot{y}$) on x and y, respectively, for the three different values of the other variable. This enabled them to trace the approximate shape of the isoclines and, finally, to show that a closed trajectory is indeed produced.

A full mathematical model of processes occurring on anodically passivated electrodes was proposed by Franck and FitzHugh.[13] The case of iron polarized to a constant potential was specifically considered, but similar arguments can be applied to other processes of this class.

Fundamental features of the model are: (i) the *a priori* assumed discontinuity in the kinetics; and (ii) recognition of the influence of H^+ ions on the shape of the polarization curve (position of the Flade potential). In this way, a coupling term is introduced between the processes of metal dissolution and passive film formation, and is an essential feature of a mechanism leading to oscillations.

The model was based on the following assumptions:

(i) An oxide film is formed on the electrode when $0 < x$, and dissolves when $x < 0$, where x is defined as $x = E_p - E_x$. Here, E_x denotes the Flade potential under the existing conditions, and E_p the electrode potential (assumed constant). The rate of film formation is proportional to the free area of the electrode $(1 - y)$ and the rate of its dissolution to the coverage y.

(ii) At the uncovered parts of the electrode, dissolution of iron occurs when $x < 0$.

(iii) Partial currents of both reactions depend linearly on the potential.

(iv) The Flade potential depends linearly on the hydrogen ion concentration.

(v) Hydrogen ions are the only carriers of charge in the electrolyte.

(vi) The distribution of H^+-ion concentration across the diffusion layer is linear at all times.

(vii) The rate of production of H^+ ions, which accompanies formation of the oxide, is neglected in comparison with transport fluxes.

These assumptions lead to the following equations:

$$\dot{x} = K_1 - K_2 x - K_3 y + K_4 x y^*, \qquad \dot{y} = K_5 x y^* \qquad (22)$$

where

$$\begin{aligned} y^* &= y & \text{for} \quad x < 0 \\ y^* &= 1 - y & \text{for} \quad 0 < x \end{aligned} \qquad (23)$$

and K's are compound constants containing the proportionality factors in the expressions for diffusional and migration fluxes of H^+, coefficients in the relation between the difference of potential x and H^+-ion concentration, and constants in the equations for the polarization lines of iron dissolution and oxide formation.

The model is thus represented by two sets each composed of two equations. Which of the sets is "operative" depends on the actual position of Flade potential in relation to the potential of the electrode.

By means of an analog computer with a relay which was feeding the multiplyer (generator of xy^* terms) either with y or with $1 - y$ values, depending on the sign of x, a stable limit cycle was generated for a particular set of numerical values of the constants. The cycle was not formed, however, when y^* was always taken as $y^* = y$, or always as $y^* = 1 - y$, independently of the sign of x. This shows clearly that a discontinuous change in the film formation–decomposition kinetics is an essential feature in determining the periodic behavior of the model. The phase-plane portrait with a limit cycle and unstable focus is shown in Fig. 9 for the following values of the constants: $K_1 = 1250$ mV sec^{-1}, $K_2 = 2$ sec^{-1}, $K_3 = 2000$ mV sec^{-1}, $K_4 = 20$ sec^{-1}, $K_5 = 1$ mV sec^{-1}.

III. Models of Electrochemical Oscillators

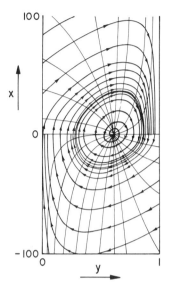

Figure 9. Phase-plane portrait of the simplified Franck–FitzHugh model.[13]

The model fails to represent adequately the real system in regions far outside the limit cycle. If the representative point is placed there (initial conditions), its trajectory, instead of winding up on to the limit cycle, will tend to infinity. The authors were able to improve their model by introducing a logarithmic, instead of a linear dependence of the Flade potential E_x on H^+-ion concentration, and taking into account charge transport by both SO_4^{2-} and Fe^{3+} ions. The improved model gave oscillations similar in shape to the current oscillations observed experimentally (Fig. 11). The phase portrait of the new model is shown in Fig. 10.

An attempt to explain periodic changes of the rest potential of platinum in a solution of 0.1 N hydrazine in 1 N KOH was made by Szpak.[55] The following mechanism was considered.

N_2H_4 (or a species "R" in general) is adsorbed on Pt from the solution and undergoes electrochemical oxidation in two stages with intermediates X_1 and X_2. X_2 can either give the final product or react with H_2O or OH^- producing H, which in turn can react with X_1, giving R. All reactions are considered as irreversible. Any

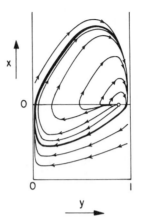

Figure 10. Phase-plane portrait of the improved Franck–FitzHugh model.[13]

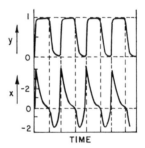

Figure 11. Oscillations exhibited by the improved Franck–FitzHugh model.[13]

encounter of species X_1 with free site S is regarded as favorable for either the appearance of another site S or an intermediate X_2, originating from X_1 by partial oxidation.

The above assumptions were regarded as being equivalent to the scheme

$$R + X_1 \xrightarrow{k_1} 2X_1$$
$$S + X_1 \xrightarrow{k_2} 2S \qquad (24)$$
$$R + S \xrightarrow{k_3} X_2$$

III. Models of Electrochemical Oscillators

from which the following kinetic equations arise:

$$\dot{x} = rk_1 x - k_2 xy, \qquad \dot{y} = k_2 xy - rk_3 y \tag{25}$$

where x, y, and r denote concentrations of X_1, S, and R, respectively. If r is treated as constant, the equations become identical with those of the well-known Volterra[3a]–Lotka[74b] system. They have two equilibrium points, a saddle at the origin $(0, 0)$ and a center at the coordinates $(rk_3/k_2, rk_1/k_2)$, around which closed trajectories exist. Typical trajectories were traced, using a graphical method of construction, for a few selected values of the kinetic constants.

Experiments with a rotating ring-disc electrode show clearly that the potential depends on the speed of rotation and that oscillations appear only when the speed is sufficiently high. When hydrogen was produced on the ring (cathodically polarized), onset of oscillations occurred at lower rotation speeds. The frequency of potential oscillations depended linearly on the square root of rotation speed (Fig. 12). This observation, in particular, can be taken as a good argument for the correctness of the model. Thus in the case of oscillations of sufficiently small amplitude around the center, the approximate dependence of frequency on r can be easily derived.

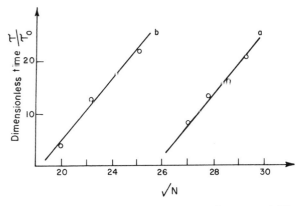

Figure 12. Effect of electrode rotation rate on frequency of diffusion-coupled oscillations.[55] (a) Pt/0.1 M N_2H_4 + 1 M KOH; (b) Pt/0.1 M N_2H_4 + 1 M KOH + H_2.

By making the center the origin of coordinates and linearizing the equations around it, one obtains

$$\dot{\bar{x}} = k_1 r \bar{y}, \qquad \dot{\bar{y}} = -k_3 r \bar{x} \qquad (26)$$

which is equivalent to one second-order equation

$$\ddot{\bar{x}} + k_1 k_3 r^2 \bar{x} = 0 \qquad (27)$$

This represents a harmonic oscillator of frequency

$$f = (r/2\pi)(k_1 k_3)^{1/2} \qquad (28)$$

which is a first approximation to the frequency of the original nonlinear system. It is seen, then, that this frequency should be proportional to r, or to the square root of the rotation speed of the electrode.

However, the relation of the model to the experimentally observed oscillations of potential is difficult to understand. Since in all stages of the process, transfer of charge is involved, the rate constants k_1, k_2, and k_3 depend on potential. Hence, the equations are not autonomous and cannot be integrated and treated as if the k's were independent of time, except under potentiostatic conditions. The role of transport effects at a rotating-disc electrode was also investigated by Wojtowicz et al.[42] in the case of ethylene oxidation at Pt, which manifests oscillations.

It may be added here that models based on equations of the Volterra–Lotka type[3a,74b] have been criticized from a more general point of view. The criticism arises from the standpoint that a differential equation which is to describe adequately a physical phenomenon, but can never represent the real system exactly, should at least be stable in the sense that some small distortion of its functional form would not change the character of its solutions significantly. This, however, is not the case with the equations which, as in the present case, have a center as an equilibrium point. It is relevant to quote Frank-Kamenetskii[71]: "Together with Salnikov[75] we have examined the influence of small additional terms on the behavior of the (Volterra–Lotka) system. We found that all terms proportional to the zeroth and first powers of x and y can lead only to damping of oscillations. On the other hand, the positive square terms, i.e., terms proportional to x^2 and y^2, can lead, under certain conditions, to the conversion of a center into an unstable focus, which results in building up of the oscillations." Minorsky[64] gives

III. Models of Electrochemical Oscillators

the following remark on the same subject:

> "...we have seen... that the center is a very special singularity and that the least change of the form of the differential equation destroys it and converts it into a neighboring weak focus for which the trajectories cease to be closed. In this way, a simple harmonic motion which, since the time of Galileo, has been assumed as a pattern of a stationary motion, turns out to be the most difficult to justify mathematically and impossible to produce experimentally. It appears thus that the singular point of the 'center' type does not correspond to any physical reality and is merely a mathematical concept separating the regions of convergent and divergent trajectories. The latter, on the other hand, have a definite physical meaning."

Nicolis[76] has considered the "noisiness" of all conservative models and remarked:

> "We see that each trajectory... has a different period depending... on initial conditions. The Volterra-Lotka model has therefore a continuous spectrum of frequencies associated with the existence of infinitely many periodic trajectories. This is a very important point, as it implies the lack of asymptotic orbital stability, that is, the lack of decay of fluctuations. As a result of infinitesimal fluctuations, the system will continuously switch to orbits with different frequencies and there will be no average, 'preferred' orbit. Intuitively, oscillations of this type (in fact, oscillations for all conservative systems) may only describe 'noise' type of effects rather than (true) periodic effects characterized by sharply defined amplitudes and frequencies."

In connection with Szpak's model, and especially with the model of Franck and FitzHugh,[13] a question arises under what conditions it is acceptable to consider the diffusion as stationary, in spite of the constantly changing rate of the surface reaction. This problem has been handled by Szpak and Rice[77] in the following way.

They assumed that a regime of oscillatory kinetics imposes on the diffusion process a condition of the form

$$(\partial c/\partial x)_{x=0} = a - b\cos(\omega t) \tag{29}$$

at the boundary $x = 0$ (interface). Taking for the other boundary condition (solution side of the diffusion layer of thickness δ), $c(\delta) = $ const, they obtained a solution of the diffusion equation and were able to show that the concentration wave that is developed in the diffusion layer is strongly damped if the frequency of oscillations is sufficiently high, and δ sufficiently small. For example, with a diffusion constant $D = 10^{-5}$ cm^2 sec^{-1}, a frequency $f = 10^{-2}$ Hz, and a diffusion layer thickness $\delta = 10^{-2}$ cm, damping is already so strong that the concentration profile resembles that of the steady state. In other words, the concentration at $x = 0$ could, with only a small error, be calculated from

$$D[c(\delta) - c(0)]/\delta = D(\partial c/\partial x)_{x=0} \tag{30}$$

where the right-hand side is a time-dependent boundary condition. (Concerning the concentration wave in the case of a finite δ, see also Rosebrugh and Lash-Miller.[78])

An explanation of oscillations observed by Wojtowicz et al.[42] during anodic oxidation of formate was attempted in terms of two alternative models. Both models assumed that electrosorption processes

$$\text{HCOO}^- + \text{Pt} \rightarrow \text{HCOO·Pt} + e^-$$
$$\text{OH}^- + \text{Pt} \rightleftarrows \text{HO·Pt} + e^- \tag{31}$$

were followed by the reaction between the two adsorbed species in which an autocatalytic effect was involved, in the sense that the greater the extent of desorption, the faster was the stripping process. Desorption of CO_2 was considered very fast.

Case 1. At a sufficiently high potential, the electrode is covered with a two-dimensional film of oxygen species with "holes" which are partially occupied by HCOO˙ and partially contain free sites. The reaction was considered to occur at the bifunctional (OH sites and Pt) *boundaries* of the holes, so that its rate would be proportional to the circumference. Assuming potentiostatic conditions and writing x and y for the coverages by formate and hydroxyl species, the following set of equations results:

$$\dot{x} = A(1 - x - y) - (1 - y)^{1/2}$$
$$\dot{y} = B(1 - x - y) - Cy - (1 - y)^{1/2} \tag{32}$$

with the time scale and A, B, C depending on the rate constants.

Case 2. A free site is needed for the occurrence of the reaction. Its rate is thus proportional to the product of the formate coverage, hydroxyl coverage, and the uncovered fraction of the total electrode surface. This gives

$$\dot{x} = A(1 - x - y) - xy(1 - x - y)$$
$$\dot{y} = B(1 - x - y) - Cy \quad xy(1 - x - y) \tag{33}$$

III. Models of Electrochemical Oscillators

It was shown that both models fulfill certain necessary criteria for the development of limit cycles. However, more detailed examination of the models[79] proved that if steady-state points are to remain inside the triangular region defined by the conditions $0 \leq x \leq 1$, $0 \leq y \leq 1$, $x + y \leq 1$, they must be stable. In other words, stable limit cycles in this region could exist only if accompanied by unstable cycles. This configuration, although in principle possible, is rather improbable.

Oscillations which, under the conditions of limited mass transport, arise at sufficiently high potentials during the oxidation of ethylene were discussed in terms of two different models.[42,80]

In Model I,[42] kinetics of oxidation through a surface process was assumed as proposed by Wroblowa et al.[81] Ethylene adsorption is considered fast and always in equilibrium. Simultaneously occurring electrosorption of oxygen-containing species by discharge from water was assumed to be relatively slow and determines[81] the overall rate (the reaction between ethylene and oxygen-containing species as well as the desorption of products is again very fast). The overall rate is then $v = k(1 - \Theta)$, where Θ denotes the coverage, which for the above assumptions, is equal to the surface concentration of ethylene species. If one-site Langmuir adsorption of ethylene is assumed (two sites may actually be involved), the rate is given by $v = k/(1 + K_L c_0)$, where K_L is the Langmuir adsorption equilibrium constant and c_0 the ethylene concentration in the solution at the electrode surface. Because in the steady state the flux of ethylene across the layer of thickness δ (with $\bar{c}$ taken as the concentration on the solution side of the layer) is $D(\bar{c} - c_0^*)/\delta$, the following relation should arise:

$$k/(1 + K_L c_0^*) = D(\bar{c} - c_0^*)/\delta \quad (34)$$

where c_0^* denotes the steady-state concentration. By solving this equation for c_0^*, it can be readily shown that stable states can be realized (c_0^* real and positive) when $[2(K_L k\delta/D)^{1/2} - 1]/K_L < \bar{c}$ and either $1/K_L < \bar{c}$ or $k\delta/D < \bar{c} < 1/K_L$. Thus, if the concentration in the bulk is too low, the diffusion layer is too thick, or the specific rate of the surface reaction is too high, the assumed kinetics can operate only during a limited period of time. Sooner or later, other factors, hitherto neglected, will become decisive and a different kinetic law must take over. For example, blocking effects would now be caused not by ethylene but by adsorbed oxygen. The current would decrease and after some period of time, the original conditions might be restored.

Model II[80] does not introduce *a priori* any discontinuity of the kinetic laws. The mechanism of the surface processes was assumed to be as proposed in Ref. 82. Both ethylene and oxygen-containing species were assumed (for simplicity) to obey Langmuir's isotherm, and the reaction between them (with rate proportional to the product of their surface concentrations) determines the overall rate, which is

$$v = k_r K K_e c_0/(1 + K + K_e c_0)^2 \tag{35}$$

where k_r is the rate constant of the chemical reaction, and K and K_e are, respectively, the Langmuir equilibrium constants for oxygen-containing species and ethylene.

Diffusion of the reactant (ethylene) across the layer of thickness δ affects the overall rate through c_0, which will no longer be considered constant. Introducing a composite constant $\beta = \delta k_r K_e/D$ and dimensionless variables $\gamma = K_e c$, $\xi = l/\delta$ (l is the diffusion coordinate), flux $\phi = v/k_r$, and time $\tau = Dt/\delta^2$, the diffusion equation

$$\partial^2 c/\partial l^2 = D\, \partial c/\partial t$$

becomes transformed into

$$\partial^2 \gamma/\partial \xi^2 = \partial \gamma/\partial \tau \tag{36}$$

and the boundary conditions appear as

$$\gamma(\tau, 1) = \bar{\gamma} = \text{const} \quad \text{and} \quad (\partial \gamma/\partial \xi)_0 = \beta K \gamma_0/(1 + K + \gamma_0)^2 \tag{37}$$

Noting that in the steady state, $\phi^* = \beta K \gamma_0^*/(1 + K + \gamma_0^*)^2$ applies as well as $\phi^* = \beta^{-1}(\bar{\gamma} - \gamma_0^*)$, the following equation in ϕ^* can be obtained:

$$\beta^2(\phi^*)^3 - 2\beta(1 + K + \bar{\gamma})(\phi^*)^2 \\ + [(1 + K + \bar{\gamma})^2 + \beta K]\phi^* - K\bar{\gamma} = 0 \tag{38}$$

It is seen that ϕ^* is a three-valued function of $\bar{\gamma}$ in a certain range of reduced concentration (it can be shown that the necessary condition for three real roots of ϕ^* is $\beta > 108$ and that they will never appear if $\bar{\gamma} \leq 8$).

III. Models of Electrochemical Oscillators

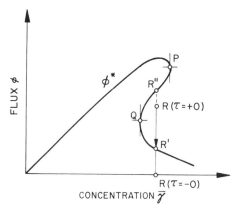

Figure 13. Steady-state flux.

A typical curve $\phi^* = f(\bar{\gamma})$ is given in Fig. 13. Solutions of the diffusion equation (for $\beta = 400$, $K = 1$) obtained by means of a digital computer have shown that the section PQ represents unstable steady states. The flux always tends to one of the two corresponding stable steady states on the upper or lower branches of the curve [e.g., if the point R represents the initial flux (at $\tau = +0$), it will move to R' and not to R''].

Change of flux in time is monotonic unless $\bar{\gamma} < \bar{\gamma}_Q$ when a single "wave" appears before the steady-state value is achieved. No oscillations are developed. ϕ^* is also a three-valued function of K under the appropriate conditions for $\bar{\gamma}$ and β (Fig. 14). As ϕ^* is

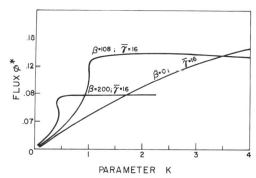

Figure 14. Influence of β on the shape of the curve $\phi = f(K)$.[80]

proportional to the current and K is potential-dependent, curves of $\phi^* = g(K)$ are directly related to the steady-state polarization curves. It is seen that an S-shaped polarization curve does not guarantee that current oscillations will arise under strictly potentiostatic conditions. If, however, K is not constant but depends on the flux (current), then oscillations are possible. Thus, we may assume the simple relation $K + R\phi = U = $ const, which may be interpreted that a voltage U composed of the true electrode potential K together with an ohmic drop through a resistance R is kept constant. Depending on the conditions of experiment, R may be, for instance, the "uncompensated resistance" between the tip of the reference electrode capillary and the test electrode, or the resistance of the whole cell with the exclusion of the test electrode. Introducing this expression into the boundary condition (37), the following is obtained:

$$\phi = \beta(U - R\phi)\gamma_0/(1 + \gamma_0 + U - R\phi)^2 \tag{39}$$

This is a cubic equation in ϕ, formally identical with (38), with R, U, and γ_0 taking the place of β, $\bar{\gamma}$, and K, so that when the appropriate conditions are fulfilled by R and U, the function $\phi = h(\gamma_0)$ will be represented by an S-shaped curve.

The behavior of the system can now be followed on the graph of this boundary condition curve (Fig. 15). The point R, which represents the state of the system as defined by (ϕ, γ_0), jumps from the position R_0 ($\tau = -0$) to R' ($\tau = +0$) and moves along the curve

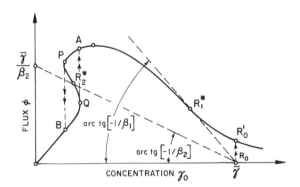

Figure 15. Boundary condition curve as a trajectory of the representative points.

III. Models of Electrochemical Oscillators

toward the steady state, i.e., to the intersection of the curve with the line $\phi = \beta^{-1}(\gamma - \gamma_0)$.

Stability of the steady state is determined by the following inequalities. If $0 < (\partial\phi/\partial\gamma_0)_{R^*}$ or $-1/\beta < (\partial\phi/\partial\gamma_0)_{R^*} < 0$, then the point R^* represents the stable steady state. If, however, $(\partial\phi/\partial\gamma_0)_{R^*} < -1/\beta$, R^* is unstable. (This has been proved, following Katchalsky and Spangler,[62] both by the method of thermodynamics of irreversible processes and by considering the fluctuations of γ_0 around R^*.) It is seen (Fig. 15), then, that the point R_2^* is unstable. This means that when $\beta = \beta_2$, the representative point R, after arriving at P, must jump down to B, then continue its movement along the curve to Q, jump up to A and return along the curve to P, etc. In effect, sustained oscillations of the discontinuous kind (relaxation oscillations) will be developed. An alternative interpretation based on the treatment of Andronov et al.[63] leads to the same result. It may happen that with appropriate values of $\bar{\gamma}$ and β, the "diffusion line" will cross the "reaction curve" at three points, two of which are stable and one unstable, or vice versa. It is clear that if the three points lie on the falling branch of the curve (Fig. 16a), the middle one is unstable, i.e., it can never be reached. On the contrary, if the three steady-state points are situated on the section PQ (Fig. 16b), the two outside ones are unstable, so that the middle one, which is, in this case, stable, can never be reached. Hence, although it is incorrect to say that section PQ is the locus of unstable steady states, it is in fact never accessible.

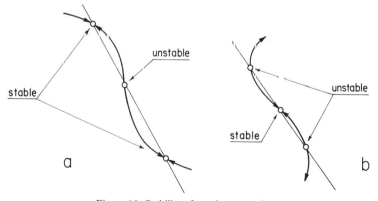

Figure 16. Stability of steady-state points.

The following general conclusion can be drawn from this model. The negative differential resistance, as determined by the S-shape of the polarization curve, evidently does not constitute a sufficient condition for oscillations. In order that oscillations can arise, the instantaneous flux (current) must be a three-valued function of instantaneous concentration. This has been realized in the model considered above by introduction of the "uncompensated resistance." However, it seems feasible that the same final result could be achieved through some kinetic factor acting as a feedback in the system of coupled surface reactions.

Katchalsky and Spangler[62] have described a model of an oscillating system composed of a membrane separating two salt solutions between which a potential difference is applied by means of two auxiliary electrodes. The membrane is formed of three layers; cation-selective (polyacid), neutral, and anion-selective (polybase). When current is passed between the two electrodes, periodic changes in the concentration of salt in the central zone develop, the resistance of the membrane changes accordingly, and the current and/or the potential difference oscillates. The following qualitative picture is given.

When the current flows across the membrane, accumulation of salt in the central zone occurs (cations flow in through the polyacid layer, anions through the polybase one). Osmotic pressure promotes the flow of solvent into the central zone and the hydrostatic pressure there increases. On the other hand, the increasing salt concentration will induce contraction of the polyelectrolyte molecules. If the point is reached where the increase of hydrostatic pressure with concentration, resulting from the polyelectrolyte configurational change, is greater than the corresponding osmotic pressure increment, the flow of solvent will be reversed. At this stage, the process becomes regenerative, with any increase in salt concentration tending to force solvent from the central region, thereby driving the concentration yet higher.[62]

Katchalsky and Spangler propose that as this regenerative process continues, salt will likewise begin to leave the neutral region because of its high concentration gradient, and will continue its flow out after the membrane has reached the point of maximum contraction. Thus, the process enters a phase of decreasing salt concentration, with consequent relaxation of the polymeric membrane matrix, and

III. Models of Electrochemical Oscillators

eventual return of the membrane to its initial state. It is clear that, as a result of the changing electrolyte content of the membrane, its conductivity properties will also vary periodically, giving rise to the externally observed oscillation.

On this basis, a mathematical model was developed as follows. Pressure p in the central zone is a function of its volume v and the salt concentration c. For its time derivative, one can thus write

$$\dot{p} = \beta\dot{v} + \gamma\dot{c} \tag{40}$$

where $\beta = \partial p/\partial v$ and $\gamma = \partial p/\partial c$. Changes in c are caused by the flow of salt and by the change of volume. If the total content of salt in the central zone is n moles, then

$$c = n/v \quad \text{and} \quad \dot{c} = (\dot{n} - c\dot{v})/v \tag{41}$$

Changes in n result from the flow of salt which may be identified with the flow of either ion, say the cation. Then

$$\dot{n} = A(J_s^\alpha - J_s^\beta) \tag{42}$$

where A denotes the area of the membrane and J_s^α and J_s^β the fluxes of the cation of the salt from the border zones α and β. For the volume change, we have

$$\dot{v} = A(J_v^\alpha - J_v^\beta) \tag{43}$$

According to Kadem and Katchalsky,[83] the fluxes can be expressed as

$$\begin{aligned} J_s^\alpha &= c(1 - \sigma^\alpha)J_v^\alpha + \omega^\alpha \Delta\pi^\alpha + t_+^\alpha I/\mathscr{F} \\ J_v &= L^\alpha(\Delta p^\alpha - \sigma^\alpha \Delta\pi^\alpha - p_E^\alpha I/\varkappa^\alpha) \end{aligned} \tag{44}$$

with the following meaning of the symbols: I is the current, t_+ the cation transference number, $\omega = D/(RT\Delta x)$ a permeability coefficient, Δx the membrane thickness, σ the selectivity coupling coefficient, L the hydraulic filtration coefficient, p_E the coefficient of electroosmotic pressure, $\varkappa$ the conductance, $\Delta p^\alpha = p^\circ - p^\alpha = -\Delta p^\beta \equiv \Delta p$, and $\Delta\pi^\alpha = \pi^\circ - \pi = 2RT(c^\circ - c) = -\Delta\pi^\beta \equiv \Delta\pi$.

Taking equal values of the coefficients ω, σ, L, and $\varkappa$ for the zones α and β, and taking into account that $p_E^\alpha = -p_E^\beta \equiv p_E$ (because of the opposite fixed charge in zones α and β), as well as assuming $t_+^\alpha = 1$ and $t_+^\beta = 0$, one obtains

$$J_v^\alpha = -J_v^\beta = J_v$$

The expressions for $\dot{v}$, $\dot{c}$, and $\dot{p}$ now become

$$\dot{v} = 2LA(\Delta p - \sigma \Delta \pi - p_E I/\varkappa)$$

$$\dot{c} = 2(\omega \Delta \pi - \sigma c J_v + I/2\mathscr{F})/d \qquad (45)$$

$$\dot{p} = 2[(\beta v - \sigma \gamma c)J_v + \gamma \omega \Delta \pi + \gamma I/2\mathscr{F}]/d$$

where $d = v/A$ is the thickness of the neutral zone.

These equations are, of course, not independent and could be further reduced to a set of two nonlinear equations in, say, c and p. However, it is simpler to determine the steady state directly from them by putting $\dot{c}^* = \dot{v}^* = \dot{p}^* = J_v^* = 0$. This gives immediately $2\omega \Delta \pi^* + I/\mathscr{F} = 0$ and $\Delta p^* = \sigma \Delta \pi^* + p_E I/\varkappa$, or

$$c^* = c^\circ + I/4RT\omega\mathscr{F}, \qquad p^* = p^\circ + I[(\sigma/2\omega\mathscr{F}) - (p_E/\varkappa)] \quad (46)$$

By linearizing the equations in the neighborhood of (c^*, p^*) and investigating the roots of the characteristic equation, it was found that the point (c^*, p^*) is a center when the following condition is fulfilled:

$$c^* = [\beta v^* + (2RT\omega/L)]/(\gamma - 2RT\sigma) \qquad (47)$$

The frequency of oscillation around the center is then

$$f = (1/\pi d)(2RTL\omega\beta v^*)^{1/2} \qquad (48)$$

By assuming certain more or less arbitrary values of the constants, the frequency and critical current density necessary for this type of oscillation in the model were calculated and appeared to be of the same order of magnitude as those experimentally observed by Shashua.[61]

In spite of that, the same criticism raised on p. 86 must apply here. Thus, the equilibrium point cannot be considered as a center. Instead, it can be an unstable focus, perhaps a weak one, and the result obtained in the paper for the frequency can be treated only as a first approximation, acceptable only for relatively short times. The real system, as always, must either tend to a limit cycle or develop quasidiscontinuous (relaxation) oscillations.

3. "Electrical" Models

The models hitherto described could, perhaps, be called "chemical," in the sense that by fixing the potential of an electrode, the kinetics

III. Models of Electrochemical Oscillators

of reactions together with those of transport processes must be solely responsible for the oscillations, including variations of the surface concentrations as well as of mass and charge fluxes. The models which will now be considered might be distinguished as "electrical." An electrode in this context is explicitly treated as just one element of the circuit, and is described either by its current–voltage characteristics, or by an equivalent electrical circuit. Naturally, the shape of the characteristic and the nature of the elements of the equivalent circuit are determined by the same factors which decide the form of equations corresponding to the "chemical" model.

As an introduction to this class of models, a paper by Franck[11] may first be considered, in which stable and unstable states of the electrode are discussed.

An electrical circuit with an electrochemical cell is considered as composed of two two-terminal elements: one of them is the electrode under consideration, while the remainder of the circuit forms the other. Each of the two elements has its own current–voltage characteristic, well-defined in the steady state. Points of intersection of the two correspond to the steady states common to both elements, i.e., to the steady states which can be realized in the overall circuit.

The characteristic of the test electrode is, of course, its polarization curve. The characteristic of the rest of the circuit is determined by the properties of its elements. In the simplest and most important case (constant external emf, all impedances real, nonpolarizable counterelectrode) the characteristic is linear. In particular, under galvanostatic or potentiostatic conditions, it is a horizontal or vertical straight line with current I as ordinate and potential E as abscissa.

If the polarization curve has a section where $dI/dE < 0$, then three steady states are possible (Fig. 17), of which two (1 and 2) are stable and one (3) is unstable.

For the case of an electrode undergoing passivation, instability of the middle steady-state point was demonstrated[12] in the following way.

The polarization curve is composed of two branches: One corresponds to the active state (zero coverage by the passive layer); the other to the passive one (full coverage). In an idealized case, they are connected by a straight vertical section at the Flade potential.

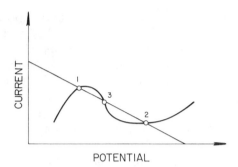

Figure 17. Steady-states as defined by the polarization curve and external circuit characteristics.

The zone between the two curves ($\Theta = 0$ and $\Theta = 1$) is filled with curves corresponding to intermediate coverages (Fig. 18). The current which flows through a partially passivated* electrode ($0 < \Theta < 1$) is

$$I = \Theta I_p + (1 - \Theta)I_a \tag{49}$$

where I_p and I_a are the relative current densities on the passivated and active parts of the electrode. Let the electrode be in the steady state S_3, i.e., at some intermediate coverage, and let it be exposed to some perturbation which will cause a small change of the potential

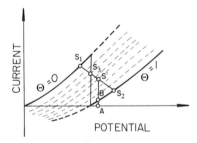

Figure 18. Instability of the mixed state (active–passive) of the electrode.

---

*Here, the initial stages of passivation are considered,[28a,28b] where a monolayer of inhibiting substance, e.g., adsorbed O species, is being built up. Complete passivation arises later in most systems when thicker oxide films are built up and only a residual anodic current passes.

toward more anodic values. The point representing the state of the electrode can move only along the linear external characteristic, so that the perturbation has brought it to, say, S'. The coverage is now larger than at S_3 and it will increase further, because a finite passivating current AB is still flowing. The point will thus move away from S_3 to the stable steady state S_2. By the same token, a cathodic perturbation of the potential will cause the system to move from S_3 to the second stable steady-state point S_1.

However, a negative differential resistance again by no means guarantees that oscillations will be generated. This has been clearly shown, for example, by Cizmadzev[84] for the case of an electrode with one degree of freedom (i.e., an electrode whose state is fully defined by one variable, such as potential), which has an N-shaped polarization curve.

Assuming a linear characteristic for the external branch of the circuit, we can write

$$dQ/dt = [(U - E)/R] - I(E) \qquad (50)$$

where U is the voltage applied, R is the resistance of the external branch, E is the potential of the electrode, Q is the charge of the electrode, and I is the Faradaic current.

Assuming further a constant electrode capacity $dQ/dE = C$, it follows that

$$C\, dE/dt = [(U - E)/R] - I(E) \qquad (51)$$

or, concisely,

$$dE/dt = F(E) \qquad (52)$$

Periodicity of $E(t)$ means that each value of E is passed (approached) alternatively from the two opposite directions. In other words, there are at least two different values of dE/dt possible, corresponding to each value of E. Hence, $F(E)$ is a multivalued function of E. The function $F(E)$ can, however, possess this property only then, when $I(E)$ is multivalued, and this is clearly not the case for the function depicted in Fig. 17.

On the other hand, if the polarization curve is S-shaped, there exists a region where $F(E)$ is three-valued, and similar reasoning for such a case would not exclude the possibility of oscillations.

The situation is different in the case of an electrode with two degrees of freedom. As an example, Czimadzev[84] considered oxidation of hydrogen, where the properties of the electrode depend on the extent of adsorption of anions. The potential E is treated as a function of the charge Q on the electrode side of the double layer and of the surface concentration Γ of the adsorbed species. The exchange current of hydrogen exceeds by 10^3 times that for adsorption of oxygen species, and it is assumed that at any value of Q, the surface concentration of hydrogen is at its equilibrium value. We thus have

$$dQ = C\, dE + \mathscr{F}\, d\Gamma \quad \text{and} \quad dQ/dt = [(U - E)/R] - I(E, \Gamma) \quad (53)$$

from which, setting $d\Gamma/dt = G(E, \Gamma)$, the following set of equations results:

$$C\, dE/dt = [(U - E)/R] - I(E, \Gamma) - G(E, \Gamma), \qquad d\Gamma/dt = G(E, \Gamma) \quad (54)$$

Investigating the character of the steady-state point S, where $(dE/dt)_S = 0$ and $(d\Gamma/dt)_S = 0$, leads to the conclusion that for S to be an unstable node or focus, the following conditions must obtain:

$$\mathscr{F}|(\partial G/\partial E)_\Gamma| > C|(\partial G/\partial \Gamma)_E| + (1/R) + (\partial I/\partial E)_\Gamma$$
$$\text{and} \quad (\partial G/\partial E)_\Gamma < 0 \quad (55)$$

where all derivatives are calculated at the point S.

It is further assumed that $(\partial G/\partial \Gamma)_E < 0$, which is always true if no attractive forces between the adsorbed species exist. In view of the condition $(\partial G/\partial E)_\Gamma < 0$, it follows that $(d\Gamma/dE)_S > 0$. The necessary condition for the periodicity to arise is hence interpreted in this way, that oscillations around S are possible only if, in its vicinity, the number of sites on which adsorption can occur increases with the potential.

If the adsorption were fast, so that no hysteresis would be exhibited, Γ would be uniquely defined by the potential, and the equations would reduce to those of the previously discussed case of $dE/dt = F(E)$, without any possibility that oscillations could arise.

Osterwald and Feller[85] observed an interesting behavior of a nickel electrode, polarized anodically in sulfuric acid. The potentiostatic polarization curve is shown in Fig. 19. Experiments were conducted in the following way. By means of a potentiostat, the

III. Models of Electrochemical Oscillators

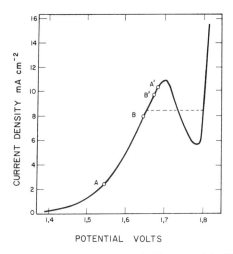

Figure 19. Polarization curve of Ni in 1 N H$_2$SO$_4$.[85]

electrode was brought to a state represented by some point (say P) on the left-hand, rising branch of the curve and the circuit was switched from potentiostatic to galvanostatic control, the latter condition being previously adjusted so that the value of the current

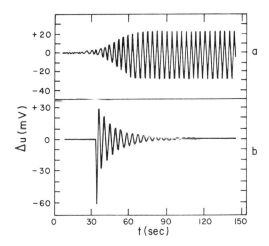

Figure 20. Oscillations of potential under galvanostatic conditions;[85] (a) $i = 2.87$, (b) $i = 2.26$ mA cm^{-2}.

did not change. The potential behaved in a different way, depending on the position of P. If $P < A$ or $A' < P$, the potential exhibited damped oscillations and tended to its original value (Fig. 20b). If $A < P < A'$, the potential either oscillated around the original value (Fig. 20a), or, if $B < P < B'$, jumped to the corresponding value on the right-hand, rising branch of the curve. It can be concluded, then, that the left-hand, rising branch did not have, under the galvanostatic conditions, a uniform character in respect to stability.

This finding was given a formal explanation in a later paper by Osterwald.[86,*] It is assumed that on an anodically polarized electrode, in parallel with the dissolution of the metal, certain processes of electrosorption of the type $A^- \to A_{ads} + e^-$ occur (cf. Ref. 28a). Their kinetics can be expressed as

$$\dot{\Gamma} = \varphi(U, \Gamma), \quad \text{or} \quad I_A = k\varphi(U, \Gamma) \tag{56}$$

and the kinetics of dissolution as

$$I_M = f(U, \Gamma) \tag{57}$$

where U is the potential, Γ the coverage, and I_A and I_M are the partial current densities of the two processes. Introducing the total current density $I = I_A + I_M$, it follows that

$$I = k\varphi(U, \Gamma) + f(U, \Gamma) \tag{58}$$

which, when solved for Γ, gives

$$\Gamma = \Psi(U, I) \tag{59}$$

Differentiating with respect to time,

$$\dot{\Gamma} = (\partial \Psi/\partial U)\dot{U} + (\partial \Psi/\partial I)\dot{I} \tag{60}$$

and substituting the last two results into (56), a relation connecting $U, \dot{U}, I,$ and $\dot{I}$ is obtained: $\Phi(U, \dot{U}, I, \dot{I}) = 0$, or, in the general case of

---

*De Levie[50] gave the following comment on Osterwald's explanation: "... Osterwald and Feller reported that no instability is encountered with Ni in H_2SO_4 when the system is controlled potentiostatically, whereas oscillations are observed under galvanostatic conditions. The proper explanation for this observation is that galvanostats have a high output impedance, in contrast to potentiostats, and it is this output resistance, rather than the control of either current or potential, which causes the system to oscillate in the region of negative immitance" (immitance = impedance or admittance).

III. Models of Electrochemical Oscillators

n electrosorption processes,

$$\Phi(U, \dot{U}, \ddot{U}, \ldots, U^{(n)}, I, \dot{I}, \ddot{I}, \ldots, I^{(n)}) = 0 \tag{61}$$

The character of equilibrium points can be established by investigating the linearized form of the equation, which, in the case of $n = 2$, is

$$a_0 \Delta I + a_1 \Delta \dot{I} + a_2 \Delta \ddot{I} = b_0 \Delta U + b_1 \Delta \dot{U} + b_2 \Delta \ddot{U} \tag{62}$$

where $a_0 = (\partial \Phi / \partial I)_s$, etc., $U = U - U_s$, $I = I - I_s$, the subscript s denoting a steady-state condition.

A second relation between U and I is given by the characteristics of the external branch of the circuit. Normally these are defined by $U = U_0 - RI$. Using this relation, either I or U can be eliminated and, for example

$$c_0 \Delta U + c_1 \Delta \dot{U} + c_2 \Delta \ddot{U} = 0 \tag{63}$$

is then obtained, where $c_n = a_n + R b_n$. Under galvanostatic conditions, the limiting $I - U$ relation corresponds to $R \to \infty$, so that $c_n = b_n$, and under potentiostatic conditions ($R = 0$), $c_n = a_n$.

The necessary and sufficient condition for the instability of an equilibrium (i.e., steady-state) point of this equation is that not all of the coefficients c_0, c_1, c_2 have the same sign. Then, and then only, is the real part of the roots of the characteristic equation $c_0 + c_1 \lambda + c_2 \lambda^2 = 0$ positive. [This is a well-known property of a linear equation of the second order and also can be proved by putting $\dot{U} = Z$, and considering the set $\dot{Z} = F(U, Z)$, $\dot{U} = Z$, as discussed on p. 69.]

It will be useful now to consider the rising branch of polarization curve (where $dI/dU > 0$), and investigate conditions under which unstable states will exist on it.

Because $(dI/dU)_s = b_0/a_0$ [from equation (62) by differentiation], it is seen that a_0 and b_0 have everywhere the same sign. Now, if under potentiostatic conditions, the rising branch represents stable steady states, then all a's are of the same sign. In these parts of the curve, where stability exists also under galvanostatic conditions, all coefficients (a's and b's) are of the same sign. If, however, a region exists which becomes unstable under galvanostatic conditions, then in this region, one of the coefficients b_0, b_1, b_2 must have a sign different from the other two. From what was said above, it cannot be b_0, so it must be either b_1 or b_2.

Writing for the roots of the characteristic equation
$$\lambda_{1,2} = (-b_1/2b_2) \mp (b_1^2 - 4b_0b_2)^{1/2} \tag{64}$$
and noting (see Fig. 20) that steady states along the whole branch are foci (stable or unstable), it is seen that everywhere the condition $b_0 b_2 > b_1^2/4$ must hold, so that b_2 is of the same sign as b_0, i.e., its sign is invariant. Consequently, b_1 must change its sign, i.e., pass through zero, at points bordering the region of instability (points A and A'). Considering finally that $c_1 = a_1 + b_1 R$, it is seen that c_1 will pass through zero when $R = -a_1/b_1$. In the general case, then, any point on this branch of the polarization curve can be made unstable by a suitable adjustment of R.

It should be understood that the above treatment does not offer a proof that oscillations must occur. What has been shown is only that part of the rising branch of the polarization curve will, under specified conditions, represent unstable steady states. This, of course, does not guarantee an oscillatory behavior of the electrode. Indeed, in the particular case of nickel in sulfuric acid, the experiments have shown that limit cycles are generated only on some sections (AB and $B'A'$) of unstable parts of the polarization curve.

Degn[58] presented a model in which diffusion and inhibition (passivation) effects and pseudopotentiostatic condition of the electrode contribute to the generation of current oscillations. The following assumptions were made.

1. The potential of the electrode is defined by electrosorption of A on free sites S, the process always being equilibrium, i.e.,
$$V = V_0 - (RT/\mathscr{F}) \ln\{[A][S]/[(AS)^+]\} \tag{65}$$
The fraction of unoccupied sites is thus given by
$$\Theta = 1/\{1 + A \exp[\mathscr{F}(V - V_0)/RT]\} \tag{66}$$
or, approximately, by
$$\Theta = k_4/(k_5 V^2 + 1) \tag{67}$$

2. Diffusion effects are accounted for by assuming that the current is proportional to Θ and to $(V - P)$, where P is the concentration polarization, i.e.,
$$I = k_1 \Theta(V - P) \tag{68}$$

III. Models of Electrochemical Oscillators

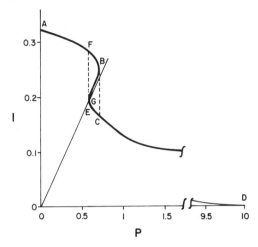

Figure 21. Current as a three-valued function of concentration polarization.[58] $E = 10$, $R = 30$, $K_2/K_3 = 3$.

Polarization increases linearly with current, while the simultaneously occurring tendency for diminution of the concentration gradient* is a first-order process:

$$dP/dt = k_2 I - k_3 P \tag{69}$$

3. The voltage U across the electrode and an ohmic resistance R in series is kept constant, i.e.,

$$U = V + RI \tag{70}$$

When it is assumed for simplicity that $k_1 = k_4 = k_5 = 1$, equations (67), (68), and (70) give

$$P = -R^2 I^3 + 2URI^2 - (E^2 + R + I)I + U \tag{71}$$

For certain values of R and U, the current I is a three-valued function of P (Fig. 21). The steady-state point is given as an intersection point of the straight line $P = (k_2/k_3)I$ [from equation (69)] and the curve given by equation (71). By suitable choice of k_2 and k_3, it can be placed on this section of the curve where $dI/dP > 0$.

*The author refers[58] to this change as "degradation" of concentration polarization.

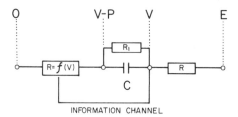

Figure 22. Circuit for simulation of oscillatory behavior.[58]

In such a case, it was concluded that the point of intersection G is inaccessible. Consequently, the representative point must cycle over the closed curve $FBCE$.

Although this statement was given without proof, it was shown by graphical methods that, for a certain set of numerical values of the constants, oscillations of current are indeed generated by the model if the jumps FG and BC (Fig. 21) are admitted. The oscillations are of the relaxation type, and the author was able to reproduce them by means of an electronic circuit, based on the scheme shown in Fig. 22. A function generator simulates the passivation effect (R_v represents $1/\Theta$ and is a function of V), a condenser simulates the concentration polarization, and the parallel resistor R_1 simulates the degradation of polarization. The following equations result:

$$I = (P/R_1) + C\, dP/dt, \qquad I = (V - P)/f(V), \qquad U = V - RI \qquad (72)$$

They correspond to (69), (67) + (68), and (70), with $k_2 = 1/C$, $k_3 = 1/CR_1$ and $f(V) = (k_5/k_1 k_4)V^2 + (1/k_1 k_4)$. The oscillations generated (for $U = 8.2\,\text{V}$, $R = 2.7\,\text{k}\Omega$, $R_1 = 3.4\,\text{k}\Omega$, $C = 25\,\mu F$) are shown in Fig. 23.

The model is given the following qualitative interpretation by Degn.[58] The true electrode potential is IR less than the externally measured value. When the external potential is kept constant, the current is not constant and consequently neither is the true electrode potential. Oscillation can be explained on this basis. If the external potential is fixed at a value which is higher than the passivation potential, there is initially no concentration polarization

III. Models of Electrochemical Oscillators

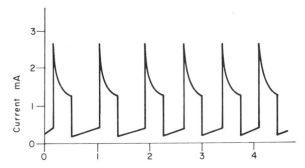

Figure 23. Oscillations obtained by means of the circuit in Fig. 22.[58]

and the current is high. At this high current, the IR term keeps the true electrode potential below the passivation potential. The current decreases with time because of increasing concentration polarization and consequently the true electrode potential will increase with time. Eventually, the true electrode potential reaches a value where passivation does take place, and the current then drops to a low value. The concentration polarization will now decrease with time. Consequently, the current will increase and the true electrode potential decrease with time. Eventually, the true electrode potential will tend to fall below the passivation potential and the electrode will become active again. Thereafter, it was considered[58] that the cycle would repeat itself.

De Levie et al.[87-89] have shown that if the charge-transfer process is characterized by a negative differential resistance in a certain region of potentials, then the impedance which, in the equivalent circuit, represents the diffusion term (Warburg impedance, see, e.g., Vetter[72]), also becomes negative.

Although this result has been proved for a few comparatively simple cases of diffusion coupled with a first-order redox reaction or electrosorption, its general applicability seems to be assured. In this connection, we may quote from Bode's conclusion[90] that, "If we do postulate ideal negative resistance elements, it follows immediately that negative elements of other types are also available. This can be shown most easily by reference to the well-known circuits shown in Fig. [24]. A simple computation shows that the input impedance Z_1 is given in either case by $Z_1 = -R^2/Z_2$."

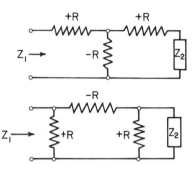

Figure 24. Negative input impedance generated by negative resistance.

Since the Warburg impedance is composed of resistance and capacitance, its negativity means that a negative capacitance has been introduced into the circuit. This, together with (a) the negative resistance, corresponding to the kinetics of the surface reactions and induced in the Warburg impedance, and (b) the double-layer capacitance (involving also possibly a pseudocapacitance due to electrosorption) constitute a set of conditions favorable for oscillation. Speaking qualitatively, there is now in the circuit, besides a dissipative element (positive resistance), also a negative resistance, which can be considered[80,89] as a source of power having an electromotive force proportional to the current flowing through the source, and a polarity determined by that current.[91] In addition, there are two reservoirs (positive and negative capacitances), between which the energy can be periodically exchanged.[92]

All this does not, however, guarantee that oscillations will in fact develop in such a circuit.

The simplest case of an electrochemical system in which oscillations may, under favorable conditions, occur has been discussed by De Levie.[50] A first-order reaction $R \rightleftarrows O + ne^-$ is coupled with the diffusion of both reactants. The equivalent circuit is given in Fig. 25(a), with the following significance of the components: R_s is the resistance of the external part of the circuit, C_{dl} is the double-layer capacitance, R_{ct} is the charge transfer resistance, and Z_W is the Warburg impedance. In this case, for the charge transfer resistance,

$$1/R_{ct} = n\mathscr{F}A[c_R'(\partial \vec{k}/\partial E) - c_O'(\partial \overleftarrow{k}/\partial E)] \tag{73}$$

III. Models of Electrochemical Oscillators

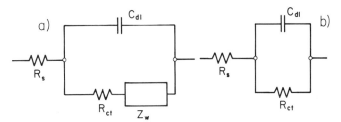

Figure 25. Equivalent circuits for the redox reaction with or without diffusion.[50]

where A is the area of the electrode surface, c' is the average concentration at the electrode surface, and E is the potential. The Warburg impedance is here[7a]

$$Z_W = R_{ct} z(j\omega)^{-1/2} \quad \text{with} \quad z = \vec{k} D_R^{-1/2} + \vec{k} D_O^{-1/2} \quad (74)$$

where the D's are the diffusion constants of R and O and $j = (-1)^{1/2}$.

The general conditions for undamped oscillations in a linear electrical circuit is that at $Z(s) = 0$, $0 \le \sigma$ and $0 < \omega$, where $Z(s)$ is the impedance function of the circuit, i.e. the impedance expressed symbolically as a function of s defined as $s = \sigma \mp \omega j$. The impedance function for the circuit of Fig. 25(a) is

$$Z(s) = R_s + \{R_{ct}(s^{1/2} + z)/[sR_{ct}C_{dl}(s^{1/2} + z) + s^{1/2}]\} \quad (75)$$

$Z(s) = 0$ has one trivial solution ($R_s = 0$, $s \to \infty$) and three non-trivial ones given by the roots of the cubic equation in $s^{1/2}$:

$$(s^{1/2})^3 + z(s^{1/2})^2 + (R_s + R_{ct}/R_s R_{ct} C_{dl})(s^{1/2}) + (z/R_s C_{dl}) = 0 \quad (76)$$

In order that conditions $0 \le \sigma$ and $0 < \omega$ are to be fulfilled in $s = \sigma \mp \omega j$, it is necessary that $|\beta| \le |\alpha|$ in $(s^{1/2}) = \alpha \mp \beta j$. Thus, by calculating the complex solutions of equation (76) in terms of $(s^{1/2})$ and solving the above inequalities, one arrives at the conditions for the R's and C_{dl} under which undamped oscillations exist. In the particular case of $\alpha = \beta$, the oscillations are sustained and harmonic.

If the diffusion rate is considered infinite, the equivalent circuit becomes that shown in Fig. 25(b), its impedance function being

$$Z(s) = (R_s R_{ct} C_{dl} s + R_s + R_{ct})/(1 + R_{ct} C_{dl} s) \quad (77)$$

[This result can also be obtained directly from equation (75) by putting $D_R = D_O = \infty$, or $z = 0$.] We thus have $Z(s) = 0$ for

$R_s = 0$, $s \to \infty$, and for $s = -(R_s + R_{ct})/R_s R_{ct} C_{dl}$. As s is always real, oscillations of a continuous character are not possible. However if $R_{ct} < 0$ and $-R_{ct} \leq R_s$, the circuit is unstable. Then, as the potential changes, the approximation $z \approx 0$ may no longer be valid, in which case, oscillatory behavior will still be observed.

The foregoing conclusions have been illustrated[50,87] by experiments on the reduction of In^{3+} on the mercury cathode in the presence of NaSCN. There is ample evidence that reduction of In^{3+} is catalyzed by SCN^-. The following mechanism has been proposed:

$$In^{3+} + 2SCN^-(ads) \xrightarrow{slow} In(SCN)_2^+(ads)$$
$$In(SCN)_2^+(ads) \quad (78)$$
$$+ 3e^- \xrightarrow{fast} In^\circ + 2SCN^-$$

SCN^- ions are immediately readsorbed, so that their concentrations in the solution and at the electrode surface remain unchanged.

The polarographic curve (Fig. 26), after the plateau (i.e., for diffusion limitation of the current by In^{3+} mass transfer), falls to a minimum, then rises again. The resulting region of negative resistance

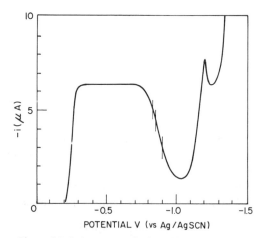

Figure 26. Polarogram of 1.2 mM $In(NO_3)_3$ + 5 M NaSCN, pH = 3.6; region of oscillations marked by vertical strokes.[50]

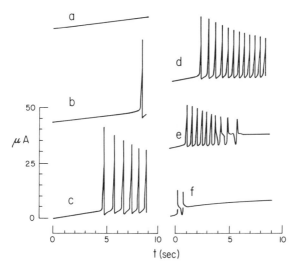

Figure 27. Current–time curves at -0.9 V. External resistance 10, 11.3, 15, 20, 30, 50 kΩ in (a)–(f), respectively.[50]

is explained by the increased desorption of SCN^- at highly negative potentials, and the subsequent increase of the current is associated with the hydrogen evolution and increase of pH of the electrolyte. In the region where $R_{ct} < 0$, oscillations arise when the external resistance in series with the cell is suitably adjusted (Fig. 27).

The following description of the events in the oscillations region was given.[50] The current, and hence the consumption of In^{3+} at the electrode, is low when the potential is near that of the minimum of the current–potential curve. Hence the interfacial concentration c_0' of indium ions increases through diffusion and natural convection. As c_0' increases, R_{ct} decreases, until the system becomes unstable [compare equation (73) where, in the present example, the term $c_R' \, \partial k/\partial E$ is negligible at the potentials considered, and the condition $-R_{ct} \leq R_s$ holds]. Then, the potential rapidly moves into the region of even more negative R_{ct}. The electrode potential can become more positive since the accompanying reduction current increases, so that the increased iR drop across R_s compensates for the decreased cell potential. This clearly shows the necessary interplay between the series resistance R_s and a negative

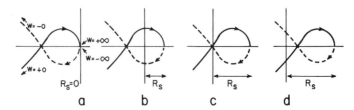

Figure 28. Application of Llewellyn criterion for the circuit of Fig. 25(a).[50]

resistance branch on the (instantaneous) $i - E$ curve. As the cell potential drops and the current increases, the interfacial concentration c_0' of indium ions rapidly decreases. This in turn diminishes the reduction current, and hence the iR_s drop, driving the cell potential back up to its original value. After a while, as c_0' has increased again, the cycle will repeat itself. It may be noted that this explanation does not require any concepts like those of "active" and "passive" states, specific adsorption sites, or competing reactions, etc.

By measuring the impedance at different frequencies, the stability of an electrochemical system may be determined. De Levie[50] proposes, for this purpose, a graphical representation on the complex plane and application of the Llewellyn criterion.[93] If the positive imaginary axis is identified with the negative ordinate,* the Llewellyn criterion is as follows: A system will be stable if the locus of $Z(j)$ encircles the origin of the complex plane in a clockwise direction as the frequency is continuously varied from $-\infty$ to $+\infty$. The curve in the range 0 to $+\infty$ is traced directly from the measurements, and the branch 0 to $-\infty$, which is symmetric (w.r.t. the real axis) in relation to the former, readily obtained.

Schematically drawn curves for the circuit of Fig. 25(a) are shown in Fig. 28. It is seen that the circuit loses its stability when R_s is increased beyond a certain critical value.

It is clear that oscillations could also find a formal explanation if the equivalent circuit contained an inductance in addition to a capacitance, and negative and positive resistances. Although Spangler and Snell's[94a] view that a chemical reaction system has

---

*In electrochemical impedance representations, it is customary to plot the *negative* imaginary axis (capacitative axis) as the positive ordinate, i.e., pointing upward, i.e., contrary to normal practice in electrical problems.

III. Models of Electrochemical Oscillators

equivalent resistive-capacitive behavior, with no means of introduction of analog inductive elements, seems to be shared by many, there is some evidence that under certain conditions, an electrode may exhibit "inductive" properties (cf. Ref. 94b).

Thus, Epelboin[95] has found that the real part of the impedance of nickel anodically polarized in $HClO_4 + CH_3COOH$ depends on the frequency and has a maximum at about 1 kHz. Gougerot and Alfieri reported similar and other related behavior with iron in $H_2SO_4 + CrO_3$[54] and Co in $HCl + CrO_3$[53]. Gerisher and Mehl[96] considered a model of the hydrogen evolution reaction based on the simultaneous occurrence of Erdey–Gruz–Volmer ($H_3O^+ + e + M \rightarrow MH$) and Heyrovsky–Horiuti type processes ($MH + H^+ + e \rightarrow H_2$) (followed by the Tafel reaction $2MH \rightarrow H_2$) and arrived at the following expression for the Faradaic admittance:

$$1/Z_F = (\mathscr{F}/RT)(\beta_v \bar{i}_v + \beta_H \bar{i}_H) - [ab/(b^2 + c_H^2 \omega^2)] + [ac_H/(b^2 + c_H^2 \omega^2)]j \qquad (79)$$

where

$$a = (\mathscr{F}/RT)(k_H - k_v)(\beta_H \bar{i}_H - \beta_v \bar{i}_v)$$

$$b = k_v + k_H + 2\Theta k_{T,0}$$

k_H and k_v are potential-dependent apparent rate constants; β_H and β_v are symmetry factors; i_H and i_v are densities of currents involved in the respective reactions; i_T is the rate of the Tafel reaction expressed in terms of the equivalent current: $i_T = 2\mathscr{F}k_T\Theta^2 = i_{T,0}\Theta^2$; Θ is the coverage by H_{ads}; and the bar indicates the constant (dc) component of the given variable.

It is seen that both real and imaginary components of the Faradaic admittance depend on the angular frequency ω, as long as $a \neq 0$. This results from the periodic changes of Θ which are not in phase with changes of the potential. Consequently, the current is out of phase with the potential, which means that the admittance has an imaginary component, and this may be either capacitative ($0 < a$) or inductive ($a < 0$). Which case will in fact be realized depends on the direction of change of Θ with the potential, and on whether the current would increase or decrease at the same time.

Schuhmann[97] considered a more general case of an electrode at which a series of processes of the type $A_i + S_i \rightarrow S_{i+1} + B_i + e$

occurred (rapid diffusion of A_i and B_i with B_i desorbing rapidly). He concluded that the Faradaic admittance could be expressed as

$$\frac{1}{Z_F} = \frac{a_{n-1}(j)^{n-1} + a_{n-2}(j)^{n-2} + \cdots + a_0}{b_{n-1}(j)^{n-1} + b_{n-2}(j)^{n-2} + \cdots + b_0'} \tag{80}$$

where the a_i and b_i are functions of the characteristic parameters of all the partial processes. The equivalent circuit will be more or less complicated and may contain equivalent inductive elements depending on the number of reactions involved, their kinetic interrelations, and the influence of potential on the various rate constants. The necessary condition is that there are at least three processes going on on the electrode, but only one of them need be a charge-transfer reaction.

In a later paper, Schuhmann[98] has shown that an equivalent circuit containing an inductive element may arise in the case of an electrode on which a reversible passivation reaction occurs accompanied by the dissolution of the metal proceeding either through some intermediate, or with participation of a catalyst.

Gougerot and Alfieri[54] observed periodic changes of the admittance of an iron electrode which closely followed oscillations of its potential. They attempted to give an explanation in terms of active and passive sites with statistically distributed, time-dependent differences of their impedances. In another paper by the same authors,[53] concerned with an oscillating cobalt cathode, changes of impedance were also observed, and an electronic model based on the theory of Schuhmann was described. Over a certain range of frequencies, it correctly simulated the observed phenomena.

To conclude, we may mention experiments in which oscillations were produced by connecting an inductive coil in series with an electrochemical cell.[99-101] Polarographic cells were used, and the cathodic process was either reduction of In^{3+} in KCNS or KBr, or reduction of Ni^{2+} in KCNS. In both cases, a negative differential resistance ρ was observed in a certain range of potentials. This, together with the double-layer capacity C, positive resistance R introduced into the circuit, and the inductance L of the coil gave the circuit (Fig. 29) to which the familiar equation applies:

$$\ddot{I} + [(1/\rho C) + (R/L)]\dot{I} + [(R + \rho)/LC\rho]I = U/LC\rho \tag{81}$$

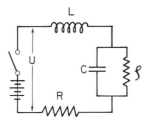

Figure 29. Inductance in series with an electrochemical cell as a generator of oscillations.

with the conditions for undamped oscillations given by:

$$(1/\rho C) + (R/L) \leq 0 \quad \text{and} \quad [(1/\rho C) - (R/L)]^2 < 4/LC$$

Agreement of the predicted behavior with that experimentally observed was shown[101] in the case of In^{3+} (values of ρ were taken from the polarographic curve and capacity from independent measurements[102]).

The oscillations in this case are of a less fundamental character, from an electrochemical point of view, since the R–C characteristics of an electrode–solution system are simply used in conjunction with an external circuit inductance which exhibits the usual characteristics of oscillation which arise in electronic circuits with RCL elements.

IV. CONCLUDING REMARKS

The overwhelming majority of the cases of electrochemical periodicity belong to the class of *relaxation* oscillations. There are often differences of several orders of magnitude in the rate of change of current or of potential within one cycle. Generally speaking, the mechanism and the kinetics of the process at these moments of short duration when the current or potential change by appreciable amounts, are unknown. Modeling of such systems is possible only by substituting the real process by a discontinuous change.

The necessary condition for these kinds of oscillations is simply the lack of any accessible, stable steady state. It is immaterial whether the unstable steady-state point is a focus or node or even a saddle point.

There is little doubt that the most common reason for instability is the coupling of diffusion with such surface processes as impart a negative differential resistance to the kinetic characteristics of the electrode.

Generation of negative resistance is usually connected with blocking effects by oxygen species, specific adsorption of anions, modifications of the energy of activation by the adsorbing species, or repulsive forces between the electrode and the discharging of catalytically active ions.

An additional factor necessary for instability and appearance of oscillatory conditions by diffusion-reaction coupling is a resistance in series with the electrode. It is always present in experimental galvanostatic arrangements, but should be virtually absent under ideal potentiostatic conditions.

In principle, there exists no fundamental reasons why kinetics of surface processes alone could not lead to oscillations. However, attempts at explaining oscillations solely in terms of kinetic equations without recourse to coupled diffusion processes have not hitherto been successful in an unambiguous way.

APPENDIX

Phase Plane. A plane the coordinates of which are the dependent variables of an autonomous system:

$$\dot{x} = P(x, y); \qquad \dot{y} = Q(x, y) \qquad (A.1)$$

Through every ordinary (nonsingular) point of the plane there passes one and only one integral curve (Cauchy theorem).

Integral Curve. A curve defined by

$$dy/dx = P(x, y)/Q(x, y) \qquad (A.2)$$

Representative Point. A point R on the phase plane which represents the instantaneous state of the system. Its motion is governed by equation (A.1).

Trajectory. An integral curve on which a motion of a point R occurs, as given by equation (A.1).

Singular Point. A point at which the direction of the tangent to the integral curves becomes indefinite, i.e., when both P and Q in equation (A.2) are zero.

Equilibrium Point. A point at which all time derivatives of the dependent variables vanish simultaneously. (This is equivalent to the steady state in kinetics.)

Center (Vortex). A singular (or equilibrium) point in the neighborhood of which the integral curves (or trajectories) are closed curves. There is a continuum of these curves.

Saddle Point. A singular (equilibrium) point reached by four integral curves (trajectories), two of which reach it as $t \to \infty$, while the other two as $t \to -\infty$. These four integral curves separate four regions containing continua of curves (resembling hyperbolae), which never reach the saddle.

Focus (Focal Point). A singular (equilibrium) point reached by integral curves (trajectories) having no definite direction. The curves are spirals winding up onto the focus or winding off from it.

Node (Nodal Point). A singular (equilibrium) point reached by the integral curves (trajectories) in such a manner that the tangents assume a strictly determined direction.

Isocline. A locus of points at which the tangents to all integral curves are inclined at the same angle.

Relaxation Oscillations. Periodic changes of a quasi-discontinuous character. Usually treated as composed of sections described by some analytical function and discontinuous jumps between them.

Warburg Impedance. An element of an electrode equivalent circuit, representing diffusion. Composed of resistance R and capacitance C (in series or in parallel) such that $RC = 1/\omega$, where ω denotes pulsation (angular frequency) of the current.

REFERENCES

[1] I. Prigogine, *Thermodynamics of Irreversible Processes*, Wiley, New York, 1955.
[2] D. Shear, *J. Theoret. Biol.* **16** (1967) 212.
[3a] V. Volterra, *Théorie mathématique de la lutte pour la vie*, Gautiers-Villars, Paris, 1931.
[3b] *Kolebatelnye processy v biologiceskich i chimiceskich sistemach* (*Oscillating Processes in Biological and Chemical Systems*), Izd. Nauka, Moscow, 1967.
[4] J. M. Douglas and N. Y. Gaitonde, *Ind. Eng. Chem. Fundamentals* **6** (1967) 265.
[5] J. M. Douglas and D. W. T. Rippin, *Chem. Eng. Sci.* **21** (1966) 305.
[6] Fechner, *Schweigg. J.* **53** (1828) 141.
[7a] E. S. Hedges and J. E. Meyers, *The Problem of Physicochemical Periodicity*, Arnold, London, 1926.
[7b] F. Flade, *Z. phys. Chem.* **76** (1911) 513.

[8a] U. F. Franck, Z. Elektrochem. **54** (1950) 540.
[8b] W. Fraenkel and H. Heinz, Z. anorg. Chem. **133** (1924) 167.
[9] K. F. Bonhoeffer and U. F. Franck, Z. Elektrochem. **55** (1951) 180.
[10] J. H. Bartlett and J. Stephenson, J. Electrochem. Soc. **99** (1952) 504.
[11] U. F. Franck, Z. phys. Chem. N.F. **3** (1954) 183.
[12] U. F. Franck, Z. Elektrochem. **62** (1958) 649.
[13] U. F. Franck and R. FitzHugh, Z. Elektrochem. **65** (1961) 156.
[14] K. F. Bonhoeffer and G. Langhammer, Z. Elektrochem. **51** (1948) 29.
[15] K. F. Bonhoeffer, E. Brauer, and G. Langhammer, Z. Elektrochem. **51** (1948) 60.
[16] K. F. Bonhoeffer and G. Langhammer, Z. Elektrochem. **51** (1948) 67.
[17] K. F. Bonhoeffer and H. Gerischer, Z. Elektrochem. **52** (1948) 149.
[18] R. S. Cooper and J. M. Bartlett, J. Electrochem. Soc. **105** (1958) 109.
[19] L. Meunier, C. R. III Réunion du CITCE **1952**, 247.
[20] L. Meunier, C. R. II Réunion du CITCE **1951**, 242.
[21] L. Meunier and G. Germain, C. R. III Réunion du CITCE **1952**, 263.
[22] F. Förster and F. Krüger, Z. Elektrochem. **33** (1927) 406.
[23] H. Lal, H. R. Thirsk, and W. F. K. Wynne-Jones, Trans. Faraday Soc. **47** (1951) 999.
[24] T. P. Hoar and J. A. S. Mowat, Nature **165** (1950) 64.
[25] A. Dmitriev and E. V. Rzhevskaya, Zh. Fiz. Chim. **35** (1961) 871.
[26] B. Pointu, Electrochim. Acta **14** (1969) 1207, 1213.
[27] M. T. Francis and W. H. Colmer, J. Electrochem. Soc. **97** (1950) 237.
[28] A. M. Shams El Din and F. M. Abd El Wahab, Electrochim. Acta **9** (1964) 883.
[28a] D. Gilroy and B. E. Conway, J. Phys. Chem. **69** (1965) 1259.
[28b] H. Wroblowa, V. Brusic, and J. O'M. Bockris, J. Phys. Chem. **75** (1971) 2823.
[29] N. Hackerman, Z. Elektrochem. **62** (1950) 632.
[30] A. R. Piggott, H. Leckie, and L. L. Shreir, Corrosion Sci. **5** (1965) 165.
[31] J. L. Rosenfeld and I. P. Danilov, Corrosion Sci. **7** (1967) 129.
[32] J. L. Rosenfeld and V. P. Maksimtchuk, Z. phys. Chem. **215** (1960) 25.
[33] Z. Szklarska-Smialowska and M. Janik-Czachor, Br. Corros. J. **4** (1969) 138.
[34] U. F. Franck and L. Meunier, Z. Naturforsch. **8b** (1953) 396.
[35] R. S. Indira, S. K. Rangarajan, and K. S. G. Doss, J. Electroanal. Chem. **21** (1969) 57.
[36] M. Thalinger and M. Volmer, Z. phys. Chem. **150** (1930) 401.
[37] G. Armstrong and J. A. V. Butler, Disc. Faraday Soc. **1** (1947) 122.
[38] D. T. Sawyer and E. T. Seo, J. Electroanal, Chem. **5** (1963) 23.
[39] E. Müller, Z. Elektrochem. **29** (1923) 264.
[40] E. Müller and G. Hindemith, Z. Elektrochem. **33** (1927) 561.
[41] E. Müller and S. Tanaka, Z. Elektrochem. **34** (1928) 256.
[42] J. Wojtowicz, N. Marincic, and B. E. Conway, J. Chem. Phys. **48** (1968) 4333.
[43] T. O. Pavela, Suomen Kemistilehti **30** (1957) 138.
[44] R. P. Buck and L. R. Griffith, J. Electrochem. Soc. **109** (1962) 1005.
[45a] J. A. Shropshire, Electrochim. Acta **12** (1967) 253.
[45b] S. Gilman, J. Phys. Chem. **67** (1963) 1898.
[46a] H. F. Hunger, J. Electrochem. Soc. **115** (1968) 492.
[46b] B. E. Conway and M. Dzieciuch, Nature **189** (1961) 914.
[47] F. Haber, Z. Elektrochem. **7** (1901) 634.
[48] A. Ya. Gochstien and A. N. Frumkin, Doklady Akad. Nauk SSSR **132** (1960) 388.
[49] A. N. Frumkin, O. A. Petrii, and N. W. Nikolaeva-Fedorovich, Doklady Akad. Nauk SSSR **133** (1961) 1158.
[50] R. De Levie, J. Electroanal. Chem. **25** (1970) 257.
[51] M. Clarke and J. A. Bernie, Electrochim. Acta **12** (1967) 205.

References

[52] M. N. Hull and F. A. Lewis, *Trans. Faraday Soc.* **64** (1968) 2472.
[53] L. Gougerot and R. Alfieri, *J. chim. phys.* **61** (1964) 843.
[54] L. Gougerot and R. Alfieri, *J. chim. phys.* **54** (1957) 514.
[55] S. Szpak, *Electrochim. Acta* **13** (1968) 483.
[56] M. L. Bhaskara Rao, *J. Electrochem. Soc.* **114** (1967) 665.
[57] L. Meunier, *C. R. III Réunion du CITCE* **1952**, 3.
[58] H. Degn, *Trans. Faraday Soc.* **64** (1968) 1348.
[59] D. Posadas, A. J. Arvia, and J. J. Podesta, *Electrochim. Acta* **16** (1971) 1041.
[60] T. Teorell, *Acta Soc. Med. Upsaliensis* **62** (1957) 60.
[61] V. E. Shashua, *Nature* **215** (1967) 846.
[62] A. Katchalsky and R. Spangler, *Quart. Rev. Biophys.* **1** (1968) 127.
[63] A. A. Andronov, A. A. Vitt, and S. E. Khaikin, *Theory of Oscillators*, Pergamon, London, 1966.
[64] N. Minorsky, *Nonlinear Oscillations*, Van Nostrand, Princeton, N.J., 1962.
[65] N. Minorsky, *Introduction to Nonlinear Mechanics*, Edwards, Ann Arbor, Mich., 1947.
[66] N. Kryloff and N. Bogoliuboff, *Introduction to Nonlinear Mechanics* (free transl. by S. Lefschetz), Princeton Univ. Press, Princeton, N.J., 1947.
[67] R. A. Struble, *Nonlinear Differential Equations*, McGraw-Hill, New York, 1962.
[68] S. Ziemba, *Vibration Analysis, Part III. Nonlinear Vibrations*, PWN, Warsaw, 1970.
[69] W. Hurewicz, *Lectures on Ordinary Differential Equations*, Wiley, New York, 1958.
[70] J. Higgins, *Ind. Eng. Chem.* **59** (1967) 18.
[71] D. A. Frank-Kamenetskii, *Diffusion and Heat Exchange in Chemical Kinetics*, Plenum Press, New York, 1969.
[72] K. J. Vetter, *Electrochemical Kinetics*, Academic, New York, 1967.
[73] M. N. Hull and F. A. Lewis, *Trans. Faraday Soc.* **64** (1968) 2472.
[74a] K. F. Bonhoeffer, *Z. Elektrochem.* **52** (1948) 24.
[74b] J. Lotka, *J. Am. Chem. Soc.* **42** (1920) 1595.
[75] D. A. Frank-Kamenetskii and J. E. Salnikov, *Zh. Fiz. Khim.* **17** (1943) 79.
[76] G. Nicolis, Stability and dissipative structures in open systems, in *Advances in Chemical Physics*, Vol. 19, Wiley–Interscience, New York, 1971.
[77] S. Szpak and E. R. Rice, *J. Chem. Phys.* **52** (1970) 6336.
[78] R. Rosebrugh and W. Lash-Miller, *J. Phys. Chem.* **14** (1910) 816.
[79] J. Wojtowicz, *Prace Naukowe Politechniki Warszawskiej, Chemia* No. 4 (1969) (Warsaw Technical University Publications, Warsaw).
[80] J. Wojtowicz and B. E. Conway, *J. Chem. Phys.* **52** (1970) 1407.
[81] H. Wroblowa, J. O'M. Bockris and B. Piersma, *J. Electroanal. Chem.* **6** (1963) 401.
[82] J. Wojtowicz, D. Gilroy, and B. E. Conway, *Electrochim. Acta* **14** (1969) 1119.
[83] O. Kadem and A. Katchalsky, *Trans. Faraday Soc.* **59** (1963) 1918.
[84] Yu. A. Cizmadzev, *Doklady Akad. Nauk SSSR* **133** (1960) 1136.
[85] J. Osterwald and H. G. Feller, *J. Electrochem. Soc.* **107** (1960) 473.
[86] J. Osterwald, *Electrochim. Acta* **7** (1962) 523.
[87] R. De Levie and A. A. Husovsky, *J. Electroanal. Chem.* **22** (1969) 29.
[88] R. De Levie and L. Pospisil, *J. Electroanal. Chem.* **22** (1969) 277.
[89] H. Moreira and R. De Levie, *J. Electroanal. Chem.* **29** (1971) 353.
[90] H. W. Bode, *Network Analysis and Feedback Amplifier Design*, Van Nostrand, Princeton, N.J., 1959, p. 187.
[91] W. H. Middendorf, *Analysis of Electric Circuits*, Wiley, New York, 1956, p. 212.
[92] W. Schwenk, *Electrochim. Acta* **5** (1961) 301.
[93] R. B. Llewellyn, *Proc. IRE* **21** (1933) 1532.

[94a] R. A. Spangler and F. M. Snell, *J. Theoret. Biol.* **16** (1967) 366.
[94b] F. Gutmann, *J. Electrochem. Soc.* **112** (1965) 94.
[95] I. Epelboin, *Compt. Rend.* **234** (1952) 950.
[96] H. Gerischer and W. Mehl, *Z. Elektrochem.* **59** (1955) 1049.
[97] D. Schuhmann, *J. chim. phys.* **60** (1963) 359.
[98] D. Schuhmann, *J. Electroanal. Chem.* **17** (1968) 45.
[99] R. Tamamushi, *J. Electroanal. Chem.* **11** (1966) 65.
[100] R. Tamamushi and K. Matsuda, *J. Electroanal. Chem.* **12** (1966) 436.
[101] B. Jakuszewski and M. Turowska, *Rocz. Chemii* **43** (1969) 2003.
[102] N. Tanaka and R. Tamamushi, in *Proc. 1st Australian Conf. on Electrochemistry*, Pergamon, Oxford, 1964, p. 248.

3

Methods and Mechanisms in Electroorganic Chemistry

A. A. Humffray

*Chemistry Department, University of Melbourne
Australia*

I. INTRODUCTION

Electroorganic chemistry, or the effecting of chemical changes in organic systems through the agency of an electric current, dates from about the time of Faraday's discovery of hydrocarbon formation during the electrolysis of aqueous acetate solutions.[1] This anodic decarboxylation reaction was later investigated for its synthetic possibilities by Kolbe[2] and others[3]; it is interesting to note that, after one and one-third centuries, it is still being actively investigated, both for the original (synthetic) purpose,[4] and with a view to obtaining information about the mechanism of this reaction, and of electrochemical reactions in general.[5] From this initial decarboxylation and radical dimerization reaction, the bounds of electroorganic chemistry have spread to include a wide range of different reaction types, ranging from complete oxidation to CO_2 of organic substrates like alkanes,[6] alcohols,[7] acids,[8-10] or even cellulose,[11] to anodic or cathodic substitution reactions, e.g., acetoxylation[12] or halide reduction,[13] to anodic or cathodic addition, e.g., methoxylation,[14] or reduction of unsaturated linkages,[15] to hydrodimerization,[16] and to formation of organometallic compounds,[17] etc. Recent reviews on oxidation,[18] electrochemical principles in electroorganic chemistry,[18a] reduction,[19] and organic electrosyntheses[20] are available.

1. Synthetic Applications of Electroorganic Chemistry

The excellent yield of ethane obtainable from acetate solutions undoubtedly encouraged many investigators, in the early days when the emphasis in organic chemistry was on synthesis, to consider electrolytic methods. Few of the latter, however, have gained widespread acceptance; some of the drawbacks to electrolytic methods, which have hindered their greater use, include the following:

(a) A long time is required for large-scale electrolysis, unless electrodes of large area are employed, since a one-electron process requires, per mole of reactant, one Faraday of charge, which at a current of 1 A requires an electrolysis duration of about 27 hr.

(b) There is an adverse influence, in prolonging the time required, of any factors which reduce the magnitude of the current attainable. A diaphragm is often essential to separate the anode and cathode in order to prevent unwanted cathodic (anodic) processes affecting products previously or simultaneously formed at the anode (cathode). The use of a diaphragm increases the resistance, resulting in a decrease in current unless the applied voltage is increased; in the latter event, the power efficiency is reduced. Deactivation of the electrode by a film of intermediates, products, or by-products is another factor which reduces the current and increases the electrolysis time. This effect is most pronounced with solid electrodes.

(c) Problems arise from the low solubility of many organic reactants in water and in similar solvents in which electrolytes can readily be dissolved to provide solutions of high electrical conductivity. Complications may arise from the mutually contradictory requirements of high solution conductivity (necessitating high supporting electrolyte concentration) and adequate solubility of organic reactants and products, often adversely affected by dissolved electrolytes ("salting out" effect).

(d) A high capital cost is associated with the need often to employ expensive electrode materials, e.g., platinum, particularly as anodes, since anodic dissolution of the metal, or oxide formation, may be the preferred anodic reaction when using electrodes of less noble metals. Another cost factor, of recent origin, is associated with the use of a potentiostat, if the greater control over product composition realizable by potentiostatic, rather than galvanostatic, electrolysis is desired.

I. Introduction

(e) Generally discouraging results are obtained as regards contamination of product by undesired by-products, when galvanostatic conditions are employed, as they frequently are for synthetic purposes, because of the much simpler equipment required for galvanostatic, as compared to potentiostatic, operation. Another factor favoring galvanostatic conditions is the greater ease in determining the charge passed during the electrolysis; in potentiostatic electrolysis, a coulometer or a current integrator is required, since the current decreases continually as the reactant concentration is reduced. This decrease in reactant concentration, resulting from reaction, necessitates an increase in potential, if a constant current is to be maintained, and a change in electrode potential can lead to the formation of undesired products, as shown below.

A major disadvantage of galvanostatic processes, associated with point (a) above, is that, in order to reduce the electrolysis time to what is considered to be a reasonable one, the current may be adjusted to a value which does not correspond to the *optimum* current density for the process involved. Most organic electrode reactions are irreversible,[18] so that overvoltage η and current density i are related through a Tafel-type relation,[21]

$$\eta = a \pm b \log i$$

for which b, the Tafel slope, usually has a value of between 59 and 118 mV,[22] although lower[23] and higher[24] values have been observed. Hence a potential change of only 50–120 mV could lead to a tenfold change in current, i.e., in the rate of reaction. Three conclusions follow from this observation:

(i) A considerably smaller, (and hence less costly) electrode would require for the same current (rate of reaction) only a small increase in potential, with only a small chance of increasing possible interference from another electrode process.

(ii) If the Coulombic efficiency with respect to the desired product is less than 100% (which is not uncommon for organic compounds, because of simultaneous occurrence of two or more electrode reactions), a small change in potential (or in current density, e.g., through changing the size of the electrode without altering the current) may significantly alter the relative rates of the electrode reactions, if their Tafel slopes are different. This is illustrated in Fig. 1 for the case of two electrode processes forming products A and B.

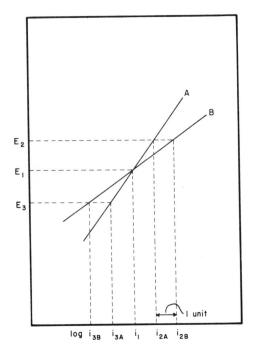

Figure 1. Effect of potential change on relative rates of two electrode processes with different Tafel slopes.

At potential E_1, or current density i_1, the Coulombic efficiency for formation of B is only 50%, while a change to potential E_2, or to current density $(i_{2A} + i_{2B})$ increases the efficiency for B formation to about 90%. In this example, a change to potential E_3 would lead to A, instead of B, becoming the predominant product. A similar discussion has been used to explain inhibition of the oxygen-evolution reaction during the Kolbe synthesis.[25]

Furthermore, if A is the desired product, and the initial (galvanostatic) conditions correspond to potential E_3 (say for 90% production of A), any process leading to a decrease in available electrode area, e.g., adsorption of a by-product, results under galvanostatic conditions, in an increase in current density and hence in potential, with a consequent decrease in the Coulombic efficiency for production of A.

I. Introduction

The potential increase toward the end of a galvanostatic electrolysis, which is caused by decreased reactant concentration, has a similar effect. Under potentiostatic conditions, a decrease in effective electrode area produces a corresponding decrease in current so as to maintain constant current density (and hence potential), and the rate of formation of A would decrease but the relative yields of A and B would remain unchanged.

(iii) Under galvanostatic conditions, if the applied current density exceeds the limiting current density[26] for the reactant (which depends on its concentration), the potential increases until another electrode process occurs (possibly solvent oxidation or reduction). The products or intermediates from this new process may then interact with the original reactant; this is especially likely if these products or intermediates are radicals. From a synthetic point of view, the yield of the final product may be adequate, but it is not inconceivable that the same product could be obtained in equally satisfactory yield at a lower potential (and hence higher power efficiency) by direct reaction at the electrode rather than by subsequent chemical reaction with the product of a prior electrode reaction. If the interest is in the mechanism of the electrode reaction, galvanostatic investigations are particularly unsatisfactory with respect to the uncertainty thus introduced.

Undaunted by all the above adverse factors, a few investigators[27,28] have persevered with attempts to find or improve electrochemical methods of synthesis; some successful attempts have been incorporated into laboratory text[29] and reference[30,31] books, but the ultimate criterion of success is undoubtedly the adoption of an electrochemical synthetic procedure as a commerical process, capable of competing with nonelectrochemical alternative methods of synthesis. Very few[4] organic electrochemical synthetic procedures yet meet this criterion, despite two distinct advantages which electrochemical methods possess compared to nonelectrochemical methods, namely:

(i) The ability for control to be exercised over the composition of the products by adjustment of the electrode potential.[32]

(ii) The consumption (in reduction processes) or production (in oxidations) of electrons only, apart from the reactant or product, respectively. The separation from, and disposal of, conjugate oxidation or reduction products derived from a chemical reducing or oxidizing agent are not necessary when the process is carried out electrochemically. This simplifies the isolation and purification of the desired product.

Haber[32] first realized the importance of potential control in determining the composition of the products from an electrode reaction. With the development of quantitative treatments for slow charge-transfer processes,[33] potential of zero charge in electrolyte solutions,[34] and adsorption processes at electrodes,[35] the fundamental significance of the electrode potential was firmly established. However, experimentally, potential control became feasible only after the first potentiostat was devised by Hickling.[36] Improvements in potentiostatic instrumentation owe much to the commercial availability of solid-state operational amplifiers, and many successful designs incorporating these have been published.[37–39]

Commercial potentiostats are also supplied by a number of firms, a recent compilation and comparison on the basis of output power, precision of potential control, response time, and cost is available.[40] At least two firms supply multipurpose electrochemical instruments which permit a wide range of measurements to be performed with a variety of techniques, without the need for auxiliary equipment; these are the PAR-170[41] and the Beckman Electroscan 30.[42]

2. Mechanistic Applications of Electroorganic Chemistry

With the growth in knowledge of organic systems, the early, predominantly synthetic bias was superseded by an approach which emphasized the explanation of the course of organic reactions in terms of mechanisms and energetics.

Electroorganic chemists, following the same trend, found that it was possible to explain the course of organic electrode reactions in terms of the same intermediates as had proved successful in systematizing the treatment of homogeneous organic reactions.

Charge-transfer reactions directly involving organic compounds can give rise to only three types of initial product, namely a carbonium ion R^+, or if the charge is localized on an atom other than C, an "onium" ion; a carbanion R^-, or if the charge is not on carbon, an anion; a radical $R^{\cdot}$. Combinations of the first two with the third, to form cation or anion radicals, are also possible. Hence transformations observed in organic electrochemistry involve largely the

I. Introduction

subsequent reactions of species of the above types, and, not surprisingly, many similarities exist to nonelectrochemical processes which proceed through similar intermediates.

Two different approaches to the investigation of organic electrode processes have developed; the first of these, termed[5] the *chemical* approach, seeks and emphasises similarities between electrode processes and homogeneous (or heterogeneous) nonelectrochemical processes, with regard to, e.g., product distributions, or relative reactivities of isomers and/or homologs, structural or steric effects, isotope effects, etc. Results are often correlated in the form of linear free-energy relationships,[43] and, as for homogeneous reactions, both experimentally based reactivity indices, e.g., Hammett substituent parameters σ, σ^+, etc.,[43] or reactivity (HOMO energies, etc.)[44] have been employed in such correlations. Polarography, in particular, lends itself to such correlations through polarographic $E_{1/2}$ values,[45] which are usually more easily evaluated than, e.g., rate or equilibrium constants.

Although organic electrode processes frequently give excellent linear free-energy correlations with rate or equilibrium data for homogeneous processes, and with theoretical measures of stability or reactivity,[46] other correlations have been sought on the basis of the special features exhibited by electrode processes. These include the heterogeneous nature of the charge-transfer process and the significance of possible adsorption of reactants, intermediates, and/or products. Successful comparisons have been made with other heterogeneous processes, e.g., gas–solid catalytic processes, and treatments of adsorption equilibria,[47] and adsorption kinetics[48] which were developed for the latter processes have been applied successfully in electrochemistry.

Some criticism[49] has been leveled at the extension of gas–solid adsorption treatments to adsorption at the solution–electrode interface because in the latter case, adsorption involves a competition between adsorbate and solvent (or other solutes) for available sites on the electrode; such competition can arise in the gas–solid case only when mixtures of gases are employed. In many cases, the electrode exerts a catalytic function (electrocatalysis)[49] analogous to heterogeneous catalysis in gas-phase reactions, and in such cases, the nature of the electrode material and the state of its surface have very significant effects on the rate of the electrode reaction, as for gas–solid heterogeneous catalytic reactions. In other electrode reactions, desorption of intermediates occurs, and subsequent chemical reactions occur either in the layer of solution immediately adjacent to the electrode, or, if the lifetime of the intermediate is high enough, in the bulk solution. The nature and surface treatment of the electrode material are of considerably less significance in these cases.

In contrast to the *chemical* approach, which emphasizes the similarities between electrode processes and homogeneous or heterogeneous reactions, as discussed above, the *electro*chemical approach concentrates on the unique features of electrode processes, i.e., on the differences between the latter and non-charge-transfer homogeneous and/or heterogeneous processes. Among these unique features are the existence of the electric double layer and the electric field at the solution–electrode interface, and the way in which these can affect the reactant concentration at the interface, where it may differ markedly from that in the bulk of the solution. Another factor taken into consideration is the effect of a net current on the concentration distribution throughout the system; this involves an analysis of the mass-transport conditions prevailing. Polar reactant molecules may be oriented in a particular manner at the interface, and this orientation may be potential-dependent. Adsorption of reactants or products is usually potential-dependent, and the quantitative measure of adsorption and evaluation of adsorption isotherms involve electrochemical techniques. The detection and measurement of some intermediates produced by charge transfer can also be performed electrochemically.[50] Complete understanding of the mechanism requires information about the magnitude of the rate constant(s) for the charge-transfer step(s), and for any associated chemical reactions, and about the electrochemical properties of any intermediate molecules, ions, or radicals—e.g., are they further oxidizable or reducible at the same potential, or at greater potentials? The chemical nature of such intermediates is often deduced from product isolation and identification studies, together with rate constant measurements on chemical reactions accompanying the charge-transfer process. Unless the reactant or the product obtained under the prevailing experimental conditions is an ion or combines with another ion present in the system, electrode processes involve, overall, the transfer of an even number of electrons. Multielectron

I. Introduction

processes usually involve a succession of single-electron-transfer steps.[51] The relative stabilities, ionization potentials, or electron affinities of reactants, intermediates, and products, and the nature and concentration of other solute or solvent species in the solution, and particularly in the double layer, give rise to a great variety of possible overall modes of reaction.

Table 1
Commonly Observed Reaction Schemes

Reaction type	Classification	Equations[a]
1. Charge transfer only	E	$O + ne \xrightarrow{k} R$
2. Consecutive charge transfers	EE	$O + n_1e \to I$
		$I + n_2e \to R$
3. Charge transfer with coupled chemical reactions		
(a) Preceding, kinetic process	CE	$Y \rightleftarrows O \xrightarrow{ne} R$
		or $Y + mH^+ \rightleftarrows O \xrightarrow{ne} R$
(b) Catalytic	EC	$O \xrightarrow{ne} R$
		$R + Z \to O$
(c) Competing	E and C	$O \xrightarrow{ne} R$
		$O + X \to Z$
(d) Subsequent (i) Simple	EC	$O \xrightarrow{ne} R \to Y$
		or $R \rightleftarrows Y$
(ii) With solute X	EC	$O \xrightarrow{ne} R$
		$R + X \to Z$
(iii) With solvent S	EC	$O \xrightarrow{ne} R$
		$R + S \to Z$
(iv) With electrode M	EC	$O \xrightarrow{ne} R$
		$R + M \to R \cdot M$
(v) Coupling	EC	$O \xrightarrow{ne} R$
		$R + O \to Z$
		or $2R \to Z$
(e) Intervening chemical reactions	ECE	$O_1 \xrightarrow{n_1e} R_1$
		$R_1 \to O_2$
		$O_2 \xrightarrow{n_2e} R_2$
4. Charge transfer with adsorbed intermediates		
(a) Adsorption following charge transfer		$O + e + M \to MR_{ads}$
		$MR_{ads} \to Z$
(b) Dissociative adsorption prior to charge transfer (anodic process)		$Y + M \to MR_{ads} + MR'_{ads}$
		$MR_{ads} \to M + O + e$
		$MR'_{ads} \to M + O' + e$

[a] O, I, and R represent electroactive components, X, Y, and Z represent electroinactive components.

The number is increased further by the possibility of reactions, especially protonation in the case of reductions and acid dissociation in the case of oxidations, preceding the charge-transfer step. Because charge-transfer steps for organic compounds are often rapid,[51] the chemical reactions accompanying the charge transfer are usually of great significance in determining the overall reaction rate. A number of commonly observed electrode reaction schemes, with or without coupled chemical reactions, is listed in Table 1, for charge transfers considered as reductions. With appropriate changes, analogous considerations apply to oxidations.

The distinction between the various possible reaction schemes is usually based on a comparison of the experimentally observed relationships between various electrochemical parameters, e.g., current and potential, current and time, reaction order, etc. and the theoretically expected relationships, which have been derived for a number of different electrochemical techniques to be discussed in Section II. The most frequently encountered examples of this approach to the elucidation of organic electrode reaction mechanisms involve techniques such as polarography, linear sweep voltammetry, cyclic voltammetry, chronopotentiometry, etc., which usually employ depolarizer concentrations of the order of millimolar, in unstirred solutions. These conditions are distinctly different from those normally employed for preparative-scale electrolyses. A further distinction is in the time scales of these diagnostic procedures compared with the time scale of electrolytic syntheses; the latter usually amounts to several minutes at least, or even hours, whereas results from the former are obtained in times seldom exceeding 1 min (polarographic drop times, for example, are usually restricted to the range of about 2–6 sec).

The short time scale of the above diagnostic procedures and the small amount of electrolysis effected leads to no significant accumulation of products in the solution, unlike preparative-scale electrolysis, so that subsequent reactions involving these products in the latter procedure may not be observed when the former techniques are employed. Chemical reactions which are relatively slow, and hence occur to an insignificant extent during the few seconds of polarographic drop life, will not affect the observed response here, but may be quite important on the longer time scale of preparative electrolysis. The higher concentration of reactant

I. Introduction

in the latter may also have a significant effect on coupled chemical reactions, particularly those of an order higher than the first, e.g., dimerizations. The continuously renewed surface of the dropping mercury electrode in polarography may behave differently[53a] from the mercury pool electrode used in macroscale electrolysis; if a solid electrode is employed, deactivation may occur on prolonged electrolysis which is not evident on the short time scales of any of the diagnostic methods. For the above reasons, mechanistic information obtained by short-time-scale procedures cannot always be considered to hold under preparative conditions. Thus, e.g., polarographic reduction of phenylmercuric salts produces two one-electron waves, the first pH-independent, the second pH-dependent, ascribed to the reactions

$$PhHgX + e \rightarrow PhHg\cdot + X^-$$

$$PhHg\cdot + H^+ + e \rightarrow PhH + Hg$$

However, unequivocal identification of the product of the second step as benzene appears not to have been achieved; macroscale controlled-potential electrolysis, using a mercury pool electrode at a potential on the plateau of the second wave, gave diphenyl-mercury as the only isolated product, presumably from the reaction[52]

$$2PhHg\cdot \rightarrow Ph_2Hg + Hg$$

In spite of such complications, many cases are known of electrochemical synthetic procedures which were developed on the basis of, e.g., polarographic measurements. A very successful example, and one of the first to have been described, was the electrochemical synthesis of 9(o-iodophenyl)9,10-dihydroacridine, by reduction of 9(o-iodophenyl)-acridine.[53] Chemical reducing agents either failed to reduce the reactant, or removed the iodine simultaneously with reduction of the acridine ring. Polarography showed two waves, separated by about 0.3 V. The first wave corresponded to reduction to the dihydro compound, the second to elimination of iodine. Controlled-potential electrolysis at a potential on the plateau of the first wave gave the desired product in good yield, at a high current efficiency.

Two of the techniques, described in Section II, which operate under conditions much more closely resembling those employed in preparative electrolyses are controlled-potential coulometry and steady-state polarization curve measurements. Measurements are normally performed in stirred solutions, although satisfactory polarization curves can be obtained in unstirred solutions for highly irreversible electrode reactions. Results from these techniques are readily applicable to preparative electrolyses, because the time scale, reactant concentration, mass-transport conditions, and extent of accumulation of products can be made very similar in both cases.

Under favorable circumstances, it may be possible to show that one of the reaction schemes in Table 1 is consistent with the experimental observations for a given reaction, or even to evaluate the rate constant for the charge-transfer step or for a coupled chemical reaction, if its value is within limits appropriate to the technique employed.

It must be stressed that organic electrode reactions often occur through complex mechanisms involving *sequences* of many steps. In such cases, the schemes listed in Table 1 may each refer only to one step in the overall reaction sequence. Electroactivity in organic compounds is often associated with the presence of functional groups containing oxygen or nitrogen, to or from which proton-transfer reactions occur rapidly.[54] In the presence of proton donors or acceptors, electron charge-transfer reactions of O- or N-containing compounds are often associated with proton transfers, thus increasing the complexity of the charge-transfer step in the overall sequence. For example, the reduction of p-nitrosophenol (I) is a commonly quoted example of the so-called ECE process. Electrochemical reduction (E) proceeds through p-hydroxylaminophenol (II), which undergoes a relatively slow chemical dehydration step (C) to form p-benzoquinoneimine (III), which is subsequently reduced (E) to p-aminophenol (IV):

I. Introduction

Compound II undergoes further reduction much less readily than does III, and II, unlike III, can be reoxidized to I. A variety of electrochemical techniques, discussed in Section II, can be used, on the basis of these reactivity differences, to measure the rate of the chemical step which converts II into III, and hence to verify this aspect of the mechanism. The two electrochemical (E) steps in the mechanism, however, are unlikely to involve simultaneous addition of two electrons and two protons, but each one much more probably involves a sequence of steps, e.g., for the first step

$$RNO \xrightarrow[(E)]{e} RNO^- \xrightarrow[(C)]{H^+} RNOH \xrightarrow[(E)]{e} RNOH^- \xrightarrow[(C)]{H^+} RNHOH$$

i.e., a mechanism designated ECEC. Other possibilities are that protonation precedes, or occurs simultaneously with, the charge-transfer steps, (or one of them), or that the two electrons are transferred in two separate steps before protonation occurs (i.e., the mechanism is EECC). Experimental evidence concerning the sequence of steps here is lacking, and is not readily obtained, because of the rapidity of the electron charge transfers and the proton transfers (the —NO, —NHOH system is polarographically reversible).[55] A scheme similar to the first one suggested above (ECEC) for RNO reduction has been proposed[56] for the reduction of aromatic hydrocarbons in the presence of proton donors, i.e.,

$$R + e \rightleftharpoons R \cdot^- \quad (E)$$
$$R \cdot^- + H^+ \rightarrow RH \cdot \quad (C)$$
$$RH \cdot + e \rightleftharpoons RH^- \quad (E)$$
$$RH^- + H^+ \rightarrow RH_2 \quad (C)$$

In the absence of proton donors, the two charge-transfer steps occur in succession,

$$R + e \rightleftharpoons R \cdot^- \quad (E)$$
$$R \cdot^- + e \rightleftharpoons R^{2-} \quad (E)$$

and both the anion radical $R \cdot^-$ and the di-anion R^{2-} may be stable in solution.

Evidently, then, the classification of the *p*-nitrosophenol reduction as an ECE process is a fairly gross simplification, since each of the E steps can itself be more complex than this overall classification indicates. The above scheme may be further complicated by hydrolysis of III to p-benzoquinone (V):

$$\underset{\text{III}}{\text{[quinone imine]}} \xrightarrow[-NH_3]{\underset{H_2O}{k'}} \underset{\text{V}}{\text{[p-benzoquinone]}}$$

This reaction is not significant during reduction of p-nitrosophenol, since III is reduced more readily than I, so that the second charge-transfer step occurs more rapidly than the chemical hydrolysis reaction. However, the rate constant k' for the hydrolysis step has been evaluated by studying the oxidation of p-aminophenol, considered as an EC process (IV $\xrightarrow{E}$ III $\xrightarrow{C}$ V)[123,154]; again the E "step" is obviously a sequence of reactions.

Any description of a reaction mechanism is incomplete without knowledge about adsorption of reactants and intermediates, including information on the type of adsorption isotherm involved,[64] so that the kinetics may be related to the species actually adsorbed at the potential concerned.[57] Information is also needed on orientation of the reacting species at the electrode surface, e.g., decarboxylation in the Kolbe reaction of a carboxylate radical adsorbed through the CO_2 group would appear to leave the the resulting alkyl radical R· too far away from the electrode for it to adsorb.[58]

3. Mass Transport Considerations

A current flow can only be maintained if unreacted electroactive material from the bulk of the solution is continually supplied to the electrode to replace that already converted. Migration, or the movement of charged electroactive species in an electric field, is seldom an important mass-transport process, since electrochemical studies are usually performed in the presence of an excess of inert supporting electrolyte, so that the transference numbers of any electroactive charged species present are reduced to negligible values. Diffusion, arising from the concentration gradient set up at the electrode–solution interface as a result of the electrode reaction, and convection, either natural, arising from thermal motion of the solution components in unstirred solution, or forced, where the process is assisted by stirring or flowing the solution, or by rotating or vibrating the electrode, then become the only significant mass-transport processes. The relationship between the various electrochemical parameters may be investigated in unstirred solution, or, particularly when the rate of the charge-transfer step itself is high, under conditions of forced convection. The most suitable form of electrode in the former case is a shielded one,[59] where a planar metal disc is sealed into a glass or plastic tube at some distance from the open end, so that the mantle limits diffusion to a direction normal to the electrode surface. Mercury pool electrodes may similarly be shielded when well-defined mass-transfer conditions are required. Transport toward electrodes of this type approximates to semi-infinite linear diffusion, and complete solution of the diffusion equations is possible.[34] Shielded electrodes of this type give best results in analytical applications, where limiting diffusion currents, proportional to concentration, are required. They are also the most suitable types to employ if the detailed shape of, e.g., a current–potential curve, or a potential–time curve at constant current, needs to be analyzed for obtaining kinetic parameters.

The dropping mercury electrode in polarography enables reproducible diffusion conditions to be established at each successive drop; in voltammetry at solid electrodes, best results are achieved when the electrodes are rotated at a constant rate (e.g., by a synchronous[60] or tachometer-controlled[61] motor). A rigorous hydrodynamic treatment is available[62] for rotating-disc electrodes, and a recent review of their application is available.[63]

II. METHODS OF INVESTIGATING ORGANIC ELECTRODE PROCESSES

Methods which are of value in the investigation of organic electrode processes, particularly from a mechanistic point of view, may be conveniently classified as either *electrochemical* or *nonelectrochemical*. In the latter category are included all those techniques that can be applied to the study of homogeneous and heterogeneous chemical reactions, i.e., techniques which are not applicable solely to electrochemical processes. Some of these techniques, for example, product isolation and identification, or reaction order determination,[64] are of equal importance and value for electrode processes and for nonelectrochemical reactions. Other information, e.g., salt and solvent effects, although often adding confirmation to an electrochemical mechanism deduced on other grounds, is usually less unequivocally diagnostic for electrode processes than for nonelectrochemical processes, because knowledge about transition states in electrode processes is much less detailed than in the latter case.

1. Electrochemical Methods

Electrochemical methods are concerned with the relationships existing between (usually pairs of) the variables potential, current or current density, time, and charge passed through the system. A large variety of techniques have been developed, depending on which of these parameters is controlled, the type of control exercised, and whether a steady-state or nonsteady-state response of the measured parameter is achieved. The most useful techniques are summarized in Table 2.

Many of these techniques have been exhaustively discussed in monographs,[65-73] or have been adequately treated in earlier volumes of this series.[8,75] The treatments which follow are therefore quite brief, and the references quoted should be consulted for details. Since most information about organic chemical aspects of electroorganic reaction processes has been gained from polarography, this technique is treated in somewhat more detail than the others.

(i) Controlled Potential

Many controlled-potential methods are available, depending on which other parameter is measured, and the type of control exercised over the potential, i.e., whether constant or varying; if varying, whether the variation is unidirectional or cyclic, single or multiple cycle, etc. Controlled-potential procedures in which the measured variable is current (or current density) are usually classed under the general title of *voltammetry*, the term polarography being best reserved for the case where a dropping mercury electrode (dme) is used.[70] Potential control is best ensured by the use of a potentiostat which may be programmed by electronic function generators to make available a wide range of potential control functions such as constant-potential, single or multiple potential step, potential ramp, and

Table 2
Electrochemical Methods of Investigating Organic Electrode Processes

Parameter controlled and type of control	Name of method	Parameter measured	Results presented as plot of:
1. Potential			
(a) Constant	Voltammetry	i	i–E
	Polarography (at dme)	i	i–E
	Potentiostatic polarization curve	i	E–log i
	Coulometry	Q or i and t	Q or i vs. t
(b) Stepped	Chronoamperometry	i as $f(t)$	i–t
(c) Varying	Linear sweep voltammetry	i as $f(t)$	i–E
	Cyclic voltammetry	i as $f(t)$	i–E
2. Current			
(a) Constant	Potentiometry ($i = 0$)	E	E–C
	Voltammetry	E	E–i
	Galvanostatic polarization curve	E	E–log i
(b) Stepped	Chronopotentiometry	E as $f(t)$	E–t or E–$t^{1/2}$
	Coulometry	Q as $f(C)$	—
	Galvanostatic charging	E as $f(t)$	E–t or E–Q
	Differential galvanostatic charging	dE/dt as $f(t)$	dE/dt vs. t
(c) Varying	Chronopotentiometry is possible with various current functions, but few applications are available		
3. Charge	Coulostatic	E	E–t
4. Concentration			
(a) Constant	Concentrostatic		
(b) Varying	(See text)	Reaction order	ln $i_{(E)}$ vs. ln c
5. Relaxation techniques	OC decay		E–t
	Repetitive square wave	Yield as $f(E)$ or as $f(t)$	
6. Miscellaneous	Impedance	Double-layer capacity or Faradaic impedance	
	Faradaic rectification		
	Oscillographic sine and square-wave methods		

cyclic functions, e.g., potential variation with time in a sinusoidal, square wave, or triangular wave manner. When varying potentials are required, the rate of change of potential is readily variable over a range of about 10^{-4} to 10^4 V sec^{-1}.[74] Three-electrode systems are usually employed, the reference electrode being terminated in a Luggin capillary placed close to the controlled electrode, to minimize iR effects.

Steady-state potentiostatic polarization curves, obtained with stationary or rotating electrodes, have been treated in detail in earlier volumes in this series[8,75] and elsewhere.[76-78] Results are analyzed in the form of Tafel plots, corresponding to the equation[21]

$$E = a' \pm b \log i$$

where E is the electrode potential corresponding to a current density i. The plus sign refers to anodic, the minus sign to cathodic processes. The slope b of the plot (Tafel slope) may be diagnostic of the mechanism of the electrode process.[22] In favorable cases, for which the standard electrode potential may be evaluated from experimental measurements, or estimated from thermodynamic data, the reversible potential for the electrode process under the conditions employed may be deduced. It is then possible to plot overvoltage η, instead of potential E, against $\log i$, where $\eta = E - E_r$, E_r being the reversible potential ($i = 0$). The Tafel slope b is unchanged by this procedure, which has the advantage that the exchange current density i_0 can also be evaluated.

The significance of the magnitude of the exchange current density has been discussed,[75-77] and Spiro[79] has pointed out the usefulness of the standard exchange current density i_0^0, i.e., the value of i_0 for unit concentration of reactants, in comparing the rates of electrode reactions.*

The advantage of steady-state polarization measurements in employing conditions similar to those involved in preparative electrolyses has already been mentioned. Measurements made at varying concentrations of reactant(s) permit *reaction orders* to be evaluated.[64] Apart, however, from measurements on the Kolbe reaction of some aliphatic carboxylic[5,24,25,80,81] and boronic[82,83] acids, reduction of aromatic ketones,[64] and fuel cell studies on

---

*Standardized double-layer conditions are also usually required, i.e., for a definite concentration of a nonspecifically adsorbed supporting electrolyte.

simple organic fuels,[6–9] the technique has not been widely applied to organic systems, although it has proved very satisfactory for elucidating electrode mechanisms for a number of simple inorganic compounds.[84–89] Until recently,[64,64a] information about the mechanisms of organic electrode reactions has come almost exclusively from investigations employing polarographic or related techniques. However, such approaches suffer important limitations in so far as basic electrode kinetic information, such as the role of adsorption of reactants, intermediates, or products, is rarely obtainable and the reaction order cannot be directly evaluated. Similarly, quantitative characterization of the kinetics of the reaction cannot be conveniently made if only a polarographic study of the reaction under diffusion-controlled conditions at low concentrations is made. Approaches which substantially extend the information obtained from polarography of organic reactions, particularly the electrode kinetic aspects, are evaluation of reaction order and its dependence on adsorption, relation of reaction order to the current–potential behavior, and evaluation of stereochemical effects where appropriate (see below).

Determinations of reaction order in relation to the adsorption isotherm and the kinetics of the reaction were made by Conway et al.[64] in the case of acetophenone reduction. In a later paper,[64a] they deduced general relations for reaction order as a function of coverage for a number of adsorption isotherms commonly employed in evaluating electrochemical adsorption. These approaches are essential for characterization of the physicochemical aspects of electroorganic reaction mechanisms and evaluation of the role of adsorption in the process.

Polarography[65,77] employing a dropping mercury electrode (dme) is usually classified as a constant-potential process because the rate of voltage change is so low that the electrode potential is effectively constant during the lifetime of a mercury drop. Currents are so small, in general, that iR effects are negligible, except in some nonaqueous solvents where R may be quite large, and a two-electrode system is satisfactory, i.e., auxiliary and reference electrodes are combined in one. Electronic iR compensators are available for use with nonaqueous solvents, but a three-electrode polarograph based on the potentiostatic principle is to be preferred. Polarography is a particularly valuable technique for preliminary application, because of the small amount of material required, the rapidity with which a polarogram can be obtained, and the number of deductions which can be made from even a single polarogram. Thus, electroactivity in the potential range investigated is immediately evident if a polarographic wave is observed. In favorable cases, polarograms (or voltammograms obtained with solid electrodes) may indicate the number of electron-transfer steps involved in the electrode reaction; these are distinguishable in the form of separate polarographic waves provided that the electron-transfer steps differ sufficiently in energy requirements ($E_{1/2}$ values differing by 150–200 mV). Diffusion current (wave height) measurements (for a given concentration) give an indication of the relative numbers of electrons transferred in each step, or, by comparison with a process of known electron number, involving a depolarizer if possible of similar molecular size and shape to that under study, the actual number of electrons involved may often be estimated. Confirmation of such estimates requires the application of the techniques of coulometry or controlled-potential electrolysis. Most investigators nowadays would regard verification by these techniques as an essential precaution, but many polarographic reports in the past have relied only on wave heights for evaluating the number of electrons involved. In some cases, the identities of the products have even been inferred solely on the basis of n values derived from wave-height measurements. In other cases, linear free-energy relationships have been used in conjunction with $E_{1/2}$ and wave-height measurements to identify the reaction occurring, and hence to indicate the nature of the products.[90] These procedures are justified on the basis that the small amount of chemical change occurring during a polarographic investigation renders product isolation and identification extremely difficult, if not impossible; the change to macroscale, controlled-potential electrolysis, while facilitating product isolation, so alters the experimental conditions that the results may not be directly comparable with polarographic results.[53a]

Experimental examination of the electrode process under diffusion-controlled conditions (e.g., in polarography), however, usually introduces complications in the interpretation of the kinetics and mechanism of the reaction unless it is a relatively reversible one; a large proportion of organic electrode reactions are, however, quite irreversible, and this necessitates careful choice of the approaches to be made in a kinetic-mechanistic study of the reaction.

By employing, as supporting electrolyte, a universal buffer solution, such as the Britton–Robinson buffer,[91] the involvement of hydrogen ions in the electrode process can readily be assessed. For an uncomplicated reduction process involving hydrogen ions, e.g.,

$$O + mH^+ + ne \rightleftharpoons R$$

if (and only if*) the reaction is polarographically *reversible*, then, at 25°C,[65]

$$E = E_{1/2} + (0.059/n) \log[(I_d - I)/I]$$

where

$$E_{1/2} = E° - (0.059/n) \log(D_O/D_R)^{1/2}(f_R/f_O) - (0.059m/n)pH$$

Hence

$$dE/d(pH) = -0.059m/n$$

n is evaluated from a plot of $\log[(I_d - I)/I]$ versus E, and m from the variation of half-wave potential, $E_{1/2}$, with pH.

If the charge transfer is polarographically irreversible, again, at 25°C,[65]

$$E = (0.059/\alpha n) \log(k°_{obs} t^{1/2}/0.76 D_0^{1/2}) + (0.059/\alpha n) \log[(I_d - I)/I]$$

or

$$E_{1/2} = (0.059/\alpha n) \log(k°_{obs} t^{1/2}/0.76 D_0^{1/2})$$

Here, $k°_{obs}$ is the rate constant for the charge transfer at the reference electrode potential for a given pH, t is the drop time, and α is the transfer coefficient.[92] If the reaction is of order m with respect to hydrogen ions, $k°_{obs}$ is a pseudo-first-order rate constant, i.e.,

$$k°_{obs} = k°[H^+]^m$$

where $k°$ is the value of $k°_{obs}$ at $[H^+] = 1$. Therefore $\log k°_{obs} = \log k° - m(pH)$ and, provided that α is pH-independent,

$$dE/d(pH) = -0.059m/\alpha n$$

---

*Few organic reactions are, in fact, really reversible, so that many reported evaluations of n are unreliable, and αn (see below) rather than n is evaluated.

In this case, analysis as above supplies values for m and αn. Since the number of protons involved changes by one when the pH passes through the pK_a value of O or R, several linear segments of different slopes may be observed in a plot of $E_{1/2}$ versus pH. From the intersections of these lines, the pK_a values of the species concerned may be evaluated.[93] A change in pH may affect the magnitude of the current, as well as $E_{1/2}$, if a kinetic process involving hydrogen ions precedes the charge-transfer step, e.g., as in the reduction of α-keto acids like pyruvic acid.[94] Thus, the un-ionized acid CH_3COCO_2H is reduced at less negative potentials than is the anion $CH_3COCO_2^-$. At a pH high enough for a significant proportion of the reactant to exist as the anion, but low enough for protonation of the anion to occur at a significant rate, then, as reduction of the un-ionized acid disturbs the ionization equilibrium, further protonation of the anion occurs, and the current due to reduction of this species increases. This current enhancement depends on the hydrogen ion concentration and on the rate constant for protonation of the anion. The latter can be evaluated from measurements of the kinetic current; detailed treatments are available,[95,96] and results obtained are generally in good agreement with those obtained by other methods, e.g., the rotating-disc method, the potentiostatic method, and chronopotentiometry.[95]

Kinetic processes other than protonation, e.g., dehydration of formaldehyde hydrate

$$CH_2(OH)_2 \rightleftharpoons CH_2O + H_2O$$

have also been studied polarographically.[97] These, and kinetic proton transfers preceding reduction, are examples of the general reaction scheme 3(a) in Table 1 and the theory applicable to polarographic studies of such reactions has been developed by Vesely and Brdicka,[97] and Koutecky.[98] The interest in such studies is usually that of measuring the rate of the chemical reaction preceding electron transfer, but no information is obtained about the mechanism of the electrochemical step. The application of polarography or related methods to kinetic studies of this type really constitutes a specialized analytical use of these methods, in which the concentration being measured depends on both the reactant concentration and the rate of the reaction preceding charge transfer.

II. Methods of Investigating Organic Electrode Processes

A qualitative inspection of the shape of the polarographic wave may give some information about the reversibility of the process, since an irreversible wave is more drawn out along the potential axis than is a reversible one. More definite conclusions can be drawn from a detailed analysis of the wave shape. Thus, for a charge-transfer process uncomplicated by a coupled chemical reaction, the slope of a plot of potential against $\log[(I_d - I)/I]$ permits a distinction into reversible (slope $2.3RT/nF$), irreversible (slope $2.3RT/\alpha nF$), and quasireversible processes (involving two segments of slopes $2.3RT/nF$ and $2.3RT/\alpha nF$, respectively). The distinction between reversible and irreversible processes is not an absolute one,[70] and the major purpose of the distinction is to assist in further treatment of the results, e.g., in evaluating the order with respect to hydrogen ion, as discussed above, or in applying a linear free-energy relationship to a series of depolarizers of regularly varying structure. For reversible processes, the $E_{1/2}$ values are closely related to $E°$ and hence to $\Delta G°$ values, and they thus have thermodynamic significance, but give no information about mechanism. For irreversible processes, however, $E_{1/2}$ values are related to $\log k$ (where k is the heterogeneous rate constant for the charge transfer process) and hence to the free energy of activation, $\Delta G^{\ddagger}$. They are also dependent, in these cases, on the value of αn, which Perrin[99] and others have interpreted, inappropriately, as the order of the reaction in electrons.[100] Thus, for irreversible processes, $E_{1/2}$ values can be used as measures of $\log k$ (and $\Delta G^{\ddagger}$) only for series of compounds which all react by the same mechanism, and for which, specifically, αn is constant throughout the series, or varies with substitution in the same way as does the appropriate substituent parameter (σ, σ^*, etc.).[43] That these are not infrequently met requirements is attested to by the large number of series to which linear free-energy relationships have successfully been applied.[46] Dryhurst and Elving[171] have discussed the correlation of molecular-orbital calculated energy levels and the energy required to add or remove an electron, and have shown that, in addition to considerations of electron-transfer reversibility, effects of adsorption and solvation energy must also be taken into account.

When chemical reactions accompany the charge-transfer process, the type of reaction scheme involved may often be determined by wave-shape analysis, or from the dependence of half-wave potential on droptime or on the depolarizer concentration.[129–134] A summary of the relevant treatments is available, with references to the original literature.[101] Dependence of the limiting current on the mercury head h can be used to distinguish diffusion-controlled waves (limiting current proportional to $h^{1/2}$) from adsorption waves (limiting current proportional to h) and kinetic waves (limiting current independent of h).[65] Analysis of current–time curves provides one of the best means of distinguishing reversible from irreversible processes. In the former, the current at any point on the wave is proportional to $t^{1/6}$; for irreversible waves, this relationship applies only at the top of the wave, where current is diffusion-controlled, but at the foot of the wave, the current is determined by the kinetics of charge transfer, and is proportional to $t^{2/3}$.[102] For satisfactory application of this criterion, special procedures are required to permit the current–time curve to be recorded on the first drop of mercury to form after the potential is applied.[103]

Relatively slow chemical reactions subsequent to charge transfer can affect the shape of the polarographic wave obtained at an electrode dropping at the normal rate (drop time 2–6 sec). If the drop time is decreased (rapid polarography) or if a streaming mercury electrode[104] is employed, the extent of occurrence of the reaction may be insufficient to have the same influence on the wave shape as in polarography under standard conditions. This situation is indicated by a difference in wave shape, or in $E_{1/2}$, when a conventional polarogram is compared with a rapid polarogram, or with a voltammetric curve obtained at a streaming mercury electrode.[105,106] This procedure is analogous to variation of scan rate in linear-sweep or cyclic voltammetry, to variation of the time between pulses in the double-step potentiostatic procedure, or the time before current reversal in reverse current chronopotentiometry, or to variation of the rotation rate in the rotating-disc and ring-disc techniques.

Information gained by the above polarographic technique may be supplemented by results from several related methods, e.g., the Kalousek commutator[107] has been used to verify that the first step in reduction of diketones is a reversible one-electron step[108]; ac polarography[109] and oscillographic polarography[110] are also useful for distinguishing reversible from irreversible processes, but few applications to organic systems have been described.

II. Methods of Investigating Organic Electrode Processes

Polarography possesses many well-known advantages as a method for investigating organic electrode processes, not the least of which are the commercial availability of reliable, sensitive polarographic instruments, and the benefit of the continuously renewed surface of the dme. The high hydrogen overvoltage on mercury renders it particularly suitable as an electrode for reduction purposes; its freedom from oxide films and ease of purification for reuse by distillation are most desirable characteristics for an electrode material. The small amount of material required, the rapidity with which results are obtained, and the deductions possible from a rapid preliminary consideration of the results (as discussed above) are advantages shared by other related techniques, e.g., cyclic voltammetry, or, to a lesser extent, chronopotentiometry. Undoubtedly, one of the major assets associated with polarography is in the extent of its accumulated literature. The cathodic electroactivity of most organic functional groups is known from studies under a variety of conditions, including the effects of positional substitution, the presence of other interacting groups, pH effects, etc. Many linear free-energy relations have been precisely established[46,172,173]; if a particular combination of conditions has not yet been explored, it is often possible to estimate the electrochemical behavior (e.g., with respect to type and sequence of steps) reasonably precisely from established results. Half-wave potentials, for linear free-energy relationships, are more easily determined than most other measures of stability or reactivity, e.g., equilibrium or rate constants, ionization potentials, electron affinities, etc. Estimation of the optimum potential for preparative reduction is considerably faster and easier than trial-and-error comparison of large-scale electrolytic procedures.

The biggest disadvantages of polarography are the limited anodic potential range available on mercury, so that oxidation processes generally have to be studied on solid noble electrodes, and the limitation, from a kinetic point of view, imposed by the necesssity of usually operating under diffusion-controlled conditions. The significance of reactant adsorption prior to charge transfer cannot always be assessed, and the possible difficulties in "scaling up" polarographic results for preparative electrolyses have been discussed above.

Voltammetry at solid electrodes has been less extensively applied than polarography, largely because of practical difficulties associated with electrode deactivation by filming. In the absence of this complication, most of the polarographic treatments mentioned are applicable to voltammograms, provided that mass-transport conditions are precisely defined; this is best achieved by rotating the electrode at a constant speed, using a synchronous or tachometer-controlled motor.[60,61] The rotating-disc and ring-disc electrodes[62,63,174] offer great promise for elucidating mechanisms in organic systems; a recent example of such an application is the verification by Adams and co-workers[111] of the ECE mechanism for oxidation of diphenylanthracene (DPA) in the presence of pyridine. In the absence of a chemical reaction removing the product of the charge transfer step, a plot of $i_L/\omega^{1/2}c$ (where i_L is the limiting current at depolariser concentration c, measured at a rotation rate ω) against $\omega^{1/2}$ is a horizontal straight line at an ordinate level indicating the number of electrons involved (since i_L is proportional to n). In an ECE process,

$$O_1 + n_1 e \xrightarrow{E} R_1 \xrightarrow[k]{C} O_2 + n_2 e \xrightarrow{E} R_2$$

if the rate of the chemical step is low, or if the rate of rotation of the electrode is high, so that R_1 is removed from the vicinity of the electrode before a significant amount of chemical change can occur, a plot of the above type indicates n_1 electrons involved. As the rate of the chemical reaction increases, or the rotation rate decreases, the apparent number of electrons involved rises to $(n_1 + n_2)$ as the second charge-transfer step becomes significant. In the absence of nucleophiles, or in the presence of pyridine at high rotation rates, DPA gives a limiting current corresponding to a one-electron process,

$$DPA \rightleftharpoons DPA^+ + e$$

At low rotation rates, in the presence of pyridine, the apparent n value increases to two, indicating the subsequent steps

$$DPA^+ + py \xrightarrow{k} DPApy^+$$
$$DPApy^+ \rightleftharpoons DPApy^{2+} + e$$

followed by

$$DPApy^{2+} + py \rightarrow DPA(py)_2^{2+}$$

where py represents pyridine. Identification of the product verified this reaction scheme.

Coulometry at constant potential, in addition to its fundamental role in establishing the number of electrons involved in an overall electrode reaction, has been developed, mainly by Bard and Meites and their co-workers,[112-114] as a powerful method of investigating mechanisms and evaluating rate constants when chemical reactions are coupled with charge transfer. An estimate of the number of electrons involved in a polarographic reduction can be made from the diffusion current, using the Ilkovic equation,[115] if the diffusion coefficient of the depolarizer is known, or by comparison of wave height with that for a process of known electron number involving a molecule of similar size and shape, so that the D values are approximately the same. Alternatively, by using a small volume of solution ($0.3 \to 0.5$ ml), a change of concentration of 50% or so may be obtained in a few hours of electrolysis at a dme, and the concentration change can be measured polarographically or, e.g., spectrophotometrically. This technique has been called millicoulometry.[116] For oxidations, a vibrating platinum electrode[117] can be successfully employed with small volumes of solution.[118] Generally, however, the number of electrons involved is determined by macrocoulometry, or controlled-potential electrolysis (cpe), using a stirred mercury pool electrode for reduction,[31] or a large-area platinum or other inert electrode for oxidations. The amount of chemical change is determined usually by isolating and measuring the product and the electric charge corresponding to this change by using a coulometer[119] or a mechanical[120] or electronic[121] integrator, or by measuring the area under a current–time curve.

The occurrence of chemical reactions preceding, following, or in parallel with the charge transfer step may give rise to nonintegral apparent n values or to apparent n values which vary with concentration of reactant or time. For the oxidation of tetraphenylborate ion in acetonitrile at a platinum electrode, Geske found that the apparent n value increased from 1.02 to almost 2 as the concentration of tetraphenylborate was reduced from 10 mM to 0.5 mM in unbuffered solution.[122] At low hydrogen ion concentration, in a buffered solution, the apparent n value was close to 2, suggesting a mechanism involving hydrogen ion in a chemical step,

$$Ph_4B^- \xrightarrow{-2e} Ph_2B^+ + PhPh$$
$$Ph_2B^+ + H_2O \xrightarrow{fast} Ph_2BOH + H^+$$
$$H^+ + Ph_4B^- \longrightarrow Ph_3B + PhH$$

Bard and Santhanam[114] have reviewed the mechanistic applications of controlled-potential coulometry.

Particularly useful for obtaining rate constants for follow-up reactions (EC mechanism) is the technique of reversal coulometry, in which, after a measured charge, insufficient for complete reaction, has been passed, the potential is switched to a value where only the product of the first electrode reaction is electroactive, e.g., a product formed by reduction is oxidized. The ratio of charges used in the two steps is unity, if no subsequent reactions remove product. However, the occurrence of chemical reactions involving the product of the first electrode process causes a decrease in the charge required for the second electrolysis, e.g., in the oxidation of p-aminophenol (discussed in Section I), the rate constant for the hydrolysis of benzoquinoneimine to benzoquinone was evaluated by oxidizing for a known time at constant potential and determining the charge required Q_0, then shifting the potential to a value at which the imine was reduced, consuming charge Q_R. From the ratio of Q_0/Q_R, a value of $1.11 \times 10^{-2} \sec^{-1}$ was obtained for the rate constant, in reasonable agreement with that found by other methods.[123]

Chronoamperometry, or the potential step method, involves measuring the current as a function of time following the application of a constant potential to a stationary electrode in a quiescent solution. If the potential is large enough for the electrode process to be diffusion-controlled, the current I_t at any time t after commencing electrolysis is given by the Cottrell equation[124]

$$I_t = nFAD^{1/2}C/\pi^{1/2}t^{1/2}$$

for an n-electron process involving a depolarizer at concentration C. A is the electrode area and D the diffusion coefficient of the depolarizer. If a current–time curve is recorded, the close relationship of this technique to controlled-potential coulometry is evident, since the charge consumed is given by the area under the curve of I versus t. In controlled-potential coulometry, however, efficient stirring is employed to minimize electrolysis time, whereas chronoamperometry is performed in quiescent solution with a shielded linear diffusion electrode. One of the main applications of chronoamperometry has been its use in evaluating electrode areas and diffusion coefficients for reactants. If the electrode potential is not high enough for an irreversible process to be diffusion-controlled,

II. Methods of Investigating Organic Electrode Processes

the current is given by an equation of the form

$$I = I_d\{\pi^{1/2}\lambda[\exp(\lambda^2)]\text{ erfc }\lambda\}$$

where $\lambda = kt^{1/2}/D^{1/2}$ and

$$\text{erfc }\lambda = 1 - (2/\pi^{1/2})\int_0^\lambda (\exp -z^2)\,dz$$

Evaluation of k, the rate constant for charge transfer, is thus possible, in principle, from these equations, but the method does not appear to have been applied for this purpose. A modification of the method has, however, been applied by Alberts and Shain[126] to the measurement of the rate constant for a chemical reaction interposed between two charge transfers, e.g., the reduction of p-nitrosophenol (ECE mechanism discussed in Section I). At short times after application of the potential, the current corresponds to that for a two-electron process forming p-hydroxylaminophenol, but as time proceeds, the concentration of benzoquinoneimine increases because of the chemical dehydration step. The second charge transfer begins to contribute significantly to the total current, which changes to that corresponding to a four-electron process. The time at which this change occurs depends on the rate constant k for the chemical step; evaluation of k is thus possible. The precautions necessary for avoiding errors in the application of chronoamperometry to measuring rate constants for follow-up reactions have been discussed by Oldham and Osteryoung.[127] Schwarz and Shain[128] have also commented on the requirements for successful applications of related methods, e.g., the shape and $E_{1/2}$ shifts of polarographic waves,[129,130] and potential–time behavior of chronopotentiometric curves[131-134], the charge transfers must be reversible and the standard potential for the charge-transfer step in the absence of kinetic complications must be known. These requirements can be avoided by the double potential step method.[128] In the first step, e.g., for a reduction process,

$$O + ne \rightarrow R$$

R is generated at a controlled rate for a short time, then the electrolysis conditions are changed so that the amount of unreacted R is measured, usually by reoxidation to O. The electrochemical reaction need not be reversible; all that is required is that some conditions can be found under which R can be oxidized to O. Often, the initial potential, before the first potential step, proves satisfactory; here, diffusion-controlled reoxidation of R gives a measure of the amount of unreacted R present, and the Faradaic current for this process is easily separated from the charging current. An EC example is provided by the reduction of azobenzene in acid solution, when the product, hydrazobenzene, undergoes rearrangement to benzidine:

$$PhN=NPh + 2H^+ + 2e \xrightarrow{E} PhNHNHPh \xrightarrow[k]{C} H_2NC_6H_4C_6H_4NH_2$$

As in the cases of the EC and ECE reaction types discussed in Section I, the first step is more complex than the simple designation E implies, and probably constitutes an ECEC process; thus,

$$Ph-\ddot{N}=\ddot{N}-Ph + e \rightleftharpoons Ph-\ddot{N}-\dot{N}-Ph^- \qquad E$$

$$Ph-\ddot{N}-\dot{N}-Ph^- + H^+ \rightarrow Ph-\overset{\overset{H}{|}}{\ddot{N}}-\dot{\ddot{N}}-Ph \qquad C$$

$$Ph-\overset{\overset{H}{|}}{\ddot{N}}-\ddot{N}-Ph + e \rightleftharpoons Ph-\overset{\overset{H}{|}}{\ddot{N}}-\ddot{N}-Ph^- \qquad E$$

$$Ph-\overset{\overset{H}{|}}{\ddot{N}}-\ddot{\dot{N}}-Ph^- + H^+ \rightarrow Ph-\overset{\overset{H}{|}}{\ddot{N}}-\overset{\overset{H}{|}}{\ddot{N}}-Ph \qquad C$$

Linear sweep voltammetry (sometimes called potential sweep chronoamperometry): In a quiescent solution at a stationary electrode, the potential is scanned linearly with time at a rate varying between several mV sec^{-1} and several hundred mV sec^{-1}, while the corresponding current is recorded. Solid electrodes give best results if shielded, but this technique appears to be less sensitive to the shape of the electrode than are many other techniques, and satisfactory results are often obtained at wire electrodes. Hanging mercury drop electrodes have been satisfactorily employed for reductions. Results, presented in the form of a plot of i versus E (or t, since E varies linearly with t), combine the features of voltammetry with those of chrono-amperometry—current is small until the potential reaches a value at which an electrode reaction occurs, then rises rapidly as for a

II. Methods of Investigating Organic Electrode Processes

voltammetric wave, since the rate of reaction is accelerated by increasing potential. The electrode reaction causes a decrease in reactant concentration near the electrode surface, and since the solution is unstirred, the current decreases with time, as in chronoamperometry. Thus a maximum or peak current is observed, which, at a given voltage sweep rate, is proportional to the reactant concentration. In a series of papers, Shain and co-workers[135] and Saveant and Vianello[136] have considered the application of this technique to most of the reaction schemes listed in Table 1.

The limitations regarding the quantitative interpretation of results obtained by the use of this procedure, and the related one of cyclic voltammetry, have been discussed by Conway et al.,[137] particularly for cases where chemisorbed intermediates are involved. One useful application, which offers promise in cases where steady-state polarization curves are difficult to measure, e.g., because of deactivation of solid electrodes by adsorption of intermediates or products, is the deduction of a Tafel slope for an irreversible process from corresponding current and potential values at the foot of the curve (currents up to 10–20% of the peak current). In this region, as for an irreversible polarographic wave, the current is kinetically controlled, and the magnitude of the current at any given potential is independent of the sweep rate. The procedure has recently been applied to inorganic[84,85] and organic[64,138] systems.

Cyclic voltammetry is an extension of the linear potential sweep method in which the direction of potential sweep is reversed periodically so that electrolysis of the products of the forward sweep occurs on sweep reversal. Like polarography, this is an excellent preliminary technique to apply, since each peak observed on the current–potential curve corresponds to a separate electrode process. Reversible processes give corresponding anodic and cathodic peak currents at peak potentials separated by $59/n$ mV at 25°C, while irreversible processes exhibit only one of these peaks. Useful information may be obtained about the occurrence of chemical reactions subsequent to charge transfer by comparing curves obtained at different scan rates (i.e., allowing different time intervals for occurrence of the subsequent reaction) or by comparing the curve obtained in the first sweep with those for subsequent sweeps. A peak, or an anodic–cathodic complementary pair of peaks, appearing only in sweeps after the first, indicate the presence in the system after the first charge transfer of an electroactive species not originally present. If the peak does not correspond simply to the reverse of the original charge transfer process (e.g., a reduction peak from the product of an initial oxidation step), then it arises from a species produced in a chemical reaction following the initial charge transfer. This type of behavior has been observed in the oxidation of aromatic amines, e.g., triphenylamine (TPA) oxidizes to a cation radical (TPA$^+$) which undergoes a chemical reaction (dimerization or coupling with more TPA) to form tetraphenyl benzidine (TPB), which is more readily oxidized (than is TPA) in two discrete one-electron steps, to form TPB$^+$ and TPB^{2+}. The first anodic sweep shows a single anodic peak, corresponding to formation of TPA$^+$; the following (and all subsequent) cathodic and anodic sweeps show two cathodic and one anodic peaks at less positive potentials than the TPA$^+$ peak (the second anodic peak corresponding to formation of TPB^{2+} is obscured by the TPA$^+$ peak). An authentic sample of TPB gave two anodic and two cathodic peaks which matched these well; the presence of TPB after electrolysis was confirmed also by EPR spectra.[139] The reaction sequence is

$$Ph_2\ddot{N}-\!\!\left\langle\bigcirc\right\rangle \underset{}{\overset{-e}{\rightleftharpoons}} Ph_2\overset{+}{\ddot{N}}-\!\!\left\langle\bigcirc\right\rangle \leftrightarrow Ph_2-\overset{+}{\ddot{N}}=\!\!\left\langle\bigcirc\right\rangle\cdot$$

$$\text{TPA} \qquad\qquad\qquad \text{TPA}^+$$

$$2\text{TPA}^+ \xrightarrow[-2H^+]{} Ph_2\ddot{N}-\!\!\left\langle\bigcirc\right\rangle\!\!-\!\!\left\langle\bigcirc\right\rangle\!\!-\ddot{N}Ph_2$$

$$\text{TPB}$$

$$\text{TPB} \overset{-e}{\rightleftharpoons} Ph_2\ddot{N}-\!\!\left\langle\bigcirc\right\rangle\!\!-\!\!\left\langle\bigcirc\right\rangle\!\!-\overset{+}{\ddot{N}}Ph_2 \xrightarrow{-e}$$

$$\text{TPB}^+$$

$$Ph_2\overset{+}{\ddot{N}}=\!\!\left\langle\bigcirc\right\rangle\!\!=\!\!\left\langle\bigcirc\right\rangle\!\!=\overset{+}{\ddot{N}}Ph_2$$

$$\text{TPB}^{2+}$$

II. Methods of Investigating Organic Electrode Processes

(ii) Controlled Current

Potentiometry, i.e., voltammetry at zero current, has found most application in organic systems for measuring concentrations, e.g., hydrogen ion concentration, using the quinhydrone electrode,[140] or for structure–stability or structure–reactivity correlations, e.g., for quinones.[141] Few organic redox systems are reversible enough for potentiometry to be useful, and since no current flows, and no net chemical change occurs, no information is obtainable from potentiometric measurements about the mechanism of the process which determines the electrode potential.

Voltammetry, in which nonzero current is controlled and the corresponding potential measured, is an alternative which usually requires somewhat simpler equipment than controlled-potential voltammetry.

Galvanostatic polarization curves often agree well with those obtained potentiostatically,[84] but, particularly where inhibition effects are observed, more information is available from potentiostatic polarization curves, hence the latter are to be preferred.[142]

Galvanostatic coulometry has advantages, already mentioned, with regard to evaluation of the charge passed simply as the product of current and time. The disadvantages of galvanostatic processes have been discussed in Section I. Recently Parker has argued in favor of galvanostatic coulometry,[143] indicating how multiple estimates of n could be obtained during a single electrolysis. By employing a comparatively high reactant concentration and an electrode of large area, and operating at a potential such that the applied current density was considerably less than the limiting current density, the potential remained constant until electrolysis was nearly complete, hence most of the disadvantages of galvanostatic procedures were avoided.

Closely related to galvanostatic coulometry is the measurement of galvanostatic charging curves for the determination of adsorption pseudocapacitance, and thus for obtaining information about electrode coverage by (electroactive) adsorbed intermediates.[50,137] At constant current, the time required to cover the electrode surface with adsorbed intermediates can be estimated from the duration of the potential arrest when electrode potential is measured as a function of time during the constant-current charging process. A potential–time trace at constant current is equivalent to a potential–charge (E–Q) trace if the current is known. Differentiation with respect to time of such a trace is equivalent to dE/dQ, i.e., to the reciprocal of the electrode capacitance, and gives better resolution of electrosorption processes. Kozlowska and Conway[144] used an operational amplifier to obtain the time derivative of the potential–time curve; this and the potential–time curve itself were displayed simultaneously on the screen of a double-beam oscilloscope. The method was applied to the detection of adsorbed intermediates in the anodic oxidation of formate.

The charge required to cover a platinum electrode with a monolayer of adsorbed hydrogen atoms, or of oxide, has been well established.[145] The degree of coverage of the electrode surface by organic reactants or intermediates can, in favorable cases, be estimated from the decrease in charge required when the uncovered portion of the electrode is covered by hydrogen or oxygen.[146] Further details on measurement of adsorbed intermediates are given in the review by Gileadi and Conway.[50]

Chronopotentiometry, or the measurement of potential as a function of time at constant current (other current functions may be used),[147–149] is related to both galvanostatic coulometry and the galvanostatic charging method. In the former of these methods, the relationship desired is that between charge consumed and amount of chemical change (i.e., concentration change); in the latter, the charge to cover an electrode completely with intermediates, or to remove the intermediates from a completely, or partly, covered electrode, or to cover the remaining free surface sites with hydrogen or oxygen is the quantity sought. In chronopotentiometry, it is the relationship between potential and time which is significant. As electrolysis proceeds, e.g.,

$$O + ne \to R$$

the concentration of O at the electrode surface decreases, and the potential changes slowly as long as both O and R are present. Provided the applied current density exceeds the limiting current density, the concentration of O at the electrode surface ultimately decreases to zero, and a potential transition occurs to a value at which another electrode process occurs. The time required before this potential change occurs is known as the transition time τ; $\tau^{1/2}$ and the square root of time $t^{1/2}$ are the quantities corresponding in

II. Methods of Investigating Organic Electrode Processes

chronopotentiometric treatments to diffusion current I_d and current I, respectively, in polarography. Thus, for a reversible process,[150]

$$E = E_{\tau/4} + 2.3(RT/nF)\log[(\tau^{1/2} - t^{1/2})/\tau^{1/2}]$$

$E_{\tau/4}$ is the potential when $t = \tau/4$, and corresponds to the half-wave potential $E_{1/2}$ in polarography. For an irreversible system,[150]

$$E = 2.3(RT/\alpha nF)\log(nFACk/I) + 2.3(RT/\alpha nF)\log[(\tau^{1/2} - t^{1/2})/\tau^{1/2}]$$

(A is the electrode area, I the applied current, and k the rate constant for the charge transfer step). Reinmuth[151] has discussed the distinction of various possible reaction schemes from a detailed analysis of the shape of potential–time curves. The method has the advantage over polarography that measurements of transition times can be made over a considerably wider range of τ values than is the case with the analogous variable, I_d, in polarography. Variation of τ, to bring it within a conveniently measurable range, is possible through variation of the applied current density, i.e., by changing I or A (or both). In spite of these advantages, fewer applications to organic systems have been reported than for polarography. Experimentally, the technique is much more sensitive to vibration, and surface-history effects for solid electrodes can markedly influence the shape and reproducibility of the potential–time curves.

An extension of the technique known as *reverse current chronopotentiometry*, is applicable to the detection of chemical reactions following charge transfer. The potential–time curve is recorded with current flow in one direction, to obtain the forward transition time τ_f. The trace is then repeated, and at any time t_f up to, but not exceeding τ_f, the current direction is reversed while potential is still recorded. A transition time τ_r for the reverse process is thus obtained. For a reversible or irreversible charge-transfer process uncomplicated by chemical reaction, Berzins and Delahay[152] showed, for current magnitudes equal in both directions, and with current reversal at time t_f, that

$$\tau_r = t_f/3$$

Geske[153] showed by this method that oxidation of cycloheptatriene in acetonitrile produced tropylium ion and a proton,

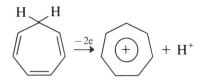

Both products were reducible, and two transitions were obtained on current reversal, one of which was independently shown to involve hydrogen ion; the ratio of forward to reverse transition times agreed satisfactorily with predictions.

If a rapid chemical reaction follows charge transfer (EC process) to produce a product which is not electroactive on current reversal, the reverse transition time is reduced to the extent that the occurrence of the chemical reaction decreases the concentration of the product of the charge transfer step. Testa and Reinmuth[154] evaluated the rate constant k for the hydrolysis of benzoquinoneimine produced on oxidation of p-aminophenol, from the decrease in τ_r.

(iii) Controlled Charge

The coulostatic or charge-step method[155] involves the injection of a controlled charge into an electrode, originally at equilibrium, and the recording of the potential–time response as equilibrium is reestablished. The differential double-layer capacity can be determined, even in the presence of the Faradaic process, and kinetic parameters for the latter may also be evaluated. The technique has not been applied to organic systems; potential applications of greatest value would be to processes occurring in aprotic solvents, where charge transfer can be studied uncomplicated by subsequent reactions, for sufficiently stable products of the initial charge-transfer step.

(iv) Controlled Concentration

A variation of the controlled-current technique was recently suggested by Oldham and Spanier[156] in which the electrolysis current is varied with time so as to control the surface concentration of electroactive species. Two examples were discussed; in the first the

II. Methods of Investigating Organic Electrode Processes 157

surface concentration was kept constant by applying a current varying with $t^{-1/2}$. This appears to be of theoretical interest only because of the practical difficulty of *initiating* such a current function. Constant surface concentration of depolarizer can also be attained by application of a suitable potential, rather than by controlling the current. This procedure has already been discussed under the heading *chronoamperometry*: As normally employed, the reactant concentration at the electrode surface is zero, but the potential can be adjusted so that the surface concentration is other than zero but still constant. Hence, if potential, rather than current, is controlled, the method would appear to be identical with controlled-potential chronoamperometry.[157] The second example, also discussed by Oldham and Spanier, for a surface concentration decreasing linearly with time, requires a current varying with $t^{1/2}$. Such a current function does seem practical, since square-root functions are readily achieved using integrated circuits,[158] but no applications of the procedure have yet been reported.

A variety of other electrochemical techniques exists, and some applications have been made mainly to rapid, uncomplicated charge-transfer processes in inorganic systems; few applications to organic systems have been described. These techniques include open-circuit potential decay measurements,[159] impedance measurements,[160] Faradaic rectification,[161] and oscillographic sine- and square-wave methods.[162]

A relaxation technique which should prove valuable for organic systems is the repetitive square-wave electrolysis procedure introduced by Wilson and Lippincott[163] and further investigated by Fleischmann *et al.*[163a] In this method, the yields of products are measured as a function of both the upper and lower potential levels of a repetitive square-wave sequence, and of the electrolysis time, i.e., pulse frequency or duty cycle. In contrast to most kinetic measurements which supply information about the rate constant for the *slow* step in a reaction sequence, this technique gives information about the *fastest* step. In the electrolysis of aqueous solutions of sodium acetate, using a train of square pulses of potential of controlled amplitude and width, it was concluded that the discharge of acetate radicals was rate-determining, and that this was followed by rapid decarboxylation to form adsorbed (cf. Ref. 5) methyl radicals, which rapidly dimerized.

$$CH_3CO_2^- \xrightarrow{r.d.s.} CH_3CO_2(ads) + e$$
$$CH_3CO_2(ads) \xrightarrow{fast} CH_3(ads) + CO_2$$
$$2CH_3(ads) \xrightarrow{fast} C_2H_6$$

Limiting values were obtained for the rate constants for the first and third steps. The discrepancy between these conclusions and those of other workers[5] that either the second step, or the adsorbed radical-anion step,

$$CH_3(ads) + CH_3CO_2^- \rightarrow C_2H_6 + CO_2 + e$$

was rate-determining, has not yet been resolved. The difficulties regarding orientation at the electrode of the adsorbed carboxylate radical have already been mentioned.[58]

Another recently developed electrochemical technique, *thin-layer electrochemistry*,[164] offers promise, once the experimental difficulties in producing a satisfactory thin-layer cell have been overcome, of wide application to organic electrode processes. Because the entire quantity of reactant is confined in a thin layer at the electrode surface (thinner than the diffusion layer arising from semiinfinite diffusion) the terms in the current–potential relationships which arise from diffusion-controlled mass transfer are eliminated or simplified. The techniques of linear sweep and cyclic voltammetry, coulometry, chronoamperometry, and chronopotentiometry can be applied. Complicating secondary reactions between products of the charge-transfer step and reactants are eliminated. In an EC process, e.g.,

$$O + ne \underset{}{\overset{E}{\rightleftharpoons}} R \overset{C}{\underset{k}{\rightarrow}} P$$

R cannot escape from the electrode into the bulk solution, so that reactions which are much slower than those accessible to conventional semiinfinite diffusion techniques can be studied. The elimination of diffusion complications leads to an increase in sensitivity in many applications, e.g., in linear sweep voltammetry, peak currents for reversible processes are proportional to n^2, instead of $n^{3/2}$ as for the conventional technique; in steady-state voltammetry, limiting currents are proportional to the diffusion coefficient D, rather than to $D^{1/2}$ as in conventional methods. Current densities employed in thin-layer cells are at least one order of magnitude smaller than those

II. Methods of Investigating Organic Electrode Processes

obtained with, e.g., conventional linear sweep voltammetry, which permits the investigation of irreversible systems which react at rates close to those of the background reactions of electrode oxidation or supporting electrolyte reduction. The applicability of the technique to organic systems has been demonstrated by evaluating rate constants for the hydrolysis of p-benzoquinoneimine produced by electrooxidation of p-aminophenol,[165] and for the benzidine rearrangement.[166] Many more applications should appear once the potentialities of the method are more widely appreciated.

Electrode reactions of organic compounds seldom involve charge transfer only, because the products of such charge-transfer reactions are usually highly reactive radical-ion species which undergo subsequent reactions with other solution components. The major activity in the study of organic electrode reactions has therefore been in the detection and characterization of these subsequent reactions. Those electrochemical techniques that produce a response proportional to concentration, and particularly those capable of rapid response, namely rotating-disc and ring-disc techniques, cyclic voltammetry, double-step chronoamperometry, controlled-potential coulometry, and reverse-current chronopotentiometry, have been successfully applied to the detection of concentration changes with time for the products of charge-transfer reactions. Such concentration changes indicate subsequent reactions, and, in favorable cases, lead to rate constants for these chemical reactions. A related situation has been studied in detail by Dubois and co-workers[167] and by others[168] in which an inorganic reagent, e.g., Br_2, is generated electrolytically, and its rate of reaction with an organic substrate is measured. Although formally an EC process, this is not classified as an electroorganic example, since the reaction is not primarily dependent on charge transfer, but can be studied just as readily with the reagent produced *in situ* by chemical means, or introduced initially. Chemical reactions preceding charge transfer, such as the protonation of acid anions or dehydration of carbonyl hydrates, are also not dependent on the occurrence of an electron charge-transfer process; their investigation by electrochemical methods relies only on the analytical capabilities of such methods. Examples are known, however, where the course of a charge-transfer reaction is modified by the occurrence of a prior chemical reaction, e.g., in the reduction of diazoacetophenone at pH above 5, the reactant is $PhCOCHN_2^+$, and ammonia and α-aminoacetophenone are produced in a six-electron process; at pH below 5, protonation forms the reactive species $PhCOCH_2N_2^+$, and reduction occurs in a two-electron process, similar to the reduction of alkyl or aryl halides, to produce acetophenone plus nitrogen.[169] Changes in steric conformation of products may also be produced by protonation prior to charge transfer; see Section V.[170]

2. Nonelectrochemical Methods

These fall into four main categories: (i) reaction order determination, (ii) product isolation, identification, and determination, (iii) detection of intermediates, and (iv) comparison with nonelectrochemical processes.

(i) Reaction Order

Vetter[78] has emphasized the importance of reaction order determination for the elucidation of electrochemical reaction mechanisms. The procedure adopted usually, in steady-state techniques, is to evaluate the order from the coefficient $[\partial(\log i)/\partial(\log C)]_E$, and several examples are available for organic systems.[5-9,24,25,64,80,81,83] Conway et al.[64,64a] showed how the reaction order determination in organic electrode reactions must be related to adsorption at the electrode and to the current–potential relation for cases where the rate constant is coverage-dependent. Applications to acetophenone reduction were investigated. Determination of order with respect to hydrogen ion from polarographic $E_{1/2}$ values has been discussed; similar results could be obtained from chronopotentiometric $E_{\tau/4}$ values. Conway and co-workers[64] have applied H_2O/D_2O isotope effect measurements to the kinetics of the proton-transfer step in the reduction of aromatic ketones. The observed kinetic isotope effect i_{H_2O}/i_{D_2O}, close to unity, indicated that proton transfer was not involved in the transition state and hence in the rate-controlling step.

(ii) Product Isolation, Identification, and Determination, and Detection of Solution Intermediates

Yield and Coulombic efficiency are of obvious importance for any synthetic organic electrochemical process. Any satisfactory analytical procedure may be employed, e.g., the product may be

II. Methods of Investigating Organic Electrode Processes

isolated and weighed, or analyzed by gas chromatography, infrared, visible, or ultraviolet spectroscopy, mass spectrometry, etc. For optically active products, polarimetry is necessary. Similar procedures may be employed for detecting intermediates in solution, provided that the lifetime of these is adequate for the method employed. Benzil can be reduced in acid or neutral solution to stilbenediol, which isomerizes to benzoin.[175] Stilbenediol exists in two isomeric forms, *cis* and *trans*, produced from the corresponding conformations of benzil,

$$\begin{array}{c} Ph-C=O \\ | \\ Ph-C=O \end{array} \xrightarrow[2H^+]{2e} \begin{array}{c} Ph-C-OH \\ \| \\ Ph-C-OH \end{array}$$

s-*cis* benzil   *cis*-stilbenediol

$$\begin{array}{c} Ph-C=O \\ | \\ O=C-Ph \end{array} \xrightarrow[2H^+]{2e} \begin{array}{c} Ph-C-OH \\ \| \\ HO-C-Ph \end{array}$$

s-*trans* benzil   *trans*-stilbenediol

Grabowski and co-workers[175] showed that the relative yields of *cis*- and *trans*-stilbenediol were markedly dependent on electrode potential, and explained this as resulting from the influence of the electric field on the conformational equilibrium between the two forms of benzil; high fields tend to favor the more polar *cis* form. Stapelfeldt and Perone[176] measured the rates of conversion of the stilbenediol isomers into benzoin both electrochemically (using linear sweep voltammetry) and spectrophotometrically (the UV spectra of the two isomers are different); good agreement was observed between rate constants obtained by both techniques.

EPR spectroscopy has been used to detect radical anions in aprotic media, e.g., that from reduction of nitrobenzene which is stabilized by charge delocalization over the ring.[177] Radical cations from oxidation of amines have been detected by the same technique.[139]

Optically transparent electrodes[178] permit the spectroscopic investigation of the layer of solution adjacent to the electrode surface; internal reflectance spectroscopy can be used for a similar purpose with opaque electrodes.[179] Tallant and Davis used infrared internal reflection to monitor concentration changes during reduction of quinone and benzil.[180] Radical anions from these reactants showed absorption bands at 1500 and 1375 cm^{-1}, respectively, the decrease in frequency being ascribed to the lower bond order in the radical anions. Absorption due to benzoquinone at 1650 cm^{-1} was negligible in the steady state, indicating a small concentration of this reactant at the electrode surface. Radical anions from benzophenone and acetophenone were more reactive, and no infrared bands attributable to them could be identified. Difficulties in using the technique were encountered from the high electrical resistance of the electrode and the need for relatively concentrated solutions, with consequent large currents; the potential was not constant over the surface of the electrode, so that precise determination of potential effects was not possible.

Methods of detecting *adsorbed* intermediates have been reviewed.[50]

(iii) Comparison with Nonelectrochemical Processes

It is possible to explain electrode reactions of organic compounds in terms of the same intermediates, carbonium ions, carbanions, and radicals as are involved in nonelectrochemical processes. Hence a comparison of product distributions from, e.g., an electrode process considered to proceed through a carbonium ion with that from homogeneous reactions believed to involve the same intermediate can supply information on the relative reactivities of such species produced by different methods. This type of comparison is discussed in Section III.

Linear free-energy relationships provide very satisfactory procedures for comparison of reactivities in homogeneous systems.[43] In such relationships, the sign of the reaction parameter ρ is often indicative of the nature of the reaction; thus, positive ρ values are observed in homogeneous reactions involving nucleophilic attack in the rate-determining step.[43] Most electrochemical reductions of aromatic compounds show positive ρ values, consistent with attack by a negatively charged reactant, namely the electron. Many aliphatic reductions show negative ρ values, perhaps indicating a chemical reaction preceding charge transfer. For the reduction of nitroparaffins, ρ changes from -0.4 at pH 2 to $+0.2$ at pH 9, but the nature of the reacting species also changes over the same pH range from RCH_2NO_2 to $RCHNO_2^-$.[181] Detailed treatments of linear free-energy relations in polarography are available.[46,99,172,173]

III. INTERMEDIATES IN ELECTROORGANIC CHEMISTRY

Most organic compounds are electrically neutral species having an even number of electrons. Transfer of a single electron to or from a neutral molecule, if unaccompanied by any simultaneous chemical change, leads to the formation of an electrically charged species having an unpaired electron (except for the uncommon case where the reactant itself is a radical). Thus a radical cation $R\cdot^+$ results from loss of a single electron, or a radical anion $R\cdot^-$ from addition of a single electron. Di-cations R^{2+} and di-anions R^{2-} result from removal or addition of two electrons in succession, although electrostatic considerations indicate that the second electron transfer occurs less readily than the first. Uncharged radicals are formed by transfer of a single electron either from an anion, e.g., in the Kolbe reaction,[5]

$$RCO_2^- \to RCO_2\cdot + e$$

or

$$RCO_2^- \to R\cdot + CO_2 + e$$

or to a cation, as in the reduction of quaternary ammonium salts,[182,183]

$$PhCH_2NEt_3^+ + e \to PhCH_2\cdot + NEt_3$$

Neutral reactants may also produce uncharged radical (adsorbed) intermediates, through dissociative chemisorption, e.g., in the electrooxidation of methanol, carbon–hydrogen bond breakage is indicated by radiochemical measurements on isotopically substituted molecules.[184,185] Electrochemical oxidation of methanol is explained in terms of an adsorbed intermediate of the type $\geqslant C-OH$,[186–188] after H dissociation. Analogous radical-type adsorbed species have been suggested as intermediates in the oxidation of many organic fuels, e.g., formic acid[8,189,190] and hydrocarbons.[8,191]

Uncharged radicals could also be produced from electrically neutral molecules if electron addition was accompanied by simultaneous anion elimination:

$$RX + e \to R\cdot + X^-$$

or if electron removal occurred simultaneously with elimination of a cation (most probably H^+), e.g.,

$$RH \to R\cdot + H^+ + e$$

The energetics of this latter type of process in the dissociative adsorption of ethylene,

$$H_2C{=}CH_2 \to \underset{\underset{M}{\cdot}}{HC}{=}\underset{\underset{M}{\cdot}}{CH} + 2H^+ + 2e$$

have been discussed by Bockris et al.,[192] who concluded that associative adsorption,

$$H_2C{=}CH_2 \to \underset{\underset{M}{\cdot}}{H_2C}{-}\underset{\underset{M}{\cdot}}{CH_2}$$

is favored on energetic grounds (cf. the infrared evidence).

A radical may be formed by anion elimination from a radical anion, produced in a prior charge-transfer step, e.g.,

$$RX + e \to RX\cdot^- \to R\cdot + X^-$$

but the subsequent (as distinct from simultaneity with charge transfer) elimination of the anion is usually considered as a property of the initially formed radical anion. Examples of this type of behavior are given below. In an analogous way, proton removal to produce an uncharged radical from an initially formed cation radical, e.g.,

$$RH \xrightarrow[-e]{} RH\cdot^+ \xrightarrow{B} R\cdot + BH^+$$

where B is a base, is not uncommonly observed reaction for cation radicals, as discussed below. Protonation of a radical anion,

$$R\cdot^- + H^+ \to RH\cdot$$

and combination of a radical cation with an anion

$$R\cdot^+ + X^- \to RX\cdot$$

are two other possible sources of radical intermediates; these also

III. Intermediates in Electroorganic Chemistry

exemplify typical properties of radical anions or radical cations, respectively.

Formation of cations R^+ or anions R^- (carbonium ions or carbanions if the charge is on carbon) requires, generally, two charge transfers and a chemical reaction. The formation of carbonium ions in the Hofer–Moest reaction,

$$RCO_2^- \to RCO_2\cdot + e$$

$$RCO_2\cdot \to R\cdot + CO_2$$

or
$$\left.\begin{array}{c} R\cdot \to R^+ + e \\ RCO_2\cdot \to R^+ + CO_2 + e \end{array}\right\} R^+ \to ROH \text{ or } RCOOR$$

and the formation of carbanions in the reduction of alkyl halides,

$$RX \xrightarrow{e} RX\cdot^- \to R\cdot + X^-$$

$$R\cdot \xrightarrow{e} R:^-$$

or

$$RX\cdot^- \xrightarrow{e} R:^- + X^-$$

are typical examples.

Since protonation of radical anions and reaction of radical cations with nucleophilic or basic reagents occurs very readily, the choice of solvent is of the utmost importance if the existence of intermediates of the above types is to be established. Thus anions or radical anions are stabilized in aprotic solvents, particularly those of a basic nature, e.g., hexamethylphosphoramide. In this solvent, radical anions from benzaldehyde have been shown to have lifetimes 500–5000 times as long as in solvents like dimethylformamide and ethylene glycol dimethyl ether.[51] Since most of the investigations which aim at detecting intermediates use techniques like cyclic voltammetry or chronopotentiometry, and operate in the millimolar concentration range for the reactant, the water content of the nonaqueous solvent is obviously of crucial importance in determining the lifetime of any reactive intermediates. The presence of water may also have a pronounced effect in anodic studies on solid electrodes, through the formation of an oxide layer on the electrode, thus altering its electrocatalytic properties. Adoption of vacuum line and dry-box techniques are essential precautions in this work.[193]

House and co-workers[194] found that the radical anion derived from 2,2,6,6-tetramethyl *trans*-4-hepten-3-one had a half-life exceeding 50 sec in dimethyl formamide and hexamethylphosphoramide. Proton donors or lithium salts when added to the solution decreased the half-life markedly and the racemic hydrodimer

$$\begin{array}{cc} Bu^t & Bu^t \\ | & | \\ H-C\!\!-\!\!-\!\!-\!\!-\!\!C-H \\ | & | \\ Bu^tCOCH_2 & CH_2COBu^t \end{array}$$

was formed. The effect of lithium cations was explained as arising from exchange of a solvent ligand of the lithium ion by the radical anion; this ion pairing lowered the repulsion between radical anions, and dimerization followed. After electrolysis in the absence of proton donors and addition of the β-deutero ketone to the electrolysis solution, EPR measurements showed the spectra of both nondeuterated and the β-deutero anion radical, confirming that electron exchange between the anion radical and the ketone was rapid.

Stabilization of cations and cation radicals is affected by the choice of solvents of low basicity and low nucleophilic reactivity, but another factor of significance is the solvating power of the solvent for cations. Thus tetra-*p*-anisylethylene is oxidized in methylene dichloride solution in two successive one-electron steps, but in acetonitrile, which has a higher solvating power for cations, a single two-electron step is observed.[195]

Reviews are available on the properties and uses of nonaqueous aprotic solvents in electrochemistry,[196-198] and the investigation of new solvent systems is an active field in electrochemistry today.[199-212] One problem associated with the use of nonaqueous solvents is the provision of a suitable reference electrode; satisfactory designs for acetonitrile, dimethylsulfoxide, propylene carbonate, and dimethylformamide have been reviewed by Butler.[213]

Apart from their effects on the lifetime and reactivity of electrogenerated anions, cations, or radicals, another important factor associated with the use of nonaqueous solvents is the extended potential range which they make available, without interference from oxygen or hydrogen evolution. In aqueous solutions, these processes often set the limits to attainable potential, although cases are known,

III. Intermediates in Electroorganic Chemistry

e.g., the Kolbe reaction, where hydrocarbons are formed anodically at high potentials in aqueous solution, the oxygen evolution reaction being almost completely inhibited.[5] In nonaqueous solvents, the potential limits are often set by the discharge potentials of the supporting electrolyte; for perchlorates and nitrates, the anodic processes occurring are[214-216]

$$ClO_4^- \rightarrow ClO_4\cdot + e$$

and

$$NO_3^- \rightarrow NO_3\cdot + e$$

and the perchlorate or nitrate radicals attack the solvent or solute. The nature of the anion of the supporting electrolyte has a pronounced effect on the product distributions from the oxidation of toluene in glacial acetic acid.[216] In the presence of acetate ion, the major products are those arising from ring substitution, namely o- and p-acetoxytoluene. With nitrate or p-toluenesulfonate as supporting electrolyte anions, side-chain substitution or side-chain coupling products predominate, including significant amounts of bibenzyl and ethylbenzene; these products suggest the intervention of benzyl and methyl radicals. The results in the presence of acetate ion can be explained by toluene oxidation followed by (or accompanied by, in a concerted process) combination of the carbonium ion and the acetate anion,

$$C_6H_5CH_3 \xrightarrow{-e} C_6H_5CH_3^+ \xrightarrow[-H^+,\,-e]{CH_3CO_2^-} CH_3CO_2C_6H_4CH_3$$

Toluenesulfonate and nitrate anions, being poorer nucleophiles than acetate ion, do not participate so readily in the nuclear substitution reaction, but are themselves oxidized to radicals, particularly at high anodic potentials. The tosylate or nitrate radicals then abstract a hydrogen atom from toluene to produce the benzyl radical, e.g.,

$$NO_3\cdot + C_6H_5CH_3 \rightarrow HNO_3 + C_6H_5CH_2\cdot$$

Methyl radicals, presumably involved in the formation of ethylbenzene, are formed from acetate as in the Kolbe reaction.[5] Benzyl acetate, which constitutes up to 48% of the products, may arise from further oxidation of a benzyl radical to the benzyl cation followed by reaction with an acetate anion,

$$C_6H_5CH_2\cdot \xrightarrow{-e} C_6H_5CH_2^+ \xrightarrow{CH_3CO_2^-} C_6H_5CH_2O_2CCH_3$$

or by combination of a benzyl radical with an acetoxy radical, the latter being formed by oxidation of acetate

$$CH_3CO_2^- \xrightarrow{-e} CH_3CO_2\cdot$$
$$C_6H_5CH_2\cdot + CH_3CO_2\cdot \longrightarrow C_6H_5CH_2O_2CCH_3$$

This type of radical combination has been proposed for the formation of methanol in the Hofer–Moest reaction of acetate ion in alkaline solution.[217] It is an alternative pathway to that involving carbonium-ion intermediates.

$$H_2O \xrightarrow{-e} \cdot OH(ads) + H^+$$
$$CH_3CO_2^- \xrightarrow{-e} CH_3\cdot(ads) + CO_2 + e$$
$$CH_3\cdot(ads) + \cdot OH(ads) \longrightarrow CH_3OH$$

Some specific examples of the general reactions outlined above are given below, followed, in Section IV, by applications depicting particular synthetic steps, and in Section V, by stereochemical applications. The examples have been chosen to be representative, rather than comprehensive, since other reviews are available.[3-5,8,18-20,46,50,51,73,99,198,218,248]

1. Anion Radicals and Anions

Aromatic hydrocarbons, particularly polycyclic ones, have supplied many examples of the general reaction types above. The aromatic π-system can act as either a source or a sink of electrons, and structural variations, which alter the extent of charge delocalization, can give rise to intermediate species of widely differing stabilities. These processes have been extensively studied by Aten and Hoijtink,[56] Peover,[198] and Adams[218] and their co-workers. Reduction of polycyclic aromatic hydrocarbons in aprotic solvents occurs in two successive one-electron steps to give the anion radical, then the di-anion,

$$R \xrightarrow{e} R\cdot^- \xrightarrow{e} R:^{2-}$$

The first electron-transfer step occurs very rapidly, and rate constants which have been measured for this step have values close to the upper limit of measurement by electrochemical techniques. The second step is much slower, as would be expected for the addition of an electron to a negatively charged species. Thus, for reduction

III. Intermediates in Electroorganic Chemistry

of anthracene in dimethyl formamide at 25°C, Aten and Hoijtink[56] quote standard rate constants $k°$ for the two charge-transfer steps (at the reversible potentials -1.95 and -2.55 V versus SCE, respectively) of >4.0 and 0.0091 cm sec^{-1}. For aryl-substituted ethylenes, the difference between the $E°$ values for the two electron-transfer steps is of the same order, about 0.5 V, but the rate constants for the two steps show much smaller differences. For dibiphenylene ethylene, the $k°$ values (at $E°$ values of -1.00 and -1.50 V versus SCE) are >6.3 and 1.7 cm sec^{-1} respectively, in dimethyl formamide at 25°C. Benzene itself and isolated double bonds are not readily reduced, but electrolytically generated electrons in solvents like aliphatic amines and hexamethylphosphoramide have been successfully employed for this purpose.[219] The question of the involvement of hydrated electrons in electrode processes has been discussed by Conway and MacKinnon,[220] who concluded that the general mechanism of electroreduction usually must involve direct electron transfer from the metal to the oxidant molecule.

Langer and Yurchak[221] have recently described a process for electrogenerative hydrogenation of the benzene ring, which requires no external applied potential. The anodic process

$$H_2 \rightarrow 2H^+ + 2e$$

is coupled with the cathodic process

$$C_6H_6 + 6H^+ + 6e \rightarrow C_6H_{12}$$

and quantitative conversion to cyclohexane occurs. Toluene is converted to methyl cyclohexane with 80% efficiency, based on hydrogen consumption.

In the presence of proton donors, the reduction of polycyclic aromatic hydrocarbons appears to be an ECEC process,

$$R \xrightarrow{e} R\cdot^- \xrightarrow{H^+} RH\cdot \xrightarrow{e} RH{:}^- \xrightarrow{H^+} RH_2$$

The second charge transfer occurs much more readily after protonation of the radical anion, as would be expected on electrostatic grounds; molecular orbital calculations also indicate that the radical RH· should have a higher electron affinity than to the parent hydrocarbon R.[222] Thus the two separate one-electron waves observed in aprotic media are replaced by a single two-electron wave in the presence of proton donors. In the case of phenanthrene, dimerization of the radical RH· competes with the second charge-transfer step, and an insoluble hydrodimer, 9,9′-dihydro-10,10′-diphenanthrene, is formed.[223] This is not a commonly observed reaction for aromatic hydrocarbons, but many examples are known of hydrodimerization involving activated alkenes.[27,224–32] Hydrodimerization of acrylonitrile appears to involve the carbanion rather than a radical coupling process, since the radical $\dot{C}H_2CH_2CN$ is reducible at the potential employed. The experimental observations are consistent with the mechanism[232]

$$CH_2CH\,CN \xrightarrow{2e} \bar{C}H_2\bar{C}H\,CN \xrightarrow{2H^+} CH_3CH_2CN$$
$$\downarrow CH_2CH\,CN$$
$$NC\bar{C}HCH_2CH_2\bar{C}H\,CN \xrightarrow{2H^+} NC(CH_2)_4CN$$
$$\downarrow CH_2CH\,CN$$
$$\text{oligomers}$$

Thus, at low pH, the yield of propionitrile, CH_3CH_2CN, increases at the expense of adiponitrile. Tetraalkylammonium salts are used as supporting electrolytes, partly to increase the water solubility of the reactant, and partly because adsorption of tetraethylammonium cations on the mercury electrode leads to the formation at the electrode surface of a "water-free" zone which discourages proton abstraction from water by the carbanion and hence promotes dimerization.[233,234]

A typical carbanion reaction (e.g., of Grignard reagents) is addition to CO_2 to form a carboxylate anion and this reaction is exhibited by electrogenerated carbanions and carbanion radicals. Thus in the presence of carbon dioxide, reduction of benzyl chloride gives phenylacetic acid,[235] stilbene gives 1,2-diphenylsuccinic acid,[236] diphenyl acetylene gives diphenylsuccinic acid, together with diphenyl maleic and diphenyl fumaric acids,[223] and naphthalene and phenanthrene produce 1,4-dicarboxy-1,4-dihydronaphthalene and 9,10-dicarboxy-9,10-dihydrophenanthrene, respectively.[223] Stereochemical aspects of these reactions are discussed in Section V.

If a functional group capable of existing as a stable anion is present in the radical anion resulting from uptake of a single electron, this group may be eliminated as an anion, leaving a radical which may dimerize or add another electron, then a proton. Halo-

III. Intermediates in Electroorganic Chemistry

genated nitroaromatic compounds give rise to examples of all these types of behavior, ranging from those where anion radicals stable enough for recording of EPR spectra are formed[117,118] to those where a halide ion is readily lost to form a neutral radical.[239–243] On reduction, *m*-nitrobenzyl chloride or bromide gives only *m*-nitro-toluene, whereas the *o* and *p* isomers add one electron in acetonitrile solution to form the radical anion which dimerizes; only a small amount of *o*- or *p*-nitrotoluene is formed. The dimer can add a further two electrons to form a di-anion, and the *o*- or *p*-nitrotoluene takes up another electron, forming an anion radical again. Delocalization of the charge on the anion radical over the aromatic ring(s) stabilizes these products, but aliphatic nitro-compounds, on reduction to form anion radicals, may lose a nitrite ion,[244]

$$RNO_2 \xrightarrow{e} RNO_2^{-} \cdot \rightarrow R \cdot + NO_2^{-}$$

The disproportionation reaction of anion radicals to form di-anions,

$$2R \cdot^{-} \rightarrow R + R:^{2-}$$

may be involved in some dimerization reactions which are ascribed to coupling of radical anions. For example, benzophenone gives a stable blue solution of the radical anion in dimethyl formamide.[245] The stability is lower in the presence of water, when a second-order reaction occurs, as measured by EPR spectroscopy.[246] This could be interpreted as the disproportionation reaction

$$2R \cdot^{-} \rightarrow R + R:^{2-} \xrightarrow{2H^+} RH_2$$

Formation of the pinacol

$$\begin{array}{cc} R_2C- & CR_2 \\ | & | \\ OH & OH \end{array}$$

on reduction of ketones could involve addition of the dianion to the carbonyl group of the reactant,

$$R_2C^{-} + R_2CO \rightarrow R_2C\!\!-\!\!CR_2 \xrightarrow{2H^+} R_2C\!\!-\!\!CR_2$$
$$\;\;|\;|\;\;\;|\;\;\;\;\;\;\;\;\;\;\;\;\;\;\;|\;\;\;\;\;|$$
$$O^- \;O^- \; O^- \;\;\;\;\;\;\;\;\;\;\;\; OH \; OH$$

instead of the radical dimerization usually postulated,[64]

$$2R_2C^{\cdot}\underset{O^-}{|} \rightarrow R_2C\underset{O^-}{|}\!\!-\!\!CR_2\underset{O^-}{|} \xrightarrow{2H^+} R_2C\underset{OH}{|}\!\!-\!\!CR_2\underset{OH}{|}$$

Mixed pinacols, obtained by reduction of two ketones together, could be explained on the same basis.[247]

2. Cation Radicals and Cations

Cation radicals and cations are generally more reactive, i.e., less stable, than anionic species, and the important stabilizing effect of solvation has already been mentioned. Structural variations in the cationic species can alter the stability quite markedly. Criteria of stability include detectability and identifiability of an EPR spectrum for the cation radical, or electrochemical observations such as the ratio of the anodic to cathodic peak currents in cyclic voltammetry (the ratio is 1 for highly stable species, less than 1 if subsequent reactions remove the anodic product before the cathodic sweep is completed). In linear sweep voltammetry, the quantity $i_p/v^{1/2}C$ (where i_p is the anodic peak current for a reactant concentration C at a voltage scan rate of v) should be constant, for a stable cationic species, over several orders of magnitude of change in v. Since i_p depends on $n^{3/2}$, the number of electrons involved n can be evaluated from the magnitude of the above quantity. Using a rotating-disc electrode, the ratio $i_L/\omega^{1/2}C$, where i_L is the limiting current for rotation rate ω, or the ratio of forward to reverse transition times in reverse current chronopotentiometry, or the ratio of forward to reverse charges in reverse current coulometry or double potential step chronoamperometry, can also give information about the occurrence of follow-up chemical reactions, and hence about the stability of the species produced in the charge-transfer step. Examples of the application of the above techniques to the detection of EC processes have been described above.[111,112,122,123,126]

Adams has reviewed the effects of structural changes on the stability of cation radicals obtained from aromatic polycyclic hydrocarbons.[218] Molecules in which a high degree of charge delocalization can occur give more stable cation radicals than do those in which the charge tends to be localized in a few positions. HMO-calculated unpaired electron distributions correlate satis-

III. Intermediates in Electroorganic Chemistry

factorily in many cases with observed stabilities, e.g., the anthracene cation radical with high electron density on positions 9 and 10 is quite unstable, whereas the perylene radical cation, where the charge is much more evenly distributed over the (larger) ring system, is quite stable, particularly in nitrobenzene solution.[249] The stabilities of the very reactive cation radicals can be improved by substitution of inert blocking groups into the reactive positions. Aromatic groups are more effective than aliphatic ones, probably because of greater charge delocalization. Thus the 9-methyl anthracene radical cation is unstable,[249] those from 9,10-dimethylanthracene and 9-phenylanthracene are moderately stable,[249] while 9,10-diphenyl anthracene gives a very stable cation radical.[249-251] Another stable cation radical is also formed from rubrene, 5,6,11,12-tetraphenyl tetracene;[249,250] if the electrode potential is cycled from an anodic value at which the cation radical is formed to a value sufficiently cathodic to produce the radical anion, or if the latter is produced simultaneously at a nearby cathode, these two species combine to form excited-state molecules, which decay with emission of light (electrochemiluminescence)[252-255]

$$R^{\cdot+} + R^{\cdot-} \rightarrow R + R^* \xrightarrow{h\nu} 2R$$

Further effects of substitution on the stabilities of cation radicals have been studied by Zweig et al.[256]

Anodic aromatic substitution, e.g., acetoxylation, discussed above, probably occurs through an ECE mechanism involving the cation radical,

$$RH \xrightarrow{-e} RH^{\cdot+} \xrightarrow{X^-} R\begin{array}{c}H\\ \diagdown\\X\end{array} \xrightarrow[-H^+]{-e} RX$$

Apart from acetoxylation,[257-264] other substitution reactions which have been investigated include bromination,[261] cyanatation (substitution by CNO),[262] cyanation (substitution by CN),[263,266-269] methoxylation,[14,262,270-274] and methylation.[257,263] An anodic aromatic substitution which appears to involve a different mechanism is thiocyanation of reactive aromatic substrates such as phenol in acetonitrile[275] and anisole in aqueous acetic acid,[276] where substitution occurs at potentials below those required to oxidize the aromatic substrate. Thus, the reaction probably involves

$$SCN^- \xrightarrow{-e} SCN \rightarrow \tfrac{1}{2}(SCN)_2$$
$$\underset{RSCN}{\overset{RH\downarrow}{}}$$

Side-chain substitution reactions of methyl-substituted aromatic hydrocarbons have been observed, e.g., formation of benzyl acetate in the acetoxylation of toluene.[216] Carbonium ions are known to react readily with nitriles, so that incorporation of the solvent in the product is not unexpected, for example, when hexamethyl benzene is oxidized in acetonitrile, and water subsequently added, side-chain acetamidation occurs, the suggested mechanism being[263]

Yields as high as 38% are obtained; two factors evidently of significance are the stabilizing effect of substitution by blocking methyl groups, discussed above, and the resonance stabilization of benzyl-type carbonium ions.[277] Polarographic and voltammetric studies indicate a reversible two-electron transfer followed by an irreversible chemical reaction,[278] but later work threw some doubt on the two-electron step.[279] Hexaethylbenzene showed a one-electron transfer followed by a slow proton loss, and it was suggested that, in the case of hexamethylbenzene, proton loss from the cation radical is so fast that the rapid sweep voltammetric technique cannot distinguish the ECE mechanism from the EE. In mixtures of acetonitrile

III. Intermediates in Electroorganic Chemistry

and acetic acid, both acetamidation and acetoxylation occurred in the presence of perchlorate anions; with BF_4^- anions, acetoxylation predominated. It was suggested that adsorption of BF_4^- at the anode and specific solvation of the ion by acetic acid increased the local concentration of the latter at the anode, hence favoring acetoxylation.[279]

Parker and Eberson[280] have pointed out that nucleophiles which are likely to react with positive carbon centers are also bases, and that proton removal from a cation radical may compete with attack on the positive carbon. The cation radical from 9,10-diphenylanthracene was found to react faster with sterically unhindered 3,5-lutidine than with the sterically hindered isomer 2,6-lutidine; reaction can occur only at the shielded 9,10 positions in the cation radical. In contrast, the 9,10-dimethylanthracene cation radical showed little difference in reactivity with the above nucleophiles, suggesting that the site of reaction was a methyl proton; thus

[Reaction scheme: 9,10-dimethylanthracene cation radical (CH₃ top, Me bottom) + N → 9-methylene-10-methylanthracene (CH₂ top, Me bottom) + NH⁺]

(cf. benzyl-type resonance stabilization).[277] Isolation of the dimer

[Structure: Me–(anthracene)–CH₂–CH₂–(anthracene)–Me dimer]

supports this explanation. Peak-height measurements in cyclic voltammetry of 9-phenylanthracene in the presence of the two lutidines showed that proton removal was much faster than nucleophilic reaction at the positive carbon in the product of the second charge transfer in the ECE mechanism, when sterically hindered nucleophiles were employed; thus

for sterically hindered nucleophiles N, $k_B \gg k_N$, while for unhindered nucleophiles, $k_B \approx k_N$.

Inclusion of substituents like OH, OCH_3, NH_2, $NHCH_3$, or $N(CH_3)_2$ increases the ease of oxidation of an aromatic molecule, because of the possibility of removing an electron from a lone pair, and accommodating a positive charge on the heteroatom. Aromatic amines have been studied in detail by Adams and co-workers.[218] A detailed reaction scheme to explain the products obtained from dimethylaniline, DMA, was proposed recently by Hand and Nelson[281] (TMB = tetramethylbenzidine):

$$PhṄMe_2 \xrightarrow{-e} Ph\overset{+}{Ṅ}Me_2 \longleftrightarrow \cdot \langle \rangle = \overset{+}{N}Me_2$$
$$(DMA) \qquad (DMA^+)$$

$$\xrightarrow[-2H^+]{\text{dimerization}} Me_2N-\langle\rangle-\langle\rangle-NMe_2 \xrightarrow{-2e} TMB^{2+}$$
$$(TMB)$$

$$DMA + DMA^+ \xrightarrow[-2H^+]{-e} PhN\begin{matrix}Me \\ \\ CH_2C_6H_5NMe_2\end{matrix}$$

III. Intermediates in Electroorganic Chemistry

$$\xrightarrow{H^+} PhNHMe + {}^+_\cdot CH_2C_6H_5NMe_2 \xrightarrow{DMA}$$

[Structure: Me_2N–C$_6$H$_4$–CH$_2$–C$_6$H$_4$–NMe$_2$]

↓ DMA$^+$

[Structure: Crystal violet – Me_2N–C$_6$H$_4$–C(=C$_6$H$_4$=$^+$NMe$_2$)–C$_6$H$_4$–NMe$_2$]

(Crystal violet)

↑ −2e

[Structure: Me_2N–C$_6$H$_4$–CH(C$_6$H$_4$–NMe$_2$)–C$_6$H$_4$–NMe$_2$]

Radical cation intermediates have been detected by EPR measurements[139,287]; introduction of blocking groups into the p position in the aromatic ring increases the stability very markedly.[282] Detailed studies of structural effects on stability are available.[283,284]

Studies on heteroatom compounds other than amines have been much less detailed, but radical cations have been detected on oxidation of nitrosobenzene,[285] Schiff bases,[286] ethers,[287] and phenols.[287] Evidence for the involvement of phenoxonium ion intermediates in the oxidation of phenols was given by Vermillion and Pearl,[289] who showed that, at low pH, a two-electron process occurred,

$$\phi OH \rightarrow \phi O^+ + 2e + H^+$$

but at high pH, where the phenol is ionized, a one-electron process was observed,

$$\phi O:^- \rightarrow \phi O\cdot + e$$

In these cases, deactivation of the electrode by film formation renders controlled-potential electrolysis and product identification difficult.

Many years earlier, Fieser[290] had discussed a reproducible electrolytic potential, called the "critical potential," which is a condition for occurrence of the reaction

$$\phi O^- \rightarrow \phi O\cdot + e$$

These critical potentials correlate well with theoretical measures of reactivity, such as charge densities.[291,292] With phenols, electrochemical polymerization occurs.

Many other examples of aromatic cationic intermediates have been discussed in a recent review of the anodic reactions of aromatic compounds.[293]

Suitably substituted aliphatic compounds give evidence for the formation of cation radicals, or even di-cations on anodic oxidation. For example, tetrakis-diethylaminoethylene (TDE) in acetonitrile undergoes two successive one-electron transfer steps, at the remarkably negative half-wave potentials of -0.75 and -0.61 V (vs aq. SCE), corresponding to the formation of the cation radical $TDE^+\cdot$ and the di-cation TDE^{2+}.[294]

$$TDE \xrightarrow{-e} TDE^+\cdot \xrightarrow{-e} TDE^{2+}$$

The latter forms a stable perchlorate $TDA^{2+}(ClO_4^-)_2$ which, on polarographic reduction, gives two one-electron waves at the same potentials as those above, indicating reversibility for the charge-transfer steps. The stability of TDE^+ was indicated by the invariance in the shape of the EPR spectrum (in dimethylformamide solution) at temperatures up to 120°C. On the basis of the half-wave potentials, the equilibrium constant for the process

$$TDE^{2+} + TDE \rightleftharpoons 2TDE^+$$

was estimated as 2.3×10^2.

III. Intermediates in Electroorganic Chemistry

Although direct evidence for the existence of cation radicals or cations in most anodic reactions of aliphatic compounds is not available, the product distribution is in many cases consistent with the involvement of cationic intermediates. The "abnormal Kolbe" products (unsaturated hydrocarbons, esters, alcohols) obtained in the Hofer–Moest oxidation of aliphatic carboxylates[5,217,295–299] and boronates[82–83,300] are readily understood in terms of carbonium-ion intermediates,

$$RCO_2^- \xrightarrow{-e} R\cdot + CO_2 \xrightarrow{-e} R^+$$

$$RB(OH)_3^- \xrightarrow{-e} R\cdot + H_3BO_3 \xrightarrow{-e} R^+$$

If R is n-butyl-$CH_3CH_2CH_2CH_2$, the major subsequent reaction pathways are

$$CH_3-CH_2-CH_2-CH_2^+ \begin{array}{l} \xrightarrow{RCO_2^-} CH_3-CH_2-CH_2-CH_2-O_2CR \\ \xrightarrow{OH^-} \\ CH_3-CH_2-CH=CH_2 \quad \text{or} \\ CH_3-CH_2-CH_2-CH_2-OH \end{array}$$

or $H_2C\overset{\displaystyle CH_2}{\underset{}{\diagup\hspace{-2pt}\diagdown}} CH-CH_3$

$$CH_3-CH_2-\overset{+}{C}H-CH_3 \begin{array}{l} \xrightarrow{OH^-} CH_3-CH=CH-CH_3 \quad \text{or} \\ CH_3-CH_2-CH-CH_3 \\ | \\ OH \\ \xrightarrow{RCO_2^-} CH_3-CH_2-CH-CH_3 \\ | \\ O_2CR \end{array}$$

The product distributions from the anodic reactions closely resemble those obtained from analogous homogeneous reactions which are interpreted as occurring through carbonium-ion intermediates. In the anodic reactions, some products are obtained which are more readily understood as arising from radical coupling reactions. Thus, n-butyl boronic acid produces substantial quantities of octane, especially in concentrated solutions.[83] The structure of R, the nature of the solvent, the concentration and pH of the solution, the presence and nature of other ions in the solution, and the nature of the electrode all have significant effects on the product distribution.[5,301]

Graphite electrodes give higher yields of carbonium-ion-type products than do platinum or iridium electrodes, and it has been suggested that the difference is due to the presence of paramagnetic centers in graphite which interact strongly with adsorbed radicals, holding them on the electrode so that a second electron transfer is possible to produce a carbonium ion.[301] The implication then is that radicals on a platinum electrode are either much more readily desorbed, and couple in solution, or that platinum catalyzes the radical combination much more effectively than does graphite, so that radical coupling predominates over further electron transfer.

The great efficiency of graphite electrodes in promoting the formation of carbonium-ion-type products was illustrated recently by the observation of acetamidation accompanying hydrocarbon formation when carboxylates are electrolyzed in acetonitrile.[302] Thus, n-pentanoic acid on electrolysis (at uncontrolled potential) yielded N-isobutylacetamide, MeCONHCMeEt, N-acetyl, N-isobutylvaleramide, BuCON(iBu)(Ac), and N-isobutylvaleramide, BuCONHiBu, as well as butenes, octane, and butanol. The formation of these products was rationalized on the basis of the scheme

$$RCO_2^- \rightarrow R^+ + CO_2 + 2e$$

$$R^+ \rightarrow (R^1)^+$$

$$(R^1)^+ + NCCH_3 \rightarrow R^1N={^+}C-CH_3 \xrightarrow{RCO_2^-} R^1N=\underset{\underset{O_2CR}{|}}{C}-CH_3$$

III. Intermediates in Electroorganic Chemistry

$$\rightarrow \begin{bmatrix} O^- \\ | \\ RC-O \\ | \quad | \\ R^1N-C-CH_3 \end{bmatrix} \rightarrow \begin{matrix} O \\ \| \\ R-C \\ | \\ R^1NCOMe \end{matrix}$$

$\downarrow H_2O$

$$\begin{matrix} O \\ \| \\ R-C \\ | \\ R^1NH \end{matrix} + HOAc$$

where R^+ is the *n*-butyl cation $CH_3-CH_2-CH_2-CH_2{}^+$ and $(R^1)^+$ is the *iso*-butyl cation $CH_3-CH_2-\overset{+}{C}H-CH_3$.

Oxidation of alkyl halides can also produce carbonium ions, as shown by the formation of rearranged products typical of carbonium ions. Thus oxidation of *neo*pentyl iodide in acetonitrile gives N-*tert*-pentylacetamide as well as N-*neo*-pentylacetamide[303]:

$$Me_3C\,CH_2I \xrightarrow{-e} Me_3C-CH_2{}^+ \rightarrow Me_2\overset{+}{C}\,Et$$

$\qquad\qquad\quad\;\; \downarrow CH_3CN \qquad\qquad\quad \downarrow CH_3CN$

$$Me_3C\,CH_2-N{=}^+C\,CH_3 \qquad Me_2EtC-N^+{=}C-CH_3$$

$\qquad\quad\; \downarrow H_2O \qquad\qquad\qquad\qquad \downarrow H_2O$

$$Me_3C\,CH_2NHCOCH_3 \qquad\; Me_2EtCNHCOCH_3$$

Formation of organometallic compounds, incorporating the cathode metal, is usually interpreted as evidence for the intervention of radical intermediates. Organo-mercury compounds are obtained in the reduction at a mercury cathode of diazonium salts,[304,306] of αβ-unsaturated carbonyl compounds, e.g., methyl-vinyl-ketone,[307]

$$\overset{O}{\diagdown\!\!\diagup\!\!\diagdown} + e + H^+ \rightarrow \cdot\overset{OH}{\diagdown\!\!\diagup\!\!\diagdown} \xrightarrow{Hg} \overset{OH}{\diagdown\!\!\diagup\!\!\diagdown}\!Hg\!\overset{OH}{\diagdown\!\!\diagup\!\!\diagdown}$$

and of organic halides.[242,308–314] Petrovich and Baizer[314] showed that 2-carbethoxy-allyl-bromide gave three polarographic waves in aqueous alcohol solution, the first two being one-electron steps, the last one a two-electron wave. The corresponding chloride and tosylate gave two two-electron waves in the same solvent, the latter wave for these compounds occurring at the same potential as the third wave for the bromide. For the bromide RBr, where

$$R \text{ is } CH_2=\underset{\underset{CH_2-}{|}}{C}-CO_2Et$$

these results are consistent with the mechanism

$$RBr + Hg \rightarrow RHgBr \xrightarrow{e} RHg\cdot + Br^- \xrightarrow{e} R^-: + Hg$$
$$\xrightarrow[2e]{3H^+} RH_3$$

and for the chloride and tosylate,

$$RX \xrightarrow[H^+]{2e} RH \xrightarrow[2H^+]{2e} RH_3$$

The authors sounded a warning about the interpretation of results in solvents like dimethylformamide, since the three compounds above all gave a single identical wave in this solvent, indicating that the reduction process observed was that for a common intermediate formed by reaction of these compounds with the solvent. This result could also serve as a warning about the possibilities for unexpected complications affecting results if a detailed study of a single compound is undertaken before the general behavior of the electroactive functional group has been established under the particular experimental conditions for a range of substituted compounds. Complications of this type, arising from interaction with the solvent prior to charge transfer, are not readily detected in any other manner.

Metals other than mercury may be incorporated into organometallic products, especially at high hydrogen-overvoltage-metals like cadmium[19] or lead.[19,315] The latter is important because of the use of tetraethyl-lead in motor fuel.[4] Linear sweep voltammetric studies at a lead electrode in acetonitrile–water mixtures showed that only the lower alkyl halides produced lead-alkyls efficiently; methyl bromide gave quantitative yields of lead-tetramethyl, with 100% current efficiency, ethyl-bromide gave 40% lead-tetraethyl plus 60% of hexaethyldilead $(Et_3Pb)_2$. Isopropyl-bromide gave only 30% of lead-alkyls, together with propene, and *tert*-butyl bromide yielded isobutene as a major product. The high efficiencies of formation of lead compounds from methyl and ethyl bromides suggest that the alkyl radical formed near the electrode either does not diffuse into the solution or reacts with the electrode faster than it can diffuse. Formation of significant amounts of $(Et_3Pb)_2$ suggests that alkylation of the electrode occurs at least to the PbR_3 stage before detachment from the surface; dimerization of PbR_3 radicals could account for $(PbEt_3)_2$ formation, or ethylation of two adjacent lead atoms on the electrode surface may be involved. In an investigation of the reduction of ethyl bromide at a lead cathode in propylene carbonate solution, Galli and Olivani found that lead-alkyls were obtained only in the presence of "onium" salts (NEt_4^+, NBu_4^+, or $SEtMe^+$). When only inorganic cations were present, the products consisted of C_2–C_4 hydrocarbons.[316] This appears to be another example of the effect of adsorbed "onium" cations inhibiting protonation, as discussed for acrylonitrile above.

The involvement of radicals in the formation of the normal Kolbe product

$$RCO_2^- \xrightarrow{-e} R\cdot + CO_2$$
$$\downarrow$$
$$\tfrac{1}{2}R_2$$

has already been mentioned. In spite of the ease with which the neophyl radical is known to rearrange,[317]

$$PhCMe_2CH_2\cdot \rightarrow PhCH_2\dot{C}Me_2$$

electrolysis of 3-phenyl-isovaleric acid $PhCMe_2CH_2CO_2H$ gave no evidence for a rearranged product.[318] To simplify product analysis, the electrolysis was conducted in the presence of acetic acid, and *tert*-amylbenzene was isolated as the sole product of mixed coupling, indicating no rearrangement of the neophyl group. The mechanism of mixed coupling is not known, however, and the product *tert*-amylbenzene could be produced without the intervention of neophyl radicals, thus

$$CH_3CO_2^- \xrightarrow{-e} CH_3CO_2\cdot \rightarrow CH_3\cdot + CO_2$$

then

$$CH_3\cdot + PhCMe_2CH_2CO_2^- \rightarrow PhCMe_2Et + CO_2 + e$$

Good yields of cross-coupled products have been obtained in this way when an acid which readily gives a Kolbe product is electrolyzed in mixture with one which does not, e.g., the diester of propane-1,2,3-tricarboxylic acid $EtO_2CCH_2CH(CO_2H)CH_2CO_2Et$ gives no Kolbe-coupled product when electrolyzed alone, but gives the mixed product $EtO_2CCH_2CH(CH_3)CH_2CO_2Et$ on oxidation in the presence of acetic acid.[319]

Instead of dimerizing, radicals generated in the presence of unsaturated compounds may add to the unsaturated linkage, e.g., oxalic acid oxidized in the presence of butadiene-1,3 in methanol gives a mixture of products formed by 1,2 and 1,4 addition of $\cdot CO_2H$ radicals to the conjugated system, followed by combination of the resulting radicals.[320] Polymerization of vinyl monomers, such as vinyl acetate, vinyl chloride, and methyl acrylate, has been initiated by radicals formed in the Kolbe reaction.[321] Evidence for the incorporation of both ethyl and $EtCO_2\cdot$ radicals in the polymer was obtained when propionic acid was electrolyzed in the presence of polymerizable monomers.[322] The other typical reactions of radicals, namely hydrogen-atom abstraction,

$$R\cdot + R'H \rightarrow RH + R'\cdot$$

and/or disproportionation

$$2R\cdot \rightarrow RH + R(-H)$$

where $R(-H)$ is an alkene, are probably responsible for the production of the monomer hydrocarbon (e.g., butane C_4H_{10}) from valeric acid $C_4H_9CO_2H$ or butaneboronic acid $C_4H_9B(OH)_2$. Disproportionation is unlikely to be the only source of the alkene, which is almost certainly produced also via a carbonium-ion mechanism, since, e.g., in the above cases, the yield of butenes exceeds that of butane.

3. Adsorption and Organic Electrode Processes

A mechanistic and general problem of great importance concerning organic electrode processes is the extent to which adsorption at the electrode significantly affects the course of the reaction. Some electroactive substances may undergo an electrode reaction when they are at the outer Helmholtz plane in the double-layer region.[323] However, adsorption at the electrode may be a necessary prerequisite to charge transfer. Adsorption of organic compounds, particularly on mercury electrodes, has been intensively studied

III. Intermediates in Electroorganic Chemistry

and reviewed.[324–327] The effects of adsorption on rates of electrode reactions have also been reviewed.[328–330] Only a limited number of investigations has yet been performed in which detailed measurements of reactant and product adsorption have been correlated with electrode kinetic studies.[57,64,192] In polarographic studies,* adsorption of reactants or products may lead to post- or pre-waves, respectively.[65] Adsorption of products or by-products may exert a retarding influence on the electrode reaction.[331] Electroinactive components may adsorb, changing the structure of the double layer, or modifying the adsorption of reactants or intermediates, and hence affecting the reaction rate.[332,333] The kinetic effects of adsorbed intermediates have been discussed by Gileadi and Conway.[334] The distinction between processes subsequent to charge transfer which occur on the electrode surface, i.e., involving adsorbed intermediates, and those occurring in the solution is not usually clear-cut. Thus, in the Kolbe reaction, the question as to whether the intermediate radicals are free or adsorbed has not yet been settled.[5,58,301] Adsorbed intermediates are expected to be more sensitive than nonadsorbed ones to the nature of the electrode material; such effects are involved in the phenomenon of electrocatalysis.[49]

In the oxidation of hydrocarbons, relationships have been observed between catalytic activity and properties of the electrode metal (e.g., d-band holes, or heat of sublimation),[335] but few applications to other organic systems have appeared. The variation of the product distribution in the oxidation of carboxylic acids when platinum is replaced by graphite has already been mentioned; this variation arises because adsorption of the intermediate affects the extent of a second charge-transfer step, rather than through a catalytic effect on the rate of a single process.[201]

That chemical steps which constitute part of overall EC or ECE electrode reactions may occur in solution is indicated by agreement between rate constants determined electrochemically and by nonelectrochemical techniques, e.g., for the dehydration of p-hydroxylaminophenol in the reduction of p-nitrosophenol, or the hydrolysis of p-benzoquinoneimine in the oxidation of p-amino-

---
*Under polarographic conditions, the low reactant concentrations used diminish the importance of adsorption which, under other conditions, however, must usually be taken into account.[18a,64]

phenol, or in bromination of reactive aromatic compounds by anodically generated bromine.[167] In other cases, chemical steps following charge transfer must occur on the electrode; the most obvious of these is the incorporation of the electrode material into an organometallic compound, e.g., the formation of lead tetramethyl, already discussed. Between these two extremes, a variety of reactant–electrode interaction effects is possible, ranging from orientation of polar molecules in the electrode field[175] through field-assisted ionization (S_N1 mechanism of halide reduction,[336,337]

$$RX \to X^- + R^+ \xrightarrow[H^+]{2e} RH)$$

to adsorption, which can range from purely physical, through associative, to dissociative chemisorption.[192] The variety of products obtained in many organic electrode reactions, particularly oxidations[5,83,217,281] may sometimes be a reflection of the simultaneous occurrence of surface and solution reactions; e.g., in the Hofer–Moest Kolbe reaction, the alkyl radicals R· may be adsorbed and dimerize on the surface to give R_2, whereas carbonium ions R^+ are much less likely to be adsorbed on a positively charged electrode, and probably undergo further reactions in solution. Recently a chronocoulometric method was suggested as a means of determining the distribution of reaction paths between the electrode surface and the solution.[338] If the charge-transfer step is fast, and the diffusion of the product of this step into the solution normally predominates over the surface reaction, it is possible by use of a potentiostatic pulse method to change the relative distribution of reaction paths by variation of the pulse length. A theory was developed which also takes into account possible adsorption of reactants and products, and makes possible evaluation of the rate of the surface reaction.

An interesting case of an organic reaction proceeding completely as a surface reaction was discovered by MacDougall et al.,[338a] who investigated the electrochemisorption of CH_3CN and other nitriles at a Pt electrode in aq. acid solutions. CH_3CN is chemisorbed at Pt in the double-layer potential region (+0.3 to +0.8 V E_H) with a cathodic transient (contrast CH_3OH) under potentiostatic conditions and then undergoes a two-stage reduction *on the Pt surface* between 0.7 and 0.0 V E_H. It is quantitatively reoxidized

IV. Types of Reaction

in the adsorbed state between 0.35 and 0.8 V in a following anodic sweep in cyclic voltammetry. Once the initial adsorption has been established, no further solution species are involved, either as reactant or product. The initial adsorption, if carried out potentiostatically in the H region at Pt, occurs with an anodic current transient due to electrochemical displacement of adsorbed H from the surface.[338a]

IV. TYPES OF REACTION
1. Substitution Reactions

(i) Cathodic

Cathodic substitution reactions, initiated by charge transfer, involve electron addition followed by elimination of a stable anion, such as a halide or nitrite anion, and combination of the resulting radical or anion (depending on whether one or two electrons were added) with a proton, or a reagent for nucleophiles like CO_2, or the electrode itself (high-hydrogen-overvoltage metals like Hg, Cd, Pb). Examples of all these types of reaction have already been discussed; some further examples are given below. The commonest example of substitution of another group by hydrogen at the cathode is in the reduction of halides. This process has been intensively studied,[13,19,337] and generalizations about structural effects on reactivity are possible.[46,99,339] Thus the ease of reduction varies with the halogen, I > Br > Cl > F, fluorides generally being reducible only when adjacent to a carbonyl group,[340,141] or in an activated trifluoromethyl group.[342-344] Allyl halides are readily reduced, whereas vinyl and aryl halides are reducible only with difficulty. On the basis of the $E_{1/2}$ values of -2.32 V for bromobenzene, -2.42 V for acetophenone, and > -2.6 V for typical alkyl chlorides, the successful specific reduction of one of three reducible groups present,

$$Br-\bigcirc-\underset{\underset{O}{\|}}{C}(CH_2)_4Cl \xrightarrow[96\% \text{ yield}]{2e} \bigcirc-\underset{\underset{O}{\|}}{C}(CH_2)_4Cl$$

was effected by controlled-potential electrolysis.[345]

Anion elimination can be materially assisted by appropriate substitution in the reactant; nitroaryl halides have already been discussed, and very ready cleavage of C—O or C—F bonds occurs in *o*- or *p*-carbomethoxy- or *p*-cyano-substituted benzyl systems, thus[344]

$$\text{p-}CR_2X\text{-C}_6H_4\text{-CO}_2Me \xrightarrow{e} [\text{radical anion}] \xrightarrow{-X^-} [\text{quinoid radical}] \leftrightarrow \text{p-}\dot{C}R_2\text{-C}_6H_4\text{-CO}_2Me \xrightarrow[H^+]{e} \text{p-}CR_2H\text{-C}_6H_4\text{-CO}_2Me$$

where R = H and X = OMe, OAc; or R = X = F.

A similar activating effect of *o*- or *p*-CN is evident in the reduction of phthalonitrile (or terephthalonitrile) which, at controlled potential, gives a radical anion stable for hours and identifiable from its EPR spectrum. At more negative potentials, another two electrons are transferred, and the EPR spectrum changes to that of benzonitrile, indicating the reaction sequence[346]

$$\text{o-}(CN)_2C_6H_4 \xrightarrow[-2.12\,V]{e} [\text{o-}(CN)_2C_6H_4]^{\cdot -} \xrightarrow[-2.76\,V]{e} \text{o-}(CN^-)_2C_6H_4 \xrightarrow{H^+} C_6H_5CN + CN^- \xrightarrow{e} [C_6H_5CN]^{\cdot -}$$

The benzonitrile radical anion is formed at a potential of −2.74 V, so the potential required to add the second electron is negative enough to add the third.

IV. Types of Reaction

Substitution of NH_2 by H is possible when an activating p-CN group is present, thus[346]

$$\underset{NH_2}{\underset{|}{C_6H_4}}\text{-CN} \xrightarrow{e} \left[\underset{NH_2}{\underset{|}{C_6H_4}}\text{-CN}\right]^{\cdot -} \longrightarrow C_6H_5\text{-CN}^{\cdot} + NH_2^-$$

$$2\, C_6H_5\text{-CN}^{\cdot} \longrightarrow NC\text{-}C_6H_4\text{-}C_6H_4\text{-}CN \xrightarrow{e} \left[NC\text{-}C_6H_4\text{-}C_6H_4\text{-}CN\right]^{\cdot -}$$

In this case, the dicyanodiphenyl radical anion was identified from its EPR spectrum.

The π-deficient pyridine nucleus shows an activating influence similar to that shown in the above examples by the p-CN group. In alkaline solution, 4-cyanopyridine is cathodically reduced to pyridine,

$$\text{4-NC-C}_5H_4N \xrightarrow[H_2O]{2e} C_5H_5N + CN^- + OH^-$$

whereas in acid solution, the CN group is reduced, not replaced,[347]

$$\text{4-NC-C}_5H_4NH^+ \xrightarrow[4H^+]{4e} \text{4-}H_2NCH_2\text{-}C_5H_4NH^+$$

A detailed study of the polarographic reduction of this compound over a wide pH range showed a current decrease in neutral solution, from a value indicative of a four-electron process to a value only-half as great as the potential became more negative. This was attributed to a change in mechanism with potential, caused by a change in orientation of the molecule at the electrode surface as the potential changed, and confirmed by isolating pyridine as product at a potential corresponding to the current minimum, and the 4-CH_2NH_2 derivative at the potential of the current maximum. At positive or slightly negative electrode potentials, the CN group apparently is close to the electrode, and is reduced; at more negative potentials, the electric field orients the molecule so that the CN group is away from the electrode, so that electron transfer to the ring can occur with expulsion of CN^-.[348]

The activating influence of an adjacent carbonyl group on the reductive replacement of F has been mentioned above.[340,341] Electrochemical reductive cleavage of other groups, e.g., C—O, C—S, and C—N, is also possible when such an activating group is present. Thus phenacyl-acetate is readily reduced,[344]

$$PhCOCH_2OAc \xrightarrow{e} [PhCOCH_2OAc]^{\cdot -}$$
$$\rightarrow OAc^- + PhCOCH_2\cdot \rightarrow PhCOCH_3$$
$$\xrightarrow[2H^+]{2e} PhCHOHCH_3$$

Recent synthetic applications of this type of activation have been to the conversion of esters into methyl ketones,[349]

$$C_6H_{13}CO_2Et + Me_2SO + NaH \rightarrow C_6H_{13}\underset{\underset{O}{\|}}{C}CH_2\underset{\underset{O}{\|}}{S}CH_3$$
$$\xrightarrow[2H^+]{2e} C_6H_{13}COCH_3$$

and to the Stevens rearrangement,[350]

$$PhCOCH_2\overset{+}{\underset{\underset{CH_2Ph}{|}}{N}}Me_2 \xrightarrow{2e}_{DMF/LiCl} PhCOCH_3 + PhCH_2NMe_2 + \left[PhCO\overset{-}{C}H\overset{+}{\underset{\underset{CH_2Ph}{|}}{N}}Me_2 \right]$$

$$\longrightarrow PhCOCHNMe_2$$
$$\qquad\qquad |$$
$$\qquad\quad CH_2Ph$$

IV. Types of Reaction

The first two products illustrate the normal reductive bond-cleavage process; in this ingenious example, the reactant itself acts as proton donor, being the strongest acid present. Proton transfer from the reactant produces the carbanionic intermediate in which the N → C shift can occur.

(ii) *Anodic*

Anodic aromatic substitution reactions, in which hydrogen is replaced by a variety of other functional groups, have been discussed already, as has halogenation with halogen generated by anodic oxidation of halide ion.[167,168] A variation of this process was recently described, in which an active iodinating agent, probably I^+ or $CH_3\overset{+}{C}=NI$ is generated anodically in acetonitrile solution in the absence of the aromatic compound, which is then added, forming iodo compounds in excellent (80–100%) yields.[351] This procedure has the advantage of not requiring the use of high temperatures in strongly acidic solvents, nor the presence of oxidizing agents, the conditions normally employed for iodination.

Groups other than hydrogen have been replaced anodically in some cases, e.g., Parker and Eberson[352] have observed nucleophilic replacement of Br from position 9 in 9,10-dibromoanthracene when this was anodically oxidized in the presence of 3,5-lutidine, thus

where N is 3,5-lutidine.

Andreades and Zahnow[268] observed replacement of an OMe group by CN during anodic cyanation of aromatic compounds,

$$\text{MeO}-\text{C}_6\text{H}_4-\text{OMe} \xrightarrow{\text{CN}^-} \text{MeO}-\text{C}_6\text{H}_4-\text{CN}$$

and Cauquis et al.[353] reported the replacement of a *tert*-butyl group when 2,4,6-tri(*tert*-butyl)aniline was oxidized in acetonitrile containing methanol or water. In the presence of water, the following scheme was postulated (where R is *tert*-butyl):

IV. Types of Reaction

Anodic formation of organometallic compounds has been observed in some cases, e.g., oxidation of acetone at a mercury electrode produces CH_3COCH_2HgOH, explained thus[354]:

$$CH_3COCH_3 + OH^- \rightleftharpoons CH_3COCH_2^- + H_2O$$

$$Hg \rightleftharpoons Hg^{2+} + 2e$$

$$CH_3COCH_2^- + Hg^{2+} \rightarrow CH_3COCH_2Hg^+$$
$$\downarrow OH^-$$
$$CH_3COCH_2HgOH$$

In this case, the anodic current simply supplies mercuric ions at a controllable rate for reaction with the organic substrate in a reactive form already present in solution; in this respect, the process has much in common with the anodic halogenation of reactive organic compounds.[167,168]

2. Addition Reactions

(i) Cathodic

Addition of hydrogen to aromatic systems has been discussed above, where the ECEC mechanism was considered, on the basis of measurements in aprotic solvents, and particularly from the influence of proton donors on the electrochemical behavior. Since hydrogen itself can be a cathodic product from protic solvents, the possibility of electrochemical hydrogenation exists, i.e., the reduction of unsaturated systems by hydrogen atoms (possibly adsorbed on the electrode) produced by prior reduction of hydrogen ions,

$$2H^+ + 2e \rightarrow 2H(ads)$$

then

$$R + 2H(ads) \rightarrow RH_2$$

This equation is not meant to imply necessarily that both hydrogen atoms are added simultaneously. Similarly, when catalytic hydrogenation occurs, using hydrogen gas and a metal catalyst in a protic solvent, the function of the catalyst may be to assist the atomization of hydrogen, or a mixed potential may be set up at the metal, where the anodic reaction

$$H_2 \rightarrow 2H^+ + 2e$$

is balanced by cathodic reactions, such as

$$R + e \rightarrow R\cdot^- \qquad R\cdot^- + H^+ \rightarrow RH\cdot$$
$$RH\cdot + e \rightarrow RH^- \qquad RH^- + H^+ \rightarrow RH_2$$

or

$$R + H^+ \rightarrow RH^+ \qquad RH^+ + e \rightarrow RH$$
$$RH + H^+ \rightarrow RH_2^+ \qquad RH_2^+ + e \rightarrow RH_2$$

Wagner has recently suggested experimental approaches which can distinguish between these possibilities.[355] Applying these methods, Takehara[356] concluded that the hydrogenation of benzoquinone involved the electrochemical mechanism, as suggested by Vetter[357] on the basis of electrochemical measurements on the cathodic reduction;

$$C_6H_4O_2 + H^+ \rightarrow C_6H_5O_2^+ \xrightarrow{e} C_6H_5O_2 \xrightarrow{H^+} C_6H_6O_2^+$$
$$\xrightarrow{e} C_6H_6O_2$$

The occurrence of this reaction path is undoubtedly favored by the rapidity with which electrochemical equilibrium is set up at the electrode for the benzoquinone–hydroquinone system. For allyl alcohol and vinyl acetate, Takehara[356] concluded that the nonelectrochemical mechanism, involving adsorbed hydrogen atoms and adsorbed substrate, was the predominant one.*

In contrast, Franklin et al.,[358] from studies on the relationship between electrode potential and fractional coverage of the electrode by adsorbed hydrogen atoms, concluded that benzoquinone was reduced by the hydrogen addition mechanism, rather than by charge transfer to the organic substrate. Their results indicated that this mechanism applied at low and high coverage of the electrode by hydrogen. For reduction of acrylonitrile, however, hydrogen addition occurred at high hydrogen coverage, but at low coverage of the electrode by hydrogen, the mechanism changed to one involving charge transfer to the organic molecule. Further work on benzoquinone is required to resolve these differences.

By comparing catalytic hydrogenation and cathodic reduction of cinnamic acid on platinized platinum, Kudryashov and Kochet-

---

*When the reactant is chemisorbed, e.g., at Pt, hydrogenation is preceded by anodic displacement of the H initially on the surface.[338a]

IV. Types of Reaction

kov[359] concluded that the two processes involved the same rate-determining step, which they suggested was the second step in each of the schemes:

$$\text{Electrochemical} \quad H^+ + e \rightarrow H(\text{ads}) \xrightarrow{R} RH$$

$$\text{Chemical} \quad \tfrac{1}{2}H_2 \rightarrow H(\text{ads}) \xrightarrow{R} RH$$

(for simplicity, only one hydrogen addition is shown).

In another study comparing hydrogenation with electron transfer, it was found that addition of ethanol to water as solvent increased the rate of electroreduction but decreased the rate of hydrogenation of maleic acid.[360]

A useful summary of the structural requirements for polarographic reducibility of multiple bonds has been given by Lund.[339] Reduction of carbon–oxygen double bonds often gives rise to α-glycols, which is taken as evidence for radical or radical anion intermediates. Hydrodimerization of acrylonitrile has been discussed above; crossed hydrocoupling of acrylonitrile with other reactants, e.g., acetone to form γ-hydroxy-γ-methyl valeronitrile, can be performed in high yield,[361]

$$Me_2CO + CH_2\!=\!CHCN \rightarrow Me_2\underset{\underset{OH}{|}}{C}CH_2CH_2CN$$

Baizer et al.[362] recently investigated the effect of potential on the product distribution in crossed hydrocoupling reactions, and concluded, for two components A and B, if A is reduced at less negative potentials than B when alone, the ratio of the yield of HABH to the self-coupled product HAAH rises as the potential becomes more negative. Thus, choice of potential permits maximization of the yield of HABH, if this crossed product is required, or alternatively A may be reduced, by potential control, to HAAH in optimum yield even in the presence of B, which might even be an impurity in A, and contamination of the product by HABH can be eliminated, or at least minimized.

For unsaturated systems of the type $RR'C\!=\!NYR''$, Lund[363] summarized the behavior on reduction thus: If Y is carbon, reduction requires two electrons per molecule and the saturated compound is the reduction product. If Y is N or O, in acid solution, four electrons per molecule are required, and the N–Y bond is split reductively; splitting of the N–Y bond generally precedes reduction of the C=N,

$$RR'C=NYR'' + H^+ \rightarrow RR'C=NYR''H^+$$
$$\rightarrow RR'\overset{+}{C}=NH_2 + HYR''$$
$$\underset{2e|2H^+}{\downarrow}$$
$$RR'\overset{+}{C}HNH_3{}^+$$

Phosphorus[364] and sulfur[365] ylides can be produced cathodically, and add to carbonyl groups, thus

$$Ph_3\overset{+}{P}CH_2R \overset{e}{\rightarrow} Ph_3\dot{P}CH_2R \rightarrow Ph_3P=CHR + H\cdot$$

$$RCHO + Ph_3P=CHR \rightarrow \underset{O}{RCH{-}CHR}$$

and

$$Me_2\overset{+}{S}CH_3 \overset{e}{\rightarrow} Me_2\dot{S}CH_3 \rightarrow Me_2S=CH_2 + H\cdot$$

$$RCHO + Me_2S=CH_2 \rightarrow \underset{O}{RCH{-}CH_2}$$

(ii) Anodic Additions

Anodic addition, e.g., of two OAc residues, or of OH plus OAc to alkenes like 1,1-diphenylethylene or stilbene, appears to involve di-cation or cation radical intermediates analogous to those discussed for anodic aromatic substitution.[258] Electrolysis of 4,4'-dimethoxy stilbene in acetonitrile containing acetic acid produced, as one product, the monomeric acetate $MeOC_6H_4CH(OH)CH$-$(OAc)C_6H_4OMe$.[366] An ECE mechanism for the formation of this product, involving a cation radical adding acetate ion, followed by further oxidation, was ruled out by considerations of linear sweep voltammetric results, i.e., the process did not involve the reactions,

$$\begin{array}{c}CHR\\||\\CHR\end{array} \xrightarrow{-e} \begin{array}{c}RCH^+\\|\\RCH\cdot\end{array} \xrightarrow{OAc^-} \begin{array}{c}RCHOAc\\|\\RCH\cdot\end{array} \xrightarrow{-e} \begin{array}{c}RCHOAc\\|\\RCH^+\end{array} \xrightarrow{H_2O} \begin{array}{c}RCHOAc\\|\\RCHOH\end{array} + H^+$$

but rather

$$\begin{array}{c}CHR\\||\\CHR\end{array} \xrightarrow{-2e} \begin{array}{c}RCH^+\\|\\RCH^+\end{array} \xrightarrow[H_2O]{HOAc} \begin{array}{c}RCHOAc\\|\\RCHOH\end{array} + 2H^+$$

IV. Types of Reaction

where R is

$$-\underset{}{\bigcirc}-\text{OMe}$$

The mechanism is thus EEC, rather than ECE. Further evidence for an EEC mechanism was obtained in the anodic addition of acetate to 9,10-dimethyl anthracene (DMA)[367]

$$\text{DMA} \rightarrow \text{DMA}^+ \rightarrow \text{DMA}^{2+} \xrightarrow{2\text{OAc}^-} \text{DMA(OAc)}_2$$

Anodic dimerization reactions have been observed for suitably substituted alkenes, e.g., 1,1-bis(dimethylamino)ethylene,[368]

$$2 \underset{\text{Me}_2\text{N}}{\overset{\text{Me}_2\text{N}}{>}}\!\!\text{C}=\text{CH}_2 \xrightarrow[-2\text{H}^+]{-2e} \underset{\text{Me}_2\text{N}}{\overset{\text{Me}_2\text{N}}{>}}\!\!\text{C}=\text{CH}-\text{CH}=\text{C}\!\!\underset{\text{NMe}_2}{\overset{\text{NMe}_2}{<}}$$

and for diphenyldiazomethane.[369] In the latter case, coulometric measurements indicated a radical chain reaction, formulated as

$$\underset{\text{Ph}}{\overset{\text{Ph}}{>}}\!\!\bar{\text{C}}\text{N}_2^+ \xrightarrow{-e} \underset{\text{Ph}}{\overset{\text{Ph}}{>}}\!\!\dot{\text{C}}\text{N}_2^+ \xrightarrow[-\text{N}_2]{+\text{I}} \underset{\text{Ph}}{\overset{\text{Ph}}{>}}\!\!\underset{\text{N}_2^+}{\overset{|}{\text{C}}}-\text{C}\!\!\underset{\text{Ph}}{\overset{\text{Ph}}{<}}$$

(I)   (II)   (III)

$$(\text{I}) + (\text{III}) \xrightarrow{-\text{N}_2} \underset{\text{Ph}}{\overset{\text{Ph}}{>}}\!\!\text{C}=\text{C}\!\!\underset{\text{Ph}}{\overset{\text{Ph}}{<}} + (\text{II})$$

(IV)

$$(\text{II}) + \text{H}_2\text{O} \xrightarrow[-\text{H}^+]{-\text{N}_2} \underset{\text{Ph}}{\overset{\text{Ph}}{>}}\!\!\dot{\text{C}}-\text{OH}$$

(V)

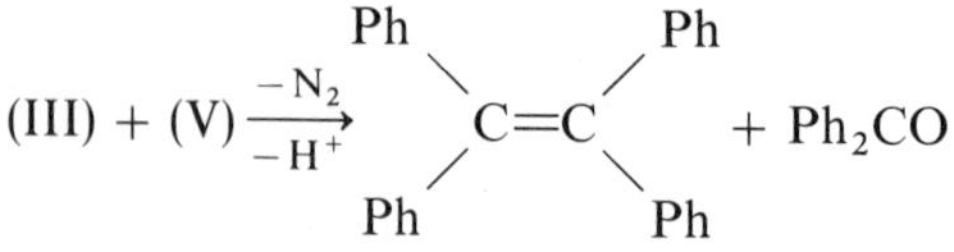

$$(III) + (V) \xrightarrow[-H^+]{-N_2} \underset{Ph}{\overset{Ph}{>}} C=C \underset{Ph}{\overset{Ph}{<}} + Ph_2CO$$

(VI)

$$(II) + (V) \longrightarrow \underset{Ph\ \ N_2^+}{\overset{Ph}{>}} C\text{---}C \underset{OH\ Ph}{\overset{Ph}{<}} \xrightarrow[-H^+]{-N_2} Ph_3CCOPh$$

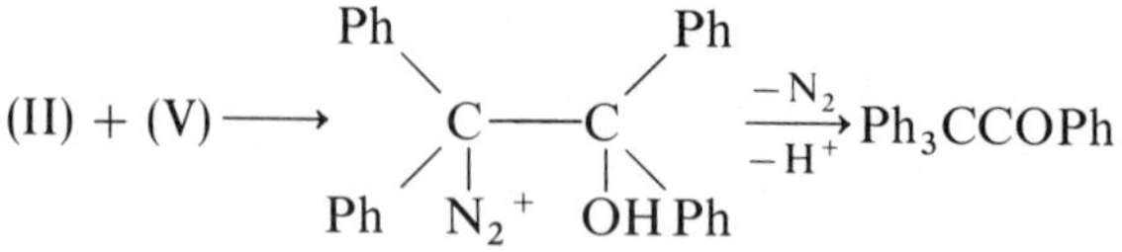

(VII)

The main product was tetraphenylethylene (80%), but the two ketones VI and VII were also isolated.

Of no apparent synthetic utility, but of interest as an example of the influence of the nature of the anion of the supporting electrolyte, is the reported polymerization of benzene in acetonitrile in the presence of tetrabutylammonium hexafluorophosphate,[370]

$$PhH \xrightarrow{-e} PhH^+ \xrightarrow{-H^+} Ph\cdot \longrightarrow polymer$$

Also, small concentrations of ClO_4^- ion have been reported[370a] to drastically modify relative yields of Kolbe/Hofer–Moest products in the electrolysis of phenylacetates.

3. Elimination Reactions

(i) Cathodic

Removal of two functional groups from adjacent carbon atoms in an organic molecule is discussed in Section V; cases in which two groups are eliminated from nonadjacent carbon atoms without replacement by hydrogen usually lead to cyclization and are discussed in Section IV.4.

(iii) Anodic

The stabilizing effect of blocking methyl groups on cations and cation radicals discussed in Section II is evident in the anodic elimination of acetyl or methyl groups from durohydroquinone

IV. Types of Reaction

acetyl esters or methyl ethers.[371] Thus,

The product was identified as duroquinone from its cyclic voltammogram.

Extending the oxidation of tropilidine to the tropylium ion, discussed in Section II, Geske[153] produced benzene by anodic oxidation of cyclohexadiene-1,3,

Dehydrogenation of a dihydrophenanthrene was observed by Ronlan and Parker,[372] during the anodic oxidation of 3,4,3',4'-tetramethoxybibenzyl. After the first oxidation sweep, a redox couple was observed on subsequent sweeps at less positive potentials than involved in the initial oxidation. Cyclic voltammetry performed on an authentic sample of 2,3,6,7-tetramethoxy-9,10-dihydrophenanthrene showed that it was the intermediate involved in the redox couple, and the product finally isolated was 2,3,6,7-tetramethoxyphenanthrene. Thus, dehydrogenation occurred in the

last step,

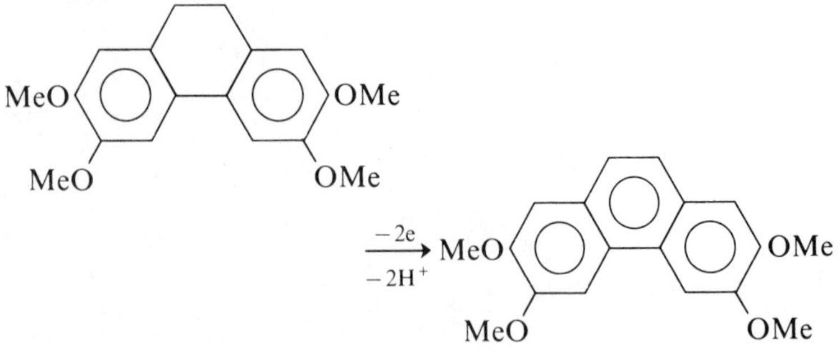

$$\text{[trimethoxy phenanthrene derivative]} \xrightarrow[-2H^+]{-2e} \text{[trimethoxy phenanthrene product]}$$

Electrolytic decarboxylation of vicinal dicarboxylic acids has been used to prepare olefins. Vicinal dicarboxylic acids are readily prepared by Diels–Alder addition of maleic anhydride to suitable dienes, so that decarboxylation of these adducts gives the products obtained by effectively adding ethylene in the Diels–Alder reaction. The synthesis of Dewar benzene provides an interesting example of this electrochemical route[373]:

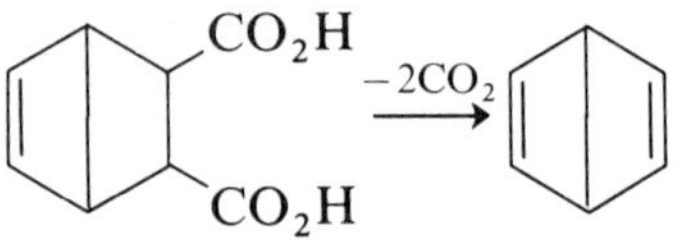

$$\text{[bicyclic diacid]} \xrightarrow{-2CO_2} \text{[Dewar benzene]}$$

4. Cyclization Reactions

(i) Cathodic

Cathodic elimination of two halogen atoms not attached to adjacent carbons may lead to cyclic products, thus[374]

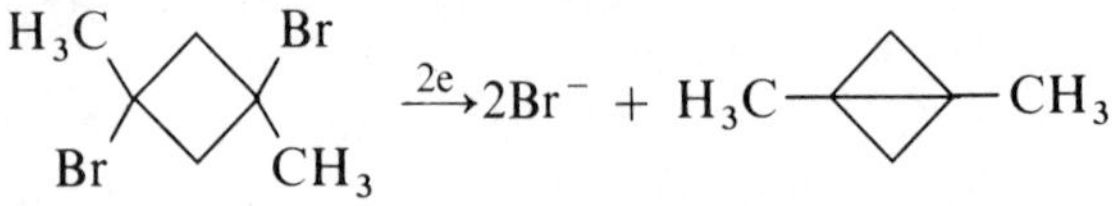

$$\text{[1,3-dibromo-1,3-dimethylcyclobutane]} \xrightarrow{2e} 2Br^- + H_3C-\diamondsuit-CH_3$$

also[373]

$$\text{[tetrakis(bromomethyl)methane]} \xrightarrow{4e} 4Br^- + \bowtie$$

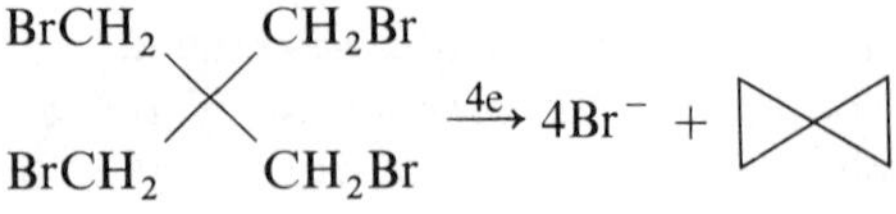

IV. Types of Reaction

1,4-dibromobutane gave 25% of cyclobutane, but 1,5-dibromopentane gave no cyclopentane. This was explained by the larger number of conformations favorable for cyclization when the dibromobutane or an intermediate therefrom was adsorbed on the electrode surface. The products could be explained on the basis of an intramolecular radical "dimerization," although other mechanisms cannot be excluded.[374] Suitable disposition, within one molecule, of two groups which, when present alone, lead to dimeric products can give rise to ring formation, e.g., the intramolecular pinacol reaction,[375]

It is interesting to note that the reaction occurred more readily in tetrahydrofuran as solvent than in acetonitrile.

Intramolecular hydrodimerization of activated alkenes has been intensively studied by Baizer et al.[376-378] as a means of producing cyclic products. Thus, diethyl cis-1,3-cyclopentane diacrylate gave diethyl trans-2,3-norbornane diacetate in 37% yield,

Ring strain probably accounts for the low yield in this case, since three-, five-, and six-membered rings are usually produced in yields exceeding 90%. The mechanism suggested[378] involves addition of a single electron to form a radical anion with simultaneous cyclization, followed by further electron uptake and protonation, e.g.,

Reduction of an organomercuric halide to a carbanion, followed by intramolecular reaction with an activated unsaturated linkage, was used by Dessy and Kandil[379] for synthesis of the acenaphthene system,

The thiocyanate group attached to an aromatic ring couples readily with an appropriately placed amino group to produce a benzothiazole derivative,[380]

In this example, electrochemical reduction is used simply to generate, in positions appropriate for cyclization, two groups which undergo a condensation reaction. Thus it differs from the previous examples, in which cyclization was achieved because of the structural disposition of an electrochemically formed intermediate which stabilizes itself by cyclization. Many variations on this method of electrochemically producing reactive groups in the correct relative positions for cyclization have been described, for example,[381]

Cyclization through lactone formation is also possible, but careful pH control is necessary to prevent hydrolysis. Ammonium phthalate can be reduced almost quantitatively to phthalide, but the pH

IV. Types of Reaction

becomes too high if sodium or potassium phthalate is used, and the yield is greatly reduced.[382]

$$\underset{CO_2^-NH_4^+}{\underset{CO_2^-NH_4^+}{\bigcirc}} \longrightarrow \underset{CO}{\underset{O}{\bigcirc}}\overset{CH_2}{\underset{}{}}$$

Two other ring-forming reagents have been produced electrochemically, but the interest to date has been solely in their novelty, and no serious attempts appear to have been made to utilize them for general synthetic purposes. These are benzyne, obtained by reduction of o-dibromobenzene, and trapped as the furan adduct, which hydrolyzed to α-naphthol,[383]

and dichlorocarbene, from reduction of carbon tetrachloride in aprotic media,[384]

$$CCl_4 + e \rightarrow Cl^- + CCl_3\cdot \xrightarrow{e} Cl^- + CCl_2$$

Dichlorocarbene formation was indicated by the production of 1,1-dichloro-2,2,3,3-tetramethyl cyclopropane when tetramethylethylene was added to the electrolysis solution,

$$Me_2C=CMe_2 + CCl_2 \rightarrow Me_2C\underset{CCl_2}{\overset{}{\diagdown\diagup}}CMe_2$$

(ii) Anodic

Attempts to form carbene or a carbene-like intermediate by anodic oxidation of disodium malonate failed to produce any evidence for its formation.[385] In the presence of cyclohexene in methanol solution, the main product isolated was cyclohexenyl methyl ether. This was explained as arising from attack on cyclohexene by radicals initially formed from malonate ion, to produce cyclohexenyl radical, which oxidized to the cation, then reacted with solvent.

Anodic cyclization has been observed during oxidation of 4,4′-dimethoxystilbene.[366] In addition to the hydroxyacetoxylation process described above, a product A of structure

was obtained, and its formation explained thus:

$$2 \begin{array}{c} RCH \\ \| \\ RCH \end{array} \rightarrow 2e + 2 \begin{array}{c} RCH^+ \\ | \\ RCH\cdot \end{array} \rightarrow \begin{array}{c} \overset{+}{R}CH \quad \overset{+}{C}HR \\ | \quad\quad | \\ RCH-CHR \end{array} \xrightarrow{AcO^-} A + H^+$$

The closely-related 3,4,3′,4′-tetramethoxybibenzyl produces a cyclic product on anodic oxidation, through the intermediate formation of a dihydrophenanthrene derivative,[372]

IV. Types of Reaction

Carbonium ions generated in the Kolbe or Hofer–Moest reaction may give rise to cyclic products, e.g., anodic oxidation of n-butyrate or of n-propaneboronate produces cyclopropane,[298,300]

$$CH_3CH_2CH_2CO_2^- \xrightarrow{-2e} CO_2 + CH_3CH_2CH_2^+ \rightarrow H^+ + \triangle$$

In other cases, cyclization could be understood either on the basis of intramolecular radical coupling or carbonium ion cyclization, and a clear distinction is not possible, e.g.,[386]

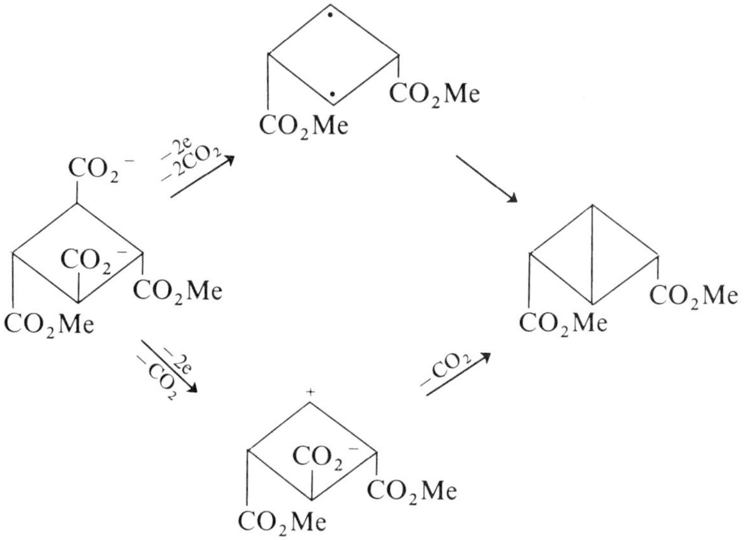

Carbonium ion intermediates are also presumably involved in the formation of cyclic products I and II below (as well as the noncyclic product III) when benzyl methyl ethanolamine was oxidized in methanol[387]:

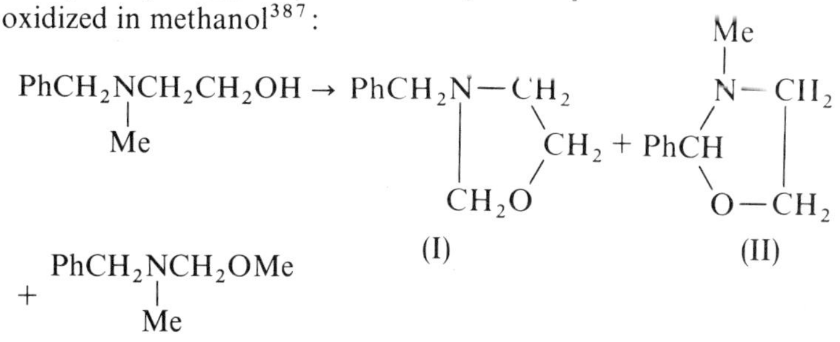

$$PhCH_2\underset{Me}{N}CH_2CH_2OH \rightarrow PhCH_2N-CH_2 \quad\quad \underset{}{\overset{Me}{N}}-CH_2$$

(I)   (II)

$$+ \quad PhCH_2\underset{Me}{N}CH_2OMe$$

(III)

As for reductive cyclization, discussed above, electrochemical introduction of a functional group into a molecule in a convenient position for cyclization with a group already present may be performed anodically, e.g.,[388]

$$\text{EtO-C}_6\text{H}_4\text{-NH}_2 + \text{SCN}^- \longrightarrow \text{EtO-C}_6\text{H}_3(\text{NH}_2)(\text{SCN}) \longrightarrow \text{benzothiazole-2-amine (EtO-substituted)}$$

Cyclization is only achieved in this way with *para*-substituted amines, since anodic thiocyanation usually occurs in the *para* position, rather than *ortho*, if the former is free.

5. Ring Opening, Expansion, or Contraction

(i) Cathodic

Controlled-potential electrolytic reduction at a mercury cathode of the cyclopropyl derivative,

$$\text{cyclopropyl-C(CH}_3\text{)=C(CN)(CO}_2\text{Et)}$$

gave no hydrodimer, but the major product was[389]

$$\text{CH}_3\text{CH}_2\text{CH}_2\text{-C(CH}_3\text{)=C(CN)(CO}_2\text{Et)}$$

Thus the cyclopropyl ring was cleaved, and the double bond left intact, although chemical reducing agents reduced the double bond.

IV. Types of Reaction

Other products obtained in smaller amounts were

$$CH_3CH_2CH_2COCH_3 \quad \text{and} \quad \triangleright\!-\!COCH_3$$

The cyclopropyl ring in the latter was not reducible, so the former must have arisen from cleavage at the double bond after reductive opening of the cyclopropyl ring.

Heterocyclic nitrogen compounds provide many examples of cathodic reduction accompanied by ring opening or ring contraction. These have been discussed by Lund,[390] and one example only, in which both processes occur, is given here:

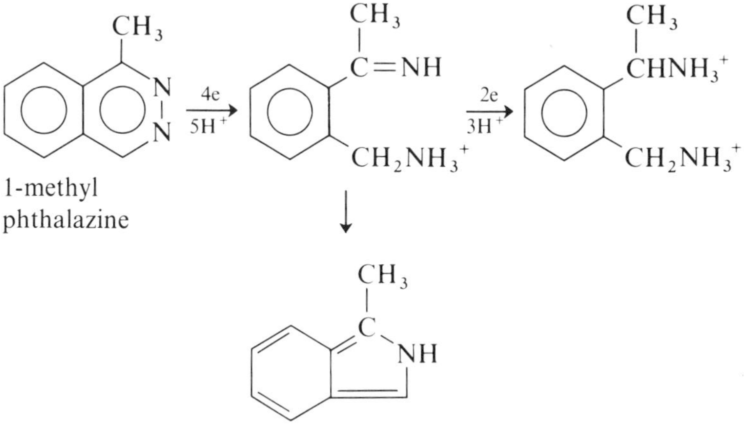

[Reaction scheme: 1-methyl phthalazine $\xrightarrow{4e, 5H^+}$ intermediate with $C=NH$ and $CH_2NH_3^+$ groups $\xrightarrow{2e, 3H^+}$ product with $CHNH_3^+$ and $CH_2NH_3^+$ groups; the intermediate also gives a cyclic product with CH_3 and NH]

1-methyl phthalazine

An example in which ring opening and ring expansion are both illustrated was supplied by Leonard et al.[391] in the reduction of 1-keto quinolizidine,

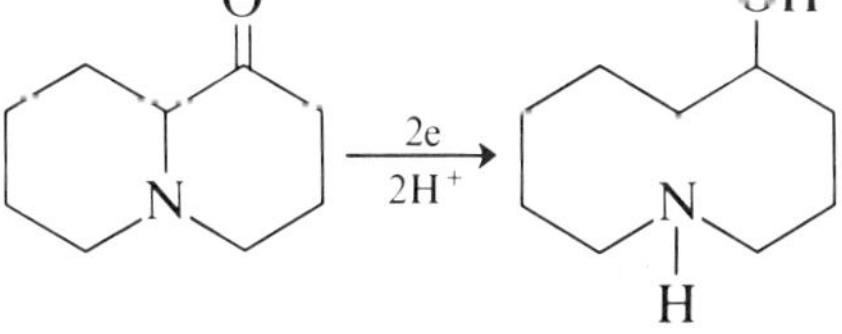

[Reaction scheme: 1-keto quinolizidine $\xrightarrow{2e, 2H^+}$ ring-expanded product with OH and NH]

(ii) Anodic

Heterocyclic compounds supply examples of anodic ring-opening processes, e.g., oxidation in acetonitrile of the α-tocopherol model compound,[392]

Once again, the stabilizing effect of blocking methyl groups is observed.

Ring expansion through the intervention of carbonium-ion intermediates has been observed during the Kolbe reaction.[297] 1-Adamantyl acetic acid in methanol on anodic oxidation, gave as the main product (72% yield) homoadamantyl methyl ether,[297] thus

Ring contraction has also been observed in oxidation of acids of appropriate structure, e.g.,[393]

V. STEREOSPECIFICITY IN ELECTRODE PROCESSES

In this section are included some electrochemical examples in which steric effects are significant in determining the conformation of the products, or in which steric considerations are used to throw some light on the reaction mechanism. Other steric effects, such as

V. Stereospecificity in Electrode Processes

those involved in steric inhibition of resonance, *ortho* effects, e.g., the difference in reductive behaviour of *o*-nitrophenol compared to the *m*-isomer, the difference in ease of replacement of axial and equatorial, or *exo*- and *endo*-halogen on reduction, have already been reviewed,[46,99] and are not dealt with further. Examples are available for each of the reaction types discussed in Section IV.

1. Substitution Reactions

Most work has been done on the replacement of halide by hydrogen, although reductive splitting of the C—O bond has also been examined stereochemically. As for homogeneous reactions of alkyl halides, electrochemical studies employing halides in which the halogen is attached to an asymmetric carbon atom should, in principle, be able to distinguish between an S_N2 type mechanism,

$$R-X \xrightarrow{2e} X^- + R^- \xrightarrow{H^+} RH$$

an S_N1,

$$R-X \to X^- + R^+ \xrightarrow[H^+]{2e} RH$$

or a radical-type mechanism,

$$RX \xrightarrow{e} R\cdot + X^- \xrightarrow{e} R^- \xrightarrow{H^+} RH$$

The first of these should lead to inversion of configuration, the latter two should both lead mainly to racemization or retention of configuration.

2-Phenyl-2-chloropropionic acid was chosen as a suitable reactant, since both the reactant and the product 2-phenyl propionic acid were known not to undergo autoracemization, and their absolute configurations were known.[175]

$$\underset{\underset{CH_3}{|}}{\overset{\overset{Cl}{|}}{Ph-C-CO_2H}} \xrightarrow[2e]{H^+} \underset{\underset{CH_3}{|}}{\overset{\overset{H}{|}}{Ph-C-CO_2H}}$$

At a mercury pool electrode, controlled-potential electrolysis at -1.4 V versus SCE ($E_{1/2} = -1.0$ V) occurs with predominant (80–90%) inversion of configuration. It is interesting to note that reduction using zinc and acetic acid also occurs with inversion, but catalytic reduction on palladium occurs with retention of configuration.[395] The electrochemical process was interpreted as occurring through two electron transfers from the electrode to the lowest vacant (antibonding) σ^* orbital, corresponding to the weakest σ bond in the molecule, the C—Cl bond. Concerted expulsion of a chloride ion leaves a carbanion of inverted configuration which protonates rapidly. The overall process thus resembles the homogeneous S_N2 mechanism.

In marked contrast, almost complete loss of optical activity occurs on electrolytic reduction of optically active O-benzoyl atrolactic acid, or its methyl ester,[396]

$$\text{Ph}-\underset{\underset{\text{CH}_3}{|}}{\overset{\overset{\text{OCOPh}}{|}}{\text{C}}}\text{CO}_2\text{H} \xrightarrow[\text{H}^+]{2e} \text{Ph}-\underset{\underset{\text{CH}_3}{|}}{\overset{\overset{\text{H}}{|}}{\text{C}}}-\text{CO}_2\text{H}$$

Reduction of this compound required a more negative electrode potential (-1.8 V versus SCE) than did 2-phenyl-2-chloropropionic acid (-1.4 V). Atrolactic acid itself was recovered unchanged under similar conditions, so that the adjacent C=O group is required to activate the carbon–oxygen bond. The authors suggested that the difference between the behavior of the O-benzoyl and the chloroacid may arise from adsorption of the leaving benzoate group but not chloride on the electrode; however, adsorption seems unlikely at a potential of -1.8 V versus SCE.

Other stereochemical investigations of halide replacement have utilized the unique and well-characterized properties of cyclopropyl systems, namely the very slow inversion of carbanions, very rapid inversion of radicals, and the rapid rearrangement of carbonium ions.[397-400] Annino et al.[397] studied compounds of the type

$$\text{Ph}_2\underset{\text{H}\quad\text{H}}{\overset{\text{R}}{\triangle}}\text{Br}$$

where R = CO_2H, CO_2Me, or Me, and found no dimeric or rearranged products, the latter tending to discount an S_N1 mechanism. For R = CO_2H, the products varied from 35% inversion (low pH) to 38% retention (high pH). For R = CO_2Me, 30–56% inversion

V. Stereospecificity in Electrode Processes

was observed, and for R = Me, 21% retention of configuration. A mechanism was proposed involving one-electron attack by the electrode from the Br side of the substrate to give a complex (substrate plus electrode) with unchanged configuration. This could give a racemate by loss of a radical into the solution, or, by uptake of another electron, a carbanion, shielded by the electrode on one side. The carbanion, on diffusion away from the electrode may invert, or react with H^+ without inversion; alternatively, resonance delocalization of charge may cause the complex to dissociate. Attack on the complex by H^+ from the less hindered side gives rise to inverted product. The authors further commented that, since cyclopropyl halides are unreactive in normal $S_N 2$ processes, results obtained by electrochemical studies on such systems may not be typical of halides in general.

The work of Annino et al.[397] was performed in aqueous ethanol solutions. Recently, similar systems were investigated in nonaqueous solvents by Webb et al.[398] employing mercury and carbon cathodes. Cyclic voltammetry and controlled-potential coulometry gave less complex results on carbon electrodes; the complications on mercury were ascribed to intervention of organomercury intermediates. 1-Bromo-1-methyl-2,2-diphenylcyclopropane gave as product 1-methyl-2,2-diphenylcyclopropane which was 25% optically pure (i.e., 63% retention) using a mercury electrode, and 47% optically pure (74% retention) using carbon. This suggested racemization prior to forming an organomercury intermediate or a carbanion. A mechanism was proposed involving a short-lived radical anion, which is directly reduced to a carbanion on carbon, accounting for the higher optical purity on this electrode material,

$$RBr \xrightarrow{e} RBr \cdot^- \xrightarrow{e} Br^- + R^- \xrightarrow{H^+} RH$$

On mercury, which is a good radical trap, formation of $R \cdot_{ads}$ (or $RHg \cdot$) competes with direct reduction to the carbanion. $R \cdot_{ads}$ or $RHg \cdot$ can abstract bromide ion from the reactant RBr,

$$RHg \cdot + RBr \rightarrow R \cdot + RHgBr$$

and racemization of $R \cdot$ accounts for the lower optical purity. Isotopic substitution of solvent acetonitrile by deuterium led to the incorporation of deuterium into the product, hence the solvent, rather than the supporting electrolyte, acts as the proton source.

The 1-iodo compound gave a product RH which was 4% optically pure (53% retention) together with R_2Hg which was racemic. This latter was probably formed thus

$$2RHg\cdot \rightarrow R_2Hg + Hg$$

RI reduces at a potential less negative than that required for RBr; at this less negative potential, the rate of reduction of $R\cdot_{ads}$ is lower, hence the concentration of $RHg\cdot$ can increase sufficiently for this reaction to become significant. The analogous compound RHgBr reduced with 100% retention of configuration.

Other stereochemical work on cyclopropyl halides has dealt with geminal dihalogenocyclopropanes.[399,400] Fry and Moore[399] studied systems of the type

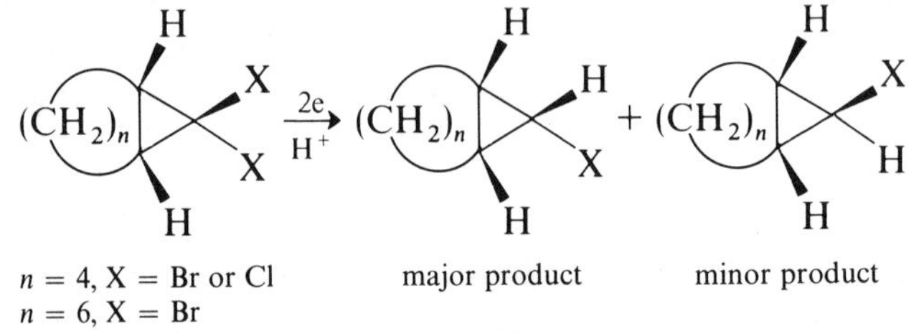

$n = 4$, $X = Br$ or Cl
$n = 6$, $X = Br$

major product minor product

and also

major minor

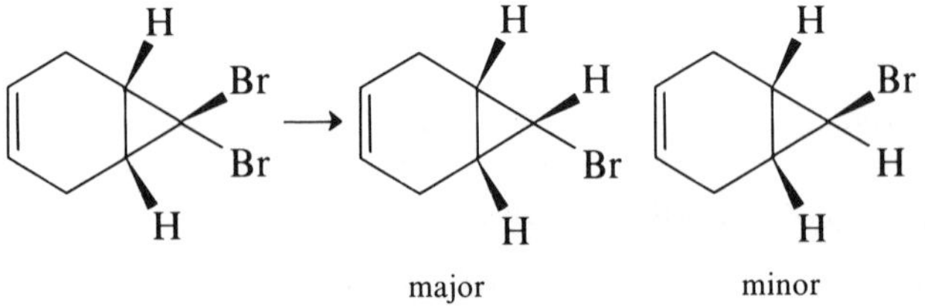

A variety of nonaqueous and partly aqueous solvent systems was employed, and in all cases, the less stable *cis* isomer was the major product; the proportion of *cis* product usually increased with increasing water content in the solvent. The predominance of *cis* product could be understood if the halide approached the electrode with the C—X bond parallel to the surface, or with X nearer to the electrode, because steric repulsions are then greater for removal of *trans*-X. If *cis*-X is removed, the carbanion diffuses out from the electrode into the solution before losing its configuration. This was

V. Stereospecificity in Electrode Processes

deduced from an experiment in a solvent consisting of 33% hexamethylphosphoramide plus 67% ethanol, which provides an aprotic region at the electrode surface,[401] and in which 100% *cis* product was obtained. Removal of *trans*-X would require subsequent rearrangement of the resulting radical or anion to a less stable *cis* isomer before protonation.

Erikson *et al.*[400] compared the reduction at a mercury cathode of the compounds

```
     Br         Cl              Cl
    /          /     and       /
   /          /               /
   Br         Cl              Br
  (I)        (II)            (III)
```

with reduction by zinc and acetic acid or lithium amalgam and ethanol. Reduction of the isomers of 7-bromo-7-chloro-[4.1.0]-bicycloheptane (III) resulted in exclusive removal of Br with predominant retention of configuration. The dichloro isomers (II) showed greater stereospecificity than the dibromo isomers (I). The results were consistent with the mechanism given above for the diphenyl cyclopropyl halides.[397] The *exo* and *endo* Br atoms in the isomers of III are reduced electrolytically with only a small change in $E_{1/2}$, suggesting no exclusive attack on the *exo* halogen in the geminal dihalide. Stereospecificity therefore depends on the carbanion intermediate; if an optically active system is used, equilibration leads to racemization,

$$\text{structure with Cl} \rightleftharpoons \text{structure with Cl}^-$$

If not optically active, the isomers do not have the same stability, which could vary with the solvent because of varying interaction with the halogen or the lone pair. If an electrode or a metal reductant can attack either halogen, the overall stereochemical result depends on the extent to which the carbanion is freed from the electrode, and also on the extent to which the above equilibrium occurs. This model is capable of explaining both solvent effects and the effect observed when Cl replaces Br.

Although unreactive, vinyl halides can be reduced to alkenes. In DMF, 3-iodohexene[402] gave hexenes in 89–92% overall yield, the *trans* isomer giving a product containing 94% *trans-* and 6% *cis*-hexene. The *cis*-iodohexene gave 70% *trans-* and 30% *cis*-hexene. Addition of phenol as a proton donor did not affect the product compositions, so that isomerization of a carbanion is unlikely. Radicals from the first electron transfer therefore rearrange, and further electron transfer and protonation follow. The partial equilibration observed implies that the isomerization rates are comparable in magnitude to the rates of reduction of radicals to carbanions. An earlier example of the same process was provided by Elving *et al.*[170] in the reduction of monobromomaleic acid, when maleic and fumaric acids were produced, together with butadiene 1,2,3,4-tetracarboxylic acid, the latter from dimerization of an intermediate radical.

2. Addition Reactions

Previous reviews[46,99] have dealt with the stereochemistry of cathodic addition of hydrogen to carbon–carbon unsaturated links, so no general discussion is given here. Two interesting recent developments will, however, be discussed. Gourley and co-workers[403] have achieved an asymmetric synthesis in an ingenious exploitation of catalytic hydrogen waves, well-known to polarographers (see review by Mairanovskii[404]). Certain organic bases, themselves polarographically inactive, cause the evolution of molecular hydrogen from a cathode and simultaneous consumption of hydrogen ions at less negative potentials than the usual waves for hydrogen discharge from the same solution in the absence of these bases. Electron transfer to the protonated form of the base produces a radical,

$$R_3\overset{+}{N}H + e \rightarrow R_3\dot{N}H$$

which does not undergo the normal reactions of cathodically produced radicals, such as further reduction to an anion, reaction with solvent, dimerization, or disproportionation, but instead produces hydrogen

$$R_3\dot{N}H \rightarrow R_3N + \tfrac{1}{2}H_2$$

V. Stereospecificity in Electrode Processes

followed by rapid protonation

$$R_3N + H^+ \to R_3\overset{+}{N}H$$

and repetition of the process.

Suitable substrates which could be reduced at a potential similar to that of the catalytic hydrogen wave to form a radical,

$$R' + H^+ + e \to R'H\cdot$$

provided the possibility for hydrogen atom transfer from $R_3\dot{N}H$ to $R'H\cdot$, thus

$$R_3\dot{N}H + R'H\cdot \to R_3N + R'H_2$$

Experiment showed that if R' was coumarin, or a substituted coumarin,

where $X = H$ or MeO, $Y = Me$, considerably enhanced yields of dihydro derivative (up to 84% yield) could be obtained (compared to 5–20% in the absence of R_3N). Dimerization was the predominant reaction in the absence of R_3N,

$$2R'H\cdot \to HR'-R'H$$

In the dimer, the carbon to which Y is attached becomes asymmetric during the reaction, and, since the dimer contains two such asymmetric carbons, both *meso* and ± forms are obtained. As long as Y is not H, the dihydro compound also contains an asymmetric carbon, and if the base R_3N is optically active, e.g., an alkaloid like yohimbine or sparteine, the ± forms of the dihydro compound are produced in unequal amounts, i.e., a product with net optical activity results. Interpretation of results is complicated by the possibility of specific orientation arising from adsorption of reactants on the electrode, but simple spatial considerations, e.g., fitting together in the transition state of $R'H\cdot$ and $R_3\dot{N}H$ so that large, medium, and small groups in the former fit opposite large, medium, and small spaces in the latter, are reasonably successful in explaining the results obtained with a range of alkaloids of varying spatial requirements. The largest optical yield achieved is 19%, and the method appears full of promise for further exploitation.

Optically active products have also been obtained in the reduction of acetophenone in the presence of large amounts of optically active amine salts.[405] Constant-current reduction at a mercury cathode and using (−) or (+) ephedrine hydrochloride or hydrobromide as supporting electrolytes yields about 40% of the optically active methyl phenyl carbinol R(+) or S(−), respectively. Also obtained is about 40% of pinacol from hydrodimerization; this is a mixture of *meso* and ± isomers. The overall current efficiency is 70–80%, and the yield of optically active product was not altered significantly at temperatures between 0 and 65°C, although the optical purity of the product decreases slightly with decreasing concentration of supporting electrolyte (0.3–1 M concentrations are satisfactory).

Cathodic addition of hydrogen to the carbonyl group of cyclic ketones has been shown to produce mixtures of axial and equatorial alcohols, in which the proportion of the less stable axial isomer is increased in the presence of acetic acid.[405] This was explained on the basis of discharge of the protonated ketone, adsorbed at the electrode through the carbonyl carbon, by means of an equatorial carbon–electrode bond

Efficient protonation aids adsorption in this sense, and also acts to preserve the configuration of the carbanion before inversion occurs. Under less efficient protonating conditions, adsorption through an equatorial oxygen–electrode bond leads to the formation of the

V. Stereospecificity in Electrode Processes

more stable equatorial alcohol,

Reduction of carbon–nitrogen double bonds in camphor oxime (R = Me) and norcamphor oxime (R = H),

at a mercury cathode in 80% ethanol, either potentiostatically or galvanostatically, produced the corresponding amines in 50–70% yields, with high degrees of stereoselectivity.[407] For example, camphor oxime gave 99% of the *exo* amine and only 1% of the more stable *endo* isomer

exo   *endo*

A similar result was obtained by reduction with lithium aluminum hydride, but sodium and ethanol gave 4% *exo* and 96% *endo*. Norcamphor oxime produced 100% *endo* amine on electrochemical reduction, and also with lithium aluminum hydride; sodium and ethanol gave 25% *endo* and 75% *exo*. The similarity in the results with LiAlH$_4$ suggests kinetic control by attack from the least-hindered side, this being *endo* for camphor, *exo* for norcamphor. Formation of an organomercury species at the electrode surface, followed by protonolysis with retention of configuration, was considered likely.

The orienting influence of the electric field at the electrode–solution interface was advanced as an explanation for the stereochemical effects observed in the reduction of benzil at a mercury cathode.[175] Benzil may exist in two conformations (cf. p. 161),

$$\underset{cis}{\begin{array}{c}Ph\diagdown\quad O\\ C\\ \|\\ C\\ Ph\diagup\quad \diagdown O\end{array}} \quad\text{and}\quad \underset{trans}{\begin{array}{c}Ph\diagdown\quad O\\ C\\ \|\\ C\\ O\diagup\quad \diagdown Ph\end{array}}$$

because of coupling between its π-electronic systems. The *trans* form is predominant at equilibrium, but in the electric field at the electrode surface, the more polar *cis* conformation is favored as the electrode potential deviates from the point of zero charge, which is close to the half-wave potential for benzil reduction. *Cis*-benzil is reduced to *cis*-stilbenediol, *trans*-benzil to *trans*-stilbenediol,

$$\underset{cis}{\begin{array}{c}Ph\diagdown\quad OH\\ C\\ \|\\ C\\ Ph\diagup\quad \diagdown OH\end{array}} \quad \underset{trans}{\begin{array}{c}Ph\diagdown\quad OH\\ C\\ \|\\ C\\ HO\diagup\quad \diagdown Ph\end{array}}$$

both of which rearrange to benzoin.

Grabowski et al.[175] using a linear anodic sweep to reoxidize the two forms of stilbenediol (*trans* at potentials less positive than for the *cis*), were able to show that the ratio *cis/trans*-stilbenediol varied with electrode potential, reaching a maximum of about 2 (in aqueous solution) at a potential of about -0.95 V (versus SCE). They argued that this ratio reflected the proportions of the two conformers of benzil in the solution, and the influence of potential, through the electric field, on these proportions. Stapelfeldt and Perone[176] studied the effect of *p*H and solvents on the reduction, and measured rate constants for the chemical transformation of the

V. Stereospecificity in Electrode Processes

two forms of stilbenediol into benzoin. Solvents had a pronounced effect, e.g., in 50% methanol at pH 7.5, the ratio cis/trans was less than 0.1, whereas at the same pH in 78% ethanol, the ratio was greater than 10.

The stereochemistry of addition of CO_2 to unsaturated systems has been studied by Wawzonek et al.[236] at uncontrolled potential and by Dietz et al.[408] at controlled potential. The latter workers formed stilbene radical anion cathodically in dimethylformamide and acetonitrile at -2.2 V versus SCE. In DMF, only meso- and dl-diphenyl succinic acids were formed, but in acetonitrile, protonation competed and some 1,2-diphenyl propionic acid was formed. Cis-stilbene gave a dl/meso ratio of about 1.5 in both solvents; trans-stilbene was studied in DMF alone and the dl/meso ratio was 2.7. Isomerization of stilbene anion radicals was shown to be slow under the experimental conditions employed, so that nonbonded interactions in the intermediate after addition of one molecule of CO_2 should determine the product distribution. If transfer of the second electron occurs from the side opposite to that of the first carboxylate group, cis-stilbene would be expected to give a carbanion with a trans arrangement of groups,

$$\underset{\ominus}{\overset{CO_2^-}{\underset{Ph}{\overset{H}{\diagup}}\underset{H}{\overset{Ph}{\diagdown}}}}$$

leading to the meso acid, whereas trans-stilbene should give the dl acid, through the carbanion

$$\underset{\ominus}{\overset{CO_2^-}{\underset{Ph}{\overset{Ph}{\diagup}}\underset{H}{\overset{H}{\diagdown}}}}$$

The radical anion may equilibrate, however, thus

$$\underset{\text{A from }trans}{\overset{CO_2^-}{\underset{Ph}{\overset{H}{\diagup}}\underset{H}{\overset{Ph}{\diagdown}}}} \rightleftharpoons \underset{\text{B from }cis}{\overset{CO_2^-}{\underset{Ph}{\overset{Ph}{\diagup}}\underset{H}{\overset{H}{\diagdown}}}}$$

Such equilibration should favor A, on steric grounds, through minimization of nonbonding repulsions. A *dl/meso* ratio higher than 1 is thus expected for both isomers of stilbene; addition of the second electron to the radical anion from *cis*-stilbene apparently is fast enough to compete with equilibration, so the *dl/meso* ratio for this isomer is smaller than for *trans*-stilbene.

The stereochemistry of the reductive dimerization (pinacolization) of aldehydes and ketones has been extensively studied by Stocker and Jenevein.[409] These authors have also compared electrochemical reduction with photochemical pinacolization of the same substrates. The reactions are both considered to proceed through intermediate formation of a ketyl radical,

$$\text{Ar}-\underset{\parallel}{\text{C}}-\text{R} \xrightarrow[\text{H}^+]{e} \text{Ar}-\underset{|}{\overset{\text{OH}}{\text{C}}}-\text{R}$$

or a ketyl radical anion in alkaline solution,

$$\text{Ar}-\underset{\parallel}{\overset{\text{O}}{\text{C}}}-\text{R} + e \rightarrow \text{Ar}-\underset{|}{\overset{\bar{\text{O}}}{\text{C}}}-\text{R}$$

In acid solution, combination of two ketyl radicals produces the pinacol; in basic media, combination of a ketyl radical and a radical anion probably predominates over reaction between two radical anions, because of charge repulsion. Considerations of nonbonded interactions and hydrogen bonding, together with the effect of *p*H on the latter, provide satisfactory explanations for the *dl/meso* ratio in the products. Minimum nonbonded interaction between like groups requires a *trans* conformation,

leading to a *meso* product. Hydrogen bonding between the two OH groups, (or between OH and O^- in basic media) requires these groups to be adjacent. This, in turn, requires either the two aryl *and* both R groups to be near-eclipsed, leading to a *meso* product,

V. Stereospecificity in Electrode Processes

or the two aryl *or* two alkyl groups to be near-eclipsed, leading to the *dl* product. On balance, these favor the *dl* product. Acetophenones show a slight predominance for formation of *dl* product, hence hydrogen bonding is significant; in basic media, as expected, the *dl/meso* ratio increases. In the case of 2-acetyl-pyridine, hydrogen bonding in the ketyl radical, and, in strong acid in its protonated form, require consideration, e.g.,

$$\underset{A}{\begin{array}{c}\text{pyridinyl-N}\cdots\text{H}-\text{O}^{-}\\|\\\text{C}-\text{CH}_3\end{array}} \xrightarrow{H^+} \underset{B}{\left[\begin{array}{c}\text{pyridinyl-N}\cdots\text{H}-\text{O}\cdots\text{H}\\|\\\text{C}-\text{CH}_3\end{array}\right]^+}$$

In neutral solution, dimerization of A is subject only to steric control and favors the *meso* product; in strongly acid media, reaction probably involves combination of A with B rather than the electrostatically less-favored dimerization of B. Hydrogen bonding in the *trans* conformation, leading to the *dl* product, is therefore possible at high acidities, so the *dl/meso* ratio increases. The protonated form B is also more readily reduced than is A, giving the carbinol, which is the major product in acid solution:

$$B + e + H^+ \rightarrow \left[\begin{array}{c}\text{pyridinyl-N}-\text{H}\cdots\text{O}-\text{H}\\|\\\text{C}-\text{CH}_3\\|\\\text{H}\end{array}\right]^+$$

Benzaldehyde in 80% alcohol[408] or in aqueous solution[410] gives a *dl/meso* ratio very close to one, but in the presence of ions which are strongly adsorbed at the interface such as I^- and tetraethyl-ammonium ion, relatively more *meso* hydrobenzoin is obtained, e.g., at 0.8 M I^-, the *dl/meso* ratio is down to 0.48.[410] The effect of these adsorbable ions suggests that the reaction occurs on, or very close to, the electrode surface, so that the presence of the ions near the reaction zone affects interradical hydrogen bonding, perhaps by competitive orientation of the OH groups, leading to more *meso* isomer. It may be significant that in this investigation, much unidentified high-molecular-weight by-product was obtained;

the total yields of hydrobenzoin were in the range 15–35% only, compared with 65–85% in 80% ethanol.[409]

Pinacolization also occurs with $\alpha,\beta,\alpha',\beta'$-unsaturated ketones,[411] in distinction from α,β-unsaturated ketones, which couple at the β position (similar to hydrodimerization of acrylonitrile). Different stereoisomeric pinacols are obtained in acid and alkaline solutions when the ketone androsta-1,4-dien-17β-ol-3-one is reduced. In acid solution, the protonated O of the C=O group faces toward the electrode, the methyl groups in positions 10,13 away from the electrode, and the OH in the radical is in the α conformation,

In alkaline solution, the O of C=O is repelled from the cathode to the β configuration, in which it also has the best chance of extracting a proton from the solvent,

Hydrodimerization of suitably substituted activated unsaturated systems can give rise to stereoisomers. For example, Wiemann and Bouguerra[412] found that hydrodimerization of ethyl cinnamate gave more *meso-* than *dl*-diethyl$\beta\beta$-diphenyl adipate, but crotonitrile gave $\beta\beta$-dimethyl adiponitrile in which the *meso/dl* ratio was one. Low yields of hydrodimer from α-phenyl cinnamic acid led Wawzonek et al.[413] to study the hydrodimerization of α-phenyl cinnamonitrile. Of the 80% yield of hydrodimerized product, approximately four-fifths was cyclic, namely the compound 1-amino-2,3,4,5-tetraphenyl-5-cyano-1-cyclopentene,

V. Stereospecificity in Electrode Processes

This could arise from a Ziegler condensation involving the anion of the expected hydrodimer,

$$\begin{array}{c} \text{Ph} \\ | \\ \text{Ph}-\text{CH}-\text{C}-\text{CN} \\ | \\ \text{Ph}-\text{CH}-\text{CH}-\text{CN} \\ | \\ \text{Ph} \end{array} \rightarrow \begin{array}{c} \text{Ph} \\ \text{Ph} \diagdown \quad | \\ \text{CH}-\text{C}-\text{CN} \\ | \quad \diagdown \\ \quad \quad \text{C}=\text{N}^{\ominus} \\ \text{CH} \diagup \\ \text{Ph} \diagup \quad | \\ \text{Ph} \end{array}$$

and may be assisted by the formation of an aprotic region at the electrode interface through adsorption of tetraalkyl-ammonium ions from the supporting electrolyte. Steric factors, involving non-bonding interactions and charge repulsions in the radical anion, indicate that the *meso* configuration,

$$\begin{array}{c} \text{H} \quad \overset{\ominus}{\downarrow} \quad \text{Ph} \\ \diagdown \quad | \quad \diagup \\ \text{Ph} \quad \overset{|}{\top} \quad \text{H} \end{array}$$

is favored for the dimeric product. If protonation occurs before dimerization, charge repulsion is absent, and steric effects alone would favor the *dl* configuration. Protonation is likely if desorption of the reactant into the bulk solution occurs. NMR measurements on the cyclic product above indicate that the hydrogens on C_3 and C_4 are *cis*, so the configuration is *meso*, supporting the view that dimerization involves the radical anion.

Schiff bases derived from benzaldehyde and *p*-substituted aromatic aldehydes, and a variety of aryl and cycloalkyl amines, undergo hydrodimerization on reduction,[414] giving *dl* and *meso* products,

$$2\text{R}-\overset{\overset{\text{H}}{|}}{\text{C}}=\text{NR}' \xrightarrow{2e,\, 2\text{H}^+} \begin{array}{c} \text{H} \\ | \quad \text{H} \\ \text{R}-\text{C}-\text{NR}' \\ | \\ \text{R}-\text{C}-\text{NR}' \\ | \quad \text{H} \\ \text{H} \end{array}$$

NMR measurements indicate a *dl/meso* ratio of one; the lack of stereospecificity is similar to that discussed for the hydrodimerization of benzaldehyde itself.[409,410] An investigation of the effect of adsorbable ions[410] would be most interesting.

3. Elimination Reactions

Elimination of two halide ions from adjacent carbon atoms in a dihalide is easier when the halogens have a *trans* conformation rather than *cis*.[415] *Trans* elimination occurs faster than protonation, as shown by the single four-electron wave obtained on polarographic reduction of *trans*-2,3-dichloroacrylonitrile to acrylonitrile, while the *cis* isomer gives two two-electron waves.[416] A further example of the preference for *trans* elimination of halide from a 1,2 dihalide is provided by the polarographic studies of Feoktistov and Markova[417] on the reduction of 1,1,2,2-tetra-bromoethane. This substance exists in two forms in the liquid phase, and in solution namely, *gauche* and *trans*,

The equations show the expected products from *trans* elimination of two bromide ions. Polarographic reduction gives two two-electron waves, the second corresponding to the reduction of *cis*-dibromoethylene. Millicoulometry at -0.5 V, -1.2 V, and -1.5 V versus SCE gave products showing no detectable amounts of the *trans* isomer, although control experiments showed that 10% of this could be detected polarographically, on the basis of the easier reduction of the *trans* isomer ($\Delta E_{1/2} = 200$ mV). The results are

V. Stereospecificity in Electrode Processes

readily understood simply on the basis of *trans* elimination from the *gauche* form, which is known to be predominant in solution. Adsorption at the electrode in a preferred orientation seems to be ruled out by the controlled-potential results, since the same authors have shown that adsorption of dibromoethylene is a maximum at -0.5 V and that desorption occurs by -1.0 V.[418]

4. Cyclization

The intramolecular pinacol reaction, previously discussed,[375]

produced both stereoisomers in approximately equal amounts, thus exhibiting no stereoselectivity. Another cyclization reaction occurs during oxidative coupling of phenolic tetrahydroisoquinolines,[419]

Since asymmetric centers are introduced at C_1 on cyclization, formation of three separable enantiomeric pairs is expected; all three were isolated, two as crystalline products, one as a glass, but the structures were not established.

5. Ring Opening

Anodic oxidation of substituted cyclopropane carboxylic acids may occur in the following ways[420]:

$$RCO_2^- \xrightarrow{-e} RCO_2\cdot \xrightarrow{-CO_2 \text{ (A)}} R\cdot \xrightarrow{-e} R^+$$
$$RCO_2^- \xrightarrow{-e(B)} \text{allylic products}$$
$$RCO_2^- \xrightarrow{-2e} RCO_2^+ \xrightarrow{-CO_2}$$

Route A yields both cyclopropyl radical and cation; the radical rapidly racemizes, while the cation rearranges nonstereospecifically to allylic products. Route B involves concerted ring opening and CO_2 loss, yielding allylic products stereospecifically. Electrolysis at uncontrolled potential, using carbon electrodes in methanol, gave stereospecific products, thus

[Reaction scheme showing phenyl-methyl cyclopropane carboxylate undergoing $-2e$ then $-CO_2$/MeOH to give Ph-C(OMe)H-CH=CH-Me plus two cyclopropyl OMe products]

[Second reaction scheme showing the other stereoisomer giving Ph-CH(OMe)-CH=CH-Me and CH=CH-CH(OMe)-CH$_3$ with Ph substituent]

Formation of the acyloxonium ion RCO_2^+ may result, in this particular example, from the cyclopropane ring strain which may obstruct decarboxylation of the cyclopropane carboxy radical. As was concluded earlier, in connection with the reduction of cyclopropyl halides, results obtained on this ring system may not be typical of the reaction under consideration.

VI. APPLICATIONS FOR COMMERCIAL SYNTHESES

The conditions under which electrochemical methods of synthesis of organic compounds are most likely to be advantageous were recently summarized by Dickinson and Ovenden[421] as being those which allow (a) a reduction in the number of steps in the preparation, (b) the use of a cheaper starting material, (c) a reduction in the number of purification stages, or (d) the use of less extreme conditions. To these may be added (e) cases like that of 9-(o-iodophenyl)dihydroacridine,[53] already discussed, for which electrochemical reduction appears to be the only suitable synthetic procedure. Many more examples will undoubtedly be discovered of systems where the benefits of determination of reaction through potential control will similarly be utilized; the effect of potential on cohydrodimer composition[362] is another example capable of greater exploitation. (f) The influence of the nature of the supporting electrolyte anion on the relative extents of side-chain and nuclear anodic substitution[216] offers many synthetic opportunities. (g) Controlled generation of an unstable reactant, e.g., thiocyanogen,[275,276] is another situation for which electrochemical procedures are particularly appropriate. (h) An attractive approach, which will undoubtedly find greater application, is that of coupling anodic and cathodic syntheses, to produce two, instead of only one, desired products. An example is the coupled cathodic synthesis of salicylaldehyde by reduction of salicylic acid in sulfate media with the anodic production of $K_2S_2O_8$.[422] In appropriate cases, it should be possible to couple two organic electrode processes in this way.

The exploitation of the stereospecificity of electrode processes, discussed in Section V, appears to offer the greatest promise for future synthetic applications. This is an active field of investigation in electrochemistry, and the incentive to achieve stereospecific syntheses is certain to maintain a high level of activity here.

ACKNOWLEDGMENT

I wish to thank Professor B. E. Conway for his encouragement to write this paper and for a critical reading of the text.

REFERENCES

[1] M. Faraday, *Pogg. Ann.* **33** (1834) 438.
[2] H. Kolbe, *Ann.* **69** (1849) 257.
[3] B. C. L. Weedon, *Advan. Org. Chem.* **1** (1960) 1; *Quart. Rev.* **6** (1952) 300.
[4] F. Beck, *Chem. Ing. Tech.* **42** (1970) 153.
[5] A. K. Vijh and B. E. Conway, *Chem. Rev.* **67** (1967) 623.
[6] G. Aranowitz and R. J. Flannery, *J. Electrochem. Soc.* **116** (1969) 938.
[7] T. O. Pavela, *Ann. Acad. Sci. Fennicae, Ser. A, Chem.* **59** (1954) 1.
[8] B. J. Piersma and E. Gileadi, in *Modern Aspects of Electrochemistry*, Vol. 4, Butterworth, London, 1966.
[9] J. W. Johnson, H. Wroblowa, and J. O'M. Bockris, *Electrochim. Acta* **9** (1964) 639.
[10] D. F. A. Koch and R. Woods, *Electrochim. Acta* **13** (1968) 2101.
[11] J. O'M. Bockris, B. J. Piersma, and E. Gileadi, *Electrochim. Acta* **9** (1964) 1329.
[12] L. Eberson, *J. Am. Chem. Soc.* **89** (1967) 4669.
[13] F. L. Lambert, *J. Org. Chem.* **31** (1966) 4184.
[14] K. Sasaki, H. Urata, K. Uneyama, and S. Nagaura, *Electrochim. Acta* **12** (1967) 137.
[15] V. Pomilio, *Z. Electrochem.* **21** (1915) 444.
[16] M. M. Baizer, *J. Org. Chem.* **31** (1966) 3487, 3885; *J. Electrochem. Soc.* **111** (1964) 215; *Tetrahedron Letters* **1963**, 973.
[17] A. Kirrmann and M. Kleine-Peter, *Bull. Soc. Chim. France* **1957**, 894.
[18] N. L. Weinberg and H. R. Weinberg, *Chem. Rev.* **68** (1968) 449.
[18a] B. E. Conway, *Principles and Approaches in the Study of Electroorganic Reactions*, Chapter 1, Electroorganic Chemistry, Ed. by N. Weinberg, in Weissberger Series on Physical Methods in Organic Chemistry, Interscience, New York (1972).
[19] F. D. Popp and H. P. Schultz, *Chem. Rev.* **62** (1962) 19.
[20] M. Fleischmann and D. Pletcher, *RIC Rev.* **2** (1969) 87; *Platinum Metals Rev.* **13** (1969) 46.
[21] R. Parsons, *Trans. Faraday Soc.* **47** (1951) 1332.
[22] B. E. Conway and E. Gileadi, *Trans. Faraday Soc.* **58** (1962) 2493.
[23] J. O'M. Bockris and A. M. Azzam, *Trans. Faraday Soc.* **48** (1952) 145.
[24] Ref. 8, p. 125.
[25] B. E. Conway and A. K. Vijh, *Z. anal. Chem.* **224** (1967) 160.
[26] G. F. A. Kortum and J. O'M. Bockris, *Textbook of Electrochemistry*, Elsevier, London, 1951, p. 405.
[27] M. M. Baizer, *J. Electrochem. Soc.* **111** (1964) 215.
[28] H. Lund, *Acta Chem. Scand.* **23** (1969) 563.
[29] A. I. Vogel, *Practical Organic Chemistry*, 3rd Ed., Longmans, London, 1967.
[30] *Organic Syntheses, Collective*, Vol. II, Ed. by A. H. Blatt, Wiley, New York, 1943, p. 574.
[31] P. E. Iversen, *J. Chem. Education* **48** (1971) 136.
[32] F. Haber, *Z. Elektrochem.* **4** (1898) 506.
[33] R. Parsons, *Trans. Faraday Soc.* **47** (1951) 1332.
[34] A. N. Frumkin, *Zh. Fiz. Khim.* **5** (1934) 240.
[35] A. N. Frumkin, *Z. Physik* **35** (1926) 792; R. S. Hansen, R. E. Minturn, and D. A. Hickson, *J. Phys. Chem.* **60** (1956) 1185.
[36] A. Hickling, *Trans. Faraday Soc.* **38** (1942) 27.
[37] J. E. Harrar, F. B. Stephens, and R. E. Pechacek, *Anal. Chem.* **34** (1962) 1036.
[38] Operational amplifiers symposium, *Anal. Chem.* **35** (1963) 1770.
[39] A. D. Goolsby and D. T. Sawyer, *Anal. Chem.* **39** (1967) 411.

References

[40] A. J. Bard and K. S. V. Santhanam, *Electroanal. Chem.* **4** (1970) 295.
[41] Princeton Applied Research Corp., Princeton, N.J.
[42] Beckman Instruments Inc., Scientific and Process Instruments Div., Fullerton, Calif.
[43] P. R. Wells, *Chem. Rev.* **63** (1963) 171; *Linear Free Energy Relationships*, Academic Press, London, 1968.
[44] A. Streitwieser, *Molecular Orbital Theory for Organic Chemists*, Wiley, New York, 1961.
[45] Ref. 26, pp. 173, 185.
[46] P. Zuman, *Prog. Phys. Organic Chem.* **5** (1967) 81.
[47] B. E. Conway and E. Gileadi, *Trans. Faraday Soc.* **58** (1962) 2493.
[48] P. Delahay and I. Tractenberg, *J. Am. Chem. Soc.* **79** (1957) 2355; W. H. Reinmuth, *Advances Anal. Chem. Instr.* **1** (1960) 241.
[49] J. O'M. Bockris and H. Wroblowa, *J. Electroanal. Chem.* **7** (1964) 428.
[50] E. Gileadi and B. E. Conway, *Modern Aspects of Electrochemistry*, Vol. III, Butterworth, London, 1964.
[51] J. M. Saveant, *J. Electroanal. Chem. Interfacial Electrochem.* **29** (1971) 87.
[52] R. Benesch and R. E. Benesch, *J. Am. Chem. Soc.* **73** (1951) 3391.
[53] J. J. Lingane, C. G. Swain, and M. Fields, *J. Am. Chem. Soc.* **65** (1943) 1348.
[53a] O. Zuman, *Disc. Faraday Soc.* **45** (1968) 61.
[54] R. P. Bell, *The Proton in Chemistry*, Methuen, London, 1959.
[55] L. Holleck and R. Schindler, *Z. Elektrochem.* **60** (1956) 1138, 1142.
[56] A. O. Aten and G. J. Hoijtink, *Advances in Polarography*, Pergamon Press, Oxford, 1961, p. 777.
[57] V. S. Bagotskii and Y. B. Vasil'ev, *Electrochim. Acta* **11** (1966) 1439.
[58] L. Eberson, *Electrochim. Acta* **12** (1967) 1473.
[59] I. M. Kolthoff and J. J. Lingane, *Polarography*, Interscience, New York, 1941, p. 23.
[60] H. A. Laitinen and I. M. Kolthoff, *J. Phys. Chem.* **45** (1941) 1079.
[61] R. H. Sonner, B. Miller, and R. E. Visco, *Anal. Chem.* **41** (1969) 1498.
[62] V. G. Levich, *Physicochemical Hydrodynamics*, Prentice Hall, Englewood Cliffs, N.J., 1962; *Acta Physicochim. USSR* **17** (1942) 257; **19** (1944) 117, 133.
[63] A. C. Riddiford, *Adv. Electrochem. Electrochem. Engng.* **4** (1966) 47.
[64] B. E. Conway, E. J. Rudd, and L. G. M. Gordon, *Disc. Faraday Soc.* **45** (1968) 87.
[64a] E. J. Rudd and B. E. Conway, *Trans. Faraday Soc.* **67** (1971) 440.
[65] J. Heyrovsky and J. Kuta, *Principles of Polarography*, Academic Press, New York, 1966.
[66] I. M. Kolthoff and J. J. Lingane, *Polarography*, Interscience, New York, 1952.
[67] L. Meites, *Polarographic Techniques*, Wiley, New York, 1965
[68] G. W. C. Milner, *The Principles and Applications of Polarography and Other Electroanalytical Processes*, Longmans Green, New York, 1957.
[69] P. Zuman and I. M. Kolthoff, *Progress in Polarography*, Interscience, New York, 1962.
[70] P. Delahay, *New Instrumental Methods in Electrochemistry*, Interscience, New York, 1954.
[71] I. M. Kolthoff and P. J. Elving, *Treatise on Analytical Chemistry*, Interscience, New York, 1963.
[72] A. Weissberger, *Techniques of Organic Chemistry: Physical Methods*, Vol. 1, Part 4, Interscience, New York, 1960.
[73] R. N. Adams, *Electrochemistry at Solid Electrodes*, Marcel Dekker, New York, 1969.
[74] J. L. Huntington and D. G. Davis, *Chem. Instr.* **2** (1969) 83.

[75] J. O'M. Bockris, *Modern Aspects of Electrochemistry*, Vol. 1, Butterworth, London and Washington, 1954, p. 180.
[76] B. E. Conway, *Theory and Principles of Electrode Processes*, The Ronald Press Co., New York, 1965.
[77] J. O'M. Bockris and A. K. N. Reddy, *Modern Electrochemistry*, Macdonald, London, 1970.
[78] K. J. Vetter, *Electrochemical Kinetics*, Academic Press, New York, 1967.
[79] M. Spiro, *Electrochim. Acta* **9** (1964) 1531.
[80] B. E. Conway and M. Dzieciuch, *Canad. J. Chem.* **41** (1963) 21.
[81] B. E. Conway and A. K. Vijh, *J. Phys. Chem.* **71** (1967) 3639.
[82] A. A. Humffray and L. F. G. Williams, *Chem. Comm.* **1965**, 616.
[83] A. A. Humffray and L. F. G. Williams, *Electrochim. Acta*, **17** (1972) 401, 409.
[84] U. Eisner and E. Gileadi, *J. Electroanal. Chem. Interfacial Electrochem.* **28** (1970) 81.
[85] J. A. Harrison and Z. A. Khan, *J. Electroanal. Chem. Interfacial Electrochem.* **28** (1970) 131.
[86] H. Gerischer and A. Mauerer, *J. Electroanal. Chem. Interfacial Electrochem.* **25** (1970) 421.
[87] J. O'M. Bockris and E. C. Potter, *J. Chem. Phys.* **20** (1952) 614.
[88] A. Damjanovic, *Modern Aspects of Electrochemistry*, Vol. 5, Plenum Press, New York, 1969.
[89] G. A. Ward and C. M. Wright, *J. Electroanal. Chem.* **8** (1964) 302.
[90] Ref. 46, p. 175.
[91] H. T. S. Britton, *Hydrogen Ions*, 4th Ed., Chapman and Hall, London, 1955, p. 360.
[92] J. O'M. Bockris, *Modern Aspects of Electrochemistry*, Vol. 1, Butterworth, London, 1954, p. 187.
[93] Ref. 65, Chapters 9, 10.
[94] R. Brdicka, *Coll. Czech. Chem. Comm.* **12** (1947) 212.
[95] H. Strehlow, in *Technique of Organic Chemistry*, Ed. by A. Weissberger, Vol. 8, Part 2, Interscience, New York, 1963, p. 799.
[96] H. W. Nurnberg, *Disc. Faraday Soc.* **39** (1965) 136.
[97] K. Vesely and R. Brdicka, *Coll. Czech. Chem. Comm.* **12** (1947) 313.
[98] J. Koutecky, *Chem. zvesti* **8** (1954) 693.
[99] C. C. Perrin, *Prog. Phys. Organic Chem.* **3** (1965) 165.
[100] T. Yokoyama and M. Enyo, *Electrochim. Acta* **15** (1971) 1921.
[101] Ref. 65, Chapter 17.
[102] Ref. 69, p. 79.
[103] J. Kuta and I. Smoler, *Progress in Polarography*, Ed. by P. Zuman and I. M. Kolthoff, Vol. 1, Interscience, New York, 1962, p. 43.
[104] J. Heyrovsky, *Chem. listy* **40** (1946) 222.
[105] W. Kemula and Z. Kublik, *Roczniki Chem.* **32** (1958) 941.
[106] J. Volke, *Chem. zvesti* **14** (1960) 807.
[107] Ref. 65, Chapter 21.
[108] S. Wawzonek, *Talanta* **12** (1965) 1232.
[109] B. Breyer and H. H. Bauer, *Alternating Current Polarography and Tensammetry*, Interscience, New York, 1963.
[110] Ref. 65, Chapter 22.
[111] G. Manning, V. D. Parker, and R. N. Adams, *J. Am. Chem. Soc.* **91** (1969) 4584.
[112] D. H. Geske and A. J. Bard, *J. Phys. Chem.* **63** (1959) 1057.
[113] L. Meites and S. A. Moros, *Anal. Chem.* **31** (1959) 23.
[114] A. J. Bard and K. S. V. Santhanam, *Electroanal. Chem.* **4** (1970) 215.

References

[115] D. Ilkovic, *Coll. Czech. Chem. Comm.* **6** (1934) 498; *J. Chim. Phys.* **35** (1938) 129.
[116] S. Bogan, L. Meites, E. Peters, and J. M. Sturtevant, *J. Am. Chem. Soc.* **73** (1951) 1584.
[117] E. D. Harris and A. J. Lindsey, *Nature* **162** (1948) 413.
[118] A. A. Humffray, unpublished work.
[119] D. A. MacInnes, *The Principles of Electrochemistry*, Reinhold, New York, 1939, pp. 29–36.
[120] J. J. Lingane and S. L. Jones, *Anal. Chem.* **22** (1950) 1220.
[121] See Ref. 38.
[122] D. H. Geske, *J. Phys. Chem.* **63** (1959) 1062.
[123] A. J. Bard and S. V. Tatwawadi, *J. Phys. Chem.* **68** (1964) 2676.
[124] F. G. Cottrell, *Z. physik. Chem.* **42** (1902) 385.
[125] Ref. 69, p. 75.
[126] G. S. Alberts and I. Shain, *Anal. Chem.* **35** (1963) 1589.
[127] K. B. Oldham and R. A. Osteryoung, *J. Electroanal. Chem.* **11** (1966) 397.
[128] W. M. Schwarz and I. Shain, *J. Phys. Chem.* **69** (1965) 30.
[129] D. M. Kern, *J. Am. Chem. Soc.* **75** (1953) 2473; **76** (1954) 1011.
[130] P. Delahay, C. C. Mattax, and T. Berzins, *J. Am. Chem. Soc.* **76** (1954) 5319.
[131] W. K. Snead and A. E. Remick, *J. Am. Chem. Soc.* **79** (1957) 6121.
[132] J. Koutecky, *Coll. Czech. Chem. Comm.* **20** (1955) 116.
[133] T. R. Rosebrugh and W. L. Miller, *J. Phys. Chem.* **14** (1910) 816.
[134] A. C. Testa and W. H. Reinmuth, *Anal. Chem.* **32** (1960) 1578.
[135] R. S. Nicholson and I. Shain, *Anal. Chem.* **36** (1964) 706; **37** (1965) 178; D. S. Polcyn and I. Shain, *Anal. Chem.* **38** (1966) 370, 376.
[136] J. M. Saveant and E. Vianello, *Electrochim. Acta* **12** (1967) 629, 1545.
[137] B. E. Conway, E. Gileadi, and H. A. Kozlowska, *J. Electrochem. Soc.* **112** (1965) 341.
[138] R. Woods, *J. Phys. Chem.* **75** (1971) 354.
[139] E. T. Seo, R. F. Nelson, J. M. Fritsch, L. S. Marcoux, D. W. Leedy, and R. N. Adams, *J. Am. Chem. Soc.* **88** (1966) 3498.
[140] G. J. Janz and D. J. G. Ives, *Reference Electrodes*, Academic Press, New York, 1961.
[141] Ref. 44, p. 251.
[142] B. E. Conway, *Progr. Reaction Kinetics* **4** (1967) 399.
[143] V. D. Parker, *Acta Chem. Scand.* **24** (1970) 2768.
[144] H. A. Kozlowska and B. E. Conway, *J. Electroanal. Chem.* **7** (1964) 109.
[145] S. Gilman, *Electroanal. Chem.* **2** (1967) Chapter 3.
[146] S. B. Brummer and K. Cahill, *J. Electroanal. Chem.* **16** (1968) 207.
[147] W. H. Reinmuth, *Anal. Chem.* **33** (1961) 322.
[148] A. C. Testa and W. H. Reinmuth, *Anal. Chem.* **33** (1961) 1320.
[149] R. W. Murray and C. N. Reilley, *J. Electroanal. Chem.* **3** (1962) 64, 182.
[150] Ref. 69, p. 187.
[151] W. H. Reinmuth, *Anal. Chem.* **32** (1960) 1514.
[152] T. Berzins and P. Delahay, *J. Am. Chem. Soc.* **75** (1953) 4205.
[153] D. H. Geske, *J. Am. Chem. Soc.* **81** (1959) 4145.
[154] A. C. Testa and W. H. Reinmuth, *Anal. Chem.* **33** (1961) 1324.
[155] P. Delahay, *J. Phys. Chem.* **66** (1962) 2204; P. Delahay and A. Aramata, *J. Phys. Chem.* **66** (1962) 2208.
[156] K. B. Oldham and J. Spanier, *J. Electroanal. Chem. Interfacial Electrochem.* **26** (1970) 331.
[157] Ref. 69, p. 52.
[158] S. D. James, *J. Electrochem. Soc.* **114** (1967) 1129.

[159] B. E. Conway and P. L. Bourgault, *Trans. Faraday Soc.* **58** (1962) 593.
[160] J. E. B. Randles, *Disc. Faraday Soc.* **1** (1947) 11.
[161] R. de Leeuwe, M. Sluyters-Rebach, and J. H. Sluyters, *Electrochim. Acta* **14** (1969) 1183.
[162] M. Sluyters-Rebach and J. H. Sluyters, *Electroanal. Chem.* **4** (1970) 1.
[163] C. L. Wilson and J. H. Lippincott, *J. Am. Chem. Soc.* **78** (1950) 4290; *J. Electrochem. Soc.* **103** (1956) 672.
[163a] M. Fleischmann, J. R. Mansfield, and W. F. K. Wynne-Jones, *J. Electroanal. Chem.* **10** (1965) 522.
[164] A. T. Hubbard and F. C. Anson, *Electroanal. Chem.* **4** (1970) 129.
[165] C. R. Christensen and F. C. Anson, *Anal. Chem.* **36** (1964) 495.
[166] D. M. Oglesby, J. D. Johnson, and C. N. Reilley, *Anal. Chem.* **38** (1966) 385.
[167] J. E. Dubois, P. Alcais, and G. Barbier, *J. Electroanal. Chem.* **8** (1964) 359.
[168] J. Janata and H. B. Mark, *Electroanal. Chem.* **3** (1969) 1.
[169] W. Gruber and H. Renner, *Monatsh.* **81** (1950) 751.
[170] P. J. Elving, I. Rosenthal, J. R. Hayes, and A. J. Martin, *Anal. Chem.* **33** (1961) 330.
[171] G. Dryhurst and P. J. Elving, *Talanta* **16** (1969) 885.
[172] R. Zahradnik and C. Parkanyi, *Talanta* **12** (1965) 1289.
[173] J. Tirouflet and E. Laviron, *Talanta* **12** (1965) 1107.
[174] V. Yu. Filinovskii, *Elektrokhimiya* **5** (1969) 635.
[175] B. Czochralska, *Chem. Physics Letters* **1** (1967) 239; Z. R. Grabowski, B. Czochralska, A. Vincenz-Chodkowska, and M. S. Balasiewicz, *Disc. Faraday Soc.* **45** (1968) 145.
[176] H. E. Stapelfeldt and S. P. Perone, *Anal. Chem.* **40** (1968) 815.
[177] D. H. Geske and A. M. Maki, *J. Am. Chem. Soc.* **82** (1960) 2671.
[178] T. Kuwana, R. K. Darlington, and P. W. Leedy, *Anal. Chem.* **36** (1964) 2023.
[179] W. N. Hansen, R. A. Osteryoung, and T. Kuwana, *Anal. Chem.* **38** (1966) 1810.
[180] D. R. Tallant and D. H. Evans, *Anal. Chem.* **41** (1969) 835.
[181] P. E. Stewart and W. A. Bonner, *Anal. Chem.* **22** (1950) 793.
[182] S. D. Ross, M. Finkelstein, and R. C. Petersen, *J. Am. Chem. Soc.* **92** (1970) 6003.
[183] M. Finkelstein, R. C. Petersen, and S. D. Ross, *Electrochim. Acta* **10** (1965) 465.
[184] R. E. Smith, H. B. Urbach, J. H. Harrison, and N. L. Hatfield, *J. Phys. Chem.* **71** (1967) 1250.
[185] V. F. Stenin, V. E. Kazarinov, and B. I. Podlovchenko, *Electrokhimiya* **5** (1969) 442.
[186] Yu. M. Vol'fkovich, Y. B. Vasil'ev, and V. S. Bagotskii, *Electrokhimiya* **5** (1969) 1462.
[187] M. W. Breiter, *Disc. Faraday Soc.* **45** (1968) 79.
[188] T. Biegler, *J. Phys. Chem.* **72** (1968) 1571.
[189] S. B. Brummer and A. C. Makrides, *J. Phys. Chem.* **68** (1964) 1448.
[190] M. W. Breiter, *J. Electroanal. Chem.* **14** (1967) 407.
[191] S. B. Brummer and M. J. Turner, *J. Phys. Chem.* **71** (1967) 3902.
[192] E. Gileadi, B. T. Rubin, and J. O'M. Bockris, *J. Phys. Chem.* **69** (1965) 3335.
[193] B. E. Conway and A. K. Vijh, *J. Phys. Chem.* **71** (1967) 3637.
[194] K. W. Bowers, R. W. Giese, J. Grimshaw, H. O. House, N. H. Kolodny, K. Kronberger, and D. K. Roe, *J. Am. Chem. Soc.* **92** (1970) 2783.
[195] A. J. Bard and J. Phelps, *J. Electroanal. Chem. Interfacial Electrochem.* **25** (1970) App. 2.
[196] C. K. Mann, *Electroanal. Chem.* **3** (1969) 57.
[197] C. K. Mann and K. K. Barnes, *Electrochemical Reactions in Nonaqueous Systems*, Marcel Dekker, New York, 1970.

References

[198] M. E. Peover, *Electroanal. Chem.* **2** (1967) 1.
[199] A. N. Dey and B. P. Sullivan, *J. Electrochem. Soc.* **117** (1970) 222.
[200] J. L. Jones, S. Adsish, R. M. Smith, and J. H. Karnes, *Anal. Chim. Acta* **49** (1970) 487.
[201] J. Courtot-Coupez and M. L'Her, *Bull. Soc. Chim. France* **1970**, 1631.
[202] J. Courtot-Coupez and M. Le Demezet, *Bull. Soc. Chim. France* **1967**, 4744.
[203] E. Kirowa-Eisner and E. Gileadi, *J. Electroanal. Chem. Interfacial Electrochem.* **25** (1970) 481.
[204] V. E. Mironov, I. M. Batyaev, and N. L. Bel'kova, *Elektrokhimiya* **6** (1970) 354.
[205] M. Breant, M. Bazouin, C. Buisson, M. Dupin and J. Lebattu, *Bull. Soc. Chim. France* **1968**, 5065.
[206] J. Courtot-Coupez and M. Le Demezet, *Bull. Chim. Soc. France* **1969**, 1033.
[207] J. Perichou and R. Buvet, *Bull. Chim. Soc. France* **1968**, 1279, 1282.
[208] G. Capobianco, G. Farnia, and F. Torzo, *Ric. Sci.* **38** (1968) 842.
[209] F. Rallo and G. Ceccaroni, *Ric. Sci.* **38** (1968) 1067.
[210] E. Tommila and J. Belinskii, *Suomen Kemistilehti* **42B** (1969) 185.
[211] J. F. Coetzee, J. M. Simon, and R. J. Bertozzi, *Anal. Chem.* **41** (1969) 766.
[212] B. L. Funt and D. G. Gray, *Canad. J. Chem.* **46** (1968) 1337.
[213] J. N. Butler, *Adv. Electrochem. Electrochem. Engng.* **7** (1970) 77.
[214] S. D. Ross, M. Finkelstein, and R. C. Petersen, *J. Am. Chem. Soc.* **89** (1967) 4088.
[215] H. Schmidt and J. Noack, *Z. anorg. allgem. Chem.* **296** (1958) 262.
[216] S. D. Ross, M. Finkelstein, and R. C. Petersen, *J. Org. Chem.* **35** (1970) 781.
[217] G. Atherton, M. Fleischmann, and F. Goodridge, *Trans. Faraday Soc.* **63** (1967) 1468.
[218] R. N. Adams, *Accounts of Chem. Res.* **2** (1969) 175.
[219] H. W. Sternberg, R. E. Markby, I. Wender, and D. M. Mohilner, *J. Am. Chem. Soc.* **91** (1969) 4191; **89** (1967) 186.
[220] B. E. Conway and D. J. MacKinnon, *J. Phys. Chem.* **74** (1969) 3663.
[221] S. H. Langer and S. Yurchak, *J. Electrochem. Soc.* **116** (1969) 1228.
[222] G. J. Hoijtink, *Rec. Trav. Chim.* **77** (1958) 555.
[223] S. Wawzonek and D. Wearring, *J. Am. Chem. Soc.* **81** (1959) 2067.
[224] M. M. Baizer and J. D. Anderson, *J. Electrochem. Soc.* **111** (1964) 226.
[225] M. M. Baizer, *J. Electrochem. Soc.* **111** (1964) 223.
[226] M. M. Baizer and J. D. Anderson, *J. Org. Chem.* **30** (1965) 3138.
[227] A. P. Tomilov, E. V. Kryukova, V. A. Klimov, and I. N. Brago, *Elecktrokhimiya* **3** (1967) 1501.
[228] A. P. Tomilov and B. L. Klynev, *Zh. Obsch. Khim.* **39** (1969) 470.
[229] T. Nonaka and K. Sugino, *J. Electrochem. Soc.* **116** (1961) 615.
[230] C. P. Andricux and J. M. Saveant, *J. Electroanal. Chem. Interfacial Electrochem.* **26** (1970) 223.
[231] S. M. Makarochkina and A. P. Tomilov, *Zh. Obsch. Khim.* **40** (1970) 676.
[232] T. Asahara, M. Seno, and T. Arai, *Bull. Chem. Soc. Japan* **42** (1969) 1316.
[233] M. M. Baizer and J. P. Petrovich, *J. Electrochem. Soc.* **114** (1967) 1023.
[234] L. G. Feoktistov, A. P. Tomilov, and I. G. Sevast'yanova, *Soviet Electrochem.* **1** (1965) 1165.
[235] S. Wawzonek, R. C. Duty, and J. H. Wagenknecht, *J. Electrochem. Soc.* **111** (1964) 74.
[236] S. Wawzonek, R. Berkey, E. W. Blaha, and M. E. Runner, *J. Electrochem. Soc.* **102** (1955) 235.
[237] T. Kitagawa, T. P. Layloff, and R. N. Adams, *Anal. Chem.* **35** (1963) 1086.
[238] T. Fujinaga, Y. Deguchi, and K. Umemoto, *Bull. Chem. Soc. Japan* **37** (1964) 822.
[239] J. G. Lawless and M. D. Hawley, *J. Electroanal. Chem.* **21** (1969) 365.

[240] G. Klopmann, *Helv. Chim. Acta* **44** (1961) 1908.
[241] A. Streitwieser, Jr. and C. Perrin, *J. Am. Chem. Soc.* **86** (1964) 4938.
[242] J. Grimshaw and J. S. Ramsey, *J. Chem. Soc. B* **1968**, 60.
[243] J. G. Lawless, D. E. Bartak, and M. D. Hawley, *J. Am. Chem. Soc.* **91** (1969) 7121.
[244] A. K. Hoffmann, W. G. Hodgson, D. L. Maricle, and W. H. Jura, *J. Am. Chem. Soc.* **86** (1964) 631.
[245] S. Wawzonek and A. Gunderson, *J. Electrochem. Soc.* **107** (1960) 537.
[246] K. Umemoto, *Bull. Chem. Soc. Japan* **40** (1967) 1058.
[247] M. J. Allen, W. G. Pieson, and J. A. Siragusa, *J. Chem. Soc.* **1961**, 2081.
[248] O. R. Brown and J. A. Harrison, *J. Electroanal. Chem. Interfacial Electrochem.* **21** (1969) 387.
[249] L. S. Marcoux, J. M. Fritsch, and R. N. Adams, *J. Am. Chem. Soc.* **89** (1967) 5766.
[250] J. Phelps, K. S. V. Santhanam, and A. J. Bard, *J. Am. Chem. Soc.* **89** (1967) 1752.
[251] M. E. Peover and B. S. White, *J. Electroanal. Chem.* **13** (1967) 93.
[252] D. M. Hercules, *Science* **145** (1964) 808.
[253] G. J. Hoijtink, *Disc. Faraday Soc.* **45** (1968) 4.
[254] J. Bader and T. Kuwana, *J. Electroanal. Chem.* **10** (1965) 104.
[255] S. W. Feldberg, *J. Am. Chem. Soc.* **88** (1966) 390.
[256] A. Zweig, A. H. Maurer, and B. G. Roberts, *J. Org. Chem.* **32** (1967) 1332.
[257] S. D. Ross, M. Finkelstein, and R. C. Petersen, *J. Org. Chem.* **86** (1964) 4139.
[258] F. D. Mango and W. H. Bonner, *J. Org. Chem.* **29** (1964) 1367.
[259] M. Leung, J. Herz, and H. W. Salzberg, *J. Org. Chem.* **30** (1965) 310.
[260] H. W. Salzberg and M. Leung, *J. Org. Chem.* **30** (1965) 2873.
[261] J. P. Millington, *J. Chem. Soc. B* **1969**, 982.
[262] V. D. Parker and B. E. Burgert, *Tetrahedron Letters* **1965**, 4065; **1968**, 2415, 3341.
[263] L. Eberson and K. Nyberg, *J. Am. Chem. Soc.* **88** (1966) 1656; *Tetrahedron Letters* **1966**, 2389.
[264] L. Eberson, *J. Am. Chem. Soc.* **89** (1967) 4669.
[265] S. Arita, Y. Takahashi, and K. Takeshita, *Kogyo Kagaku Zasshi* **72** (1969) 2289; *Chem. Abstr.* **72**, 138926.
[266] L. Eberson, *Disc. Faraday Soc.* **45** (1968) 242.
[267] K. Koyana, T. Susuki, S. Tsutsumi, *Tetrahedron Letters* **1965**, 627; **1966**, 2501; *Tetrahedron* **23** (1966) 2675.
[268] S. Andreades and E. W. Zahnow, *J. Am. Chem. Soc.* **91** (1969) 4181.
[269] K. Yoshida and T. Freno, *Bull. Chem. Soc. Japan* **42** (1969) 2411; *Chem. Comm.* **1970**, 711.
[270] N. Clauson-Klass, F. Limborg, and K. Glens, *Acta Chem. Scand.* **6** (1952) 531.
[271] A. J. Baggaley and R. Brettle, *J. Chem. Soc. C* **1968**, 969.
[272] S. D. Ross, M. Finkelstein, and J. J. Uebel, *J. Org. Chem.* **34** (1969) 1018.
[273] N. L. Weinberg and E. A. Brown, *J. Org. Chem.* **31** (1966) 4054.
[274] K. Yoshida and T. Fueno, *Bull. Chem. Soc. Japan* **42** (1969) 2411.
[275] G. Cauquis and G. Pierre, *Compt. Rend.* **C266** (1968) 883.
[276] A. A. Humffray, unpublished work.
[277] Ref. 44, Chapter 12.
[278] A. E. Coleman, H. H. Richtol, and D. A. Aikens, *J. Electroanal. Chem. Interfacial Electrochem.* **18** (1968) 165.
[279] K. Nyberg, *Chem. Comm.* **1969**, 774.
[280] V. D. Parker and L. Eberson, *Tetrahedron Letters* **33** (1969) 2839, 2843.
[281] R. Hand and R. F. Nelson, *J. Electrochem. Soc.* **117** (1970) 1353.
[282] R. F. Nelson and R. N. Adams, *J. Am. Chem. Soc.* **90** (1968) 3925.
[283] A. Zweig, J. E. Lancaster, M. T. Neglia, and W. H. Jura, *J. Am. Chem. Soc.* **86** (1964) 4130.

References

[284] B. M. Latta and R. W. Taft, *J. Am. Chem. Soc.* **89** (1967) 5172.
[285] G. Cauquis, M. Genies, H. Lemaire, A. Rassat, and J. P. Ravet, *J. Chem. Phys.* **47** (1967) 4642.
[286] P. Martinet, J. Simonet, and J. Tendil, *Compt. Rend. Ser. C.* **286** (1969) 2329.
[287] A. Zweig, W. G. Hodgson, and W. H. Jura, *J. Am. Chem. Soc.* **86** (1964) 4124.
[288] F. W. Steuber and K. Dimroth, *Ber.* **99** (1966) 258.
[289] F. J. Vermillion and I. A. Pearl, *J. Electrochem. Soc.* **111** (1964) 1392.
[290] L. F. Fieser, *J. Am. Chem. Soc.* **52** (1930) 5204.
[291] N. S. Hush, *J. Chem. Soc.* **1953**, 2375.
[292] T. Fueno, T. Lee, and H. Eyring, *J. Phys. Chem.* **63** (1959) 1940.
[293] K. Sasaki and W. J. Newby, *J. Electroanal. Chem.* **20** (1969) 137.
[294] K. Kuwata and D. H. Geske, *J. Am. Chem. Soc.* **56** (1964) 2101.
[295] L. Eberson and K. Nyberg, *Acta Chem. Scand.* **18** (1964) 1567.
[296] P. S. Skell, *J. Am. Chem. Soc.* **91** (1969) 695.
[297] J. A. Van Zorge, J. Strating, and H. Wynberg, *Rec. Trav. Chim. Pays Bas* **89** (1970) 781.
[298] W. J. Koehl, *J. Am. Chem. Soc.* **86** (1964) 4686.
[299] P. S. Skell and P. H. Reichenbacher, *J. Am. Chem. Soc.* **90** (1968) 2309.
[300] J. T. Keating and P. S. Skell, *J. Org. Chem.* **34** (1969) 1479.
[301] S. D. Ross and M. Finkelstein, *J. Org. Chem.* **34** (1969) 2923.
[302] D. L. Muck and E. R. Wilson, *J. Electrochem. Soc.* **117** (1970) 1358.
[303] L. L. Miller and A. K. Hoffmann, *J. Am. Chem. Soc.* **89** (1967) 593.
[304] E. R. Atkinson, H. H. Warren, P. I. Abell, and R. E. Wing, *J. Am. Chem. Soc.* **72** (1950) 915.
[305] J. K. Kochi, *J. Am. Chem. Soc.* **77** (1955) 3208.
[306] R. M. Elofson, *Canad. J. Chem.* **36** (1958) 1207.
[307] L. Holleck and D. Marquarding, *Naturwiss.* **49** (1962) 468.
[308] S. G. Mairanovskii and A. D. Filanova-Krasnova, *Izv. Akad. Nauk SSSR Ser. Khim.* **1967**, 1673.
[309] A. B. Ershler, G. A. Tedoradze, M. Fakhimi, and K. P. Butin, *Soviet Electrochem.* **2** (1966) 295.
[310] A. Kirrman and M. Kleine-Peter, *Bull. Soc. Chim. France* **1957**, 894.
[311] N. S. Hush and K. B. Oldham, *J. Electroanal. Chem.* **6** (1963) 34.
[312] L. B. Rogers and A. N. Diefenderfer, *J. Electrochem. Soc.* **114** (1967) 942.
[313] L. G. Feoktistov and S. I. Zhdanov, *Electrochim. Acta* **10** (1965) 647.
[314] J. P. Petrovich and M. M. Baizer, *Electrochim. Acta* **12** (1967) 1249.
[315] H. E. Ulery, *J. Electrochem. Soc.* **116** (1969) 1201.
[316] R. Galli and F. Olivani, *J. Electroanal. Chem.* **25** (1970) 331.
[317] S. Winstein and F. H. Seubold, *J. Am. Chem. Soc.* **69** (1947) 2916.
[318] H. Breederveld and E. C. Kooyman, *Rec. Trav. Chim.* **76** (1957) 297.
[319] W. Von Miller, *Z. Elektrochem.* **4** (1897) 55.
[320] A. G. Kornienko, A. V. Uvarov, L. A. Mirkind, and M. Ya. Fioshin, *Elektrokhimiya* **4** (1968) 1140.
[321] W. B. Smith and H. Gilde, *J. Am. Chem. Soc.* **82** (1960) 659.
[322] A. Kunugi, T. Shimizu, and S. Nagaura, *Bull. Chem. Soc. Japan* **43** (1970) 1298.
[323] O. Stern, *Z. Elektrochem.* **30** (1924) 508.
[324] M. A. V. Devanathan and B. V. K. S. R. A. Tilak, *Chem. Rev.* **65** (1965) 635.
[325] E. Gileadi, *J. Electroanal. Chem.* **11** (1966) 137.
[326] K. Sasaki, *Denki Kagaku Oyobi Kogyo Butsuri Kogaku* **37** (1969) 80, 148; *Chem. Abst.* **71**, 35365.
[327] R. Parsons, *J. Electroanal. Chem.* **7** (1964) 136; **8** (1964) 93.

[328] A. P. Tomilov, L. A. Fedorova, V. A. Klimov, and G. A. Tedoradze, *Elektrokhimiya* **4** (1968) 1264.
[329] J. Kuta, *Record Chem. Progr.* **29** (1968) 231.
[330] R. Parsons, *Disc. Faraday Soc.* **45** (1968) 40; *J. Electroanal. Chem.* **21** (1969) 35.
[331] D. Gilroy and B. E. Conway, *J. Phys. Chem.* **69** (1965) 1259.
[332] H. W. Nurnberg, *J. Electroanal. Chem.* **21** (1969) 99.
[333] A. Avrutskaya and M. Ya. Fioshin, *Elektrokhimiya* **2** (1966) 920.
[334] E. Gileadi and B. E. Conway, *J. Chem. Phys.* **39** (1963) 3420.
[335] A. T. Kuhn, H. Wroblowa, and J. O'M. Bockris, *Trans. Faraday Soc.* **63** (1967) 1458.
[336] P. J. Elving and B. Pullman, in *Advances in Chemical Physics* Ed. by I. Prigogine, Interscience, New York, 1961, Vol. 3, p. 1.
[337] M. Von Stackelberg and W. Stracke, *Z. Elektrochem.* **53** (1949) 118.
[338] R. Koopmann, *Ber. Bursenges* **72** (1968) 32.
[338a] B. MacDougall, B. E. Conway, and H. A. Kozlowska, *J. Electroanal. Chem.* **32** (1971) App. 15; *J. Chem. Soc., Faraday Trans.*, in press (1972).
[339] H. Lund, *Talanta* **12** (1965) 1065.
[340] P. J. Elving and J. T. Leone, *J. Am. Chem. Soc.* **79** (1957) 1546.
[341] A. M. Wilson and N. L. Allinger, *J. Am. Chem. Soc.* **83** (1961) 1999.
[342] H. Lund, *Acta Chem. Scand.* **13** (1959) 192.
[343] J. H. Stocker and R. M. Jenevein, *Chem. Comm.* **1968**, 934.
[344] J. P. Coleman, H. G. Gilde, J. P. Utley, and B. C. L. Weedon, *Chem. Comm.* **1970**, 738.
[345] A. J. Fry, A. L. Mitnick, and R. G. Reed, *J. Org. Chem.* **35** (1970) 1232.
[346] P. H. Rieger, I. Bernal, W. H. Reinmuth, and G. K. Fraenkel, *J. Am. Chem. Soc.* **55** (1963) 683.
[347] J. Volke and J. Holubek, *Coll. Czech. Chem. Comm.* **28** (1963) 1597.
[348] J. Volke and A. M. Kardos, *Coll. Czech. Chem. Comm.* **33** (1968) 2560.
[349] B. Lamm and B. Sammelson, *Acta Chem. Scand.* **23** (1969) 691.
[350] P. E. Iversen, *Tetrahedron Letters* **1971**, 55.
[351] L. L. Miller, E. P. Kujawa, and C. B. Campbell, *J. Am. Chem. Soc.* **92** (1970) 2821.
[352] V. D. Parker and L. Eberson, *Chem. Comm.* **1969**, 973.
[353] G. Cauquis, G. Fauvelot, and J. Rigaudy, *Bull. Soc. Chim. France* **1968**, 4928.
[354] M. Heyrovsky, *Coll. Czech. Chem. Comm.* **28** (1963) 26.
[355] C. Wagner, *Electrochim. Acta* **15** (1970) 987.
[356] Z. Takehara, *Electrochim. Acta* **15** (1970) 999.
[357] K. J. Vetter, *Z. Elektrochem.* **56** (1952) 797.
[358] T. C. Franklin, M. Matto, T. Itoh, and D. H. McLelland, *J. Electroanal. Chem. Interfacial Electrochem.* **27** (1970) 303.
[359] I. V. Kudryashov and V. L. Kochetkov, *Russ. J. Phys. Chem.* **43** (1969) 728.
[360] L. N. Ivanovskaya, I. V. Kudryashov, and S. A. Kilesnikova, *Elektrokhimiya* **5** (1969) 760.
[361] K. Sugino and T. Nonaka, *Electrochim. Acta* **13** (1968) 613.
[362] M. M. Baizer, J. P. Petrovich, and D. A. Tyssee, *J. Electrochem. Soc.* **117** (1970) 173.
[363] H. Lund, *Acta Chem. Scand.* **13** (1959) 249.
[364] T. Shono and M. Mitani, *J. Am. Chem. Soc.* **90** (1968) 2728.
[365] T. Shono and M. Mitani, *Tetrahedron Letters* **1969**, 687.
[366] V. D. Parker and L. Eberson, *Chem. Comm.* **1969**, 340.
[367] V. D. Parker, *Chem. Comm.* **1969**, 848.
[368] J. M. Fritsch and H. Weingarten, *J. Am. Chem. Soc.* **90** (1968) 793.
[369] W. Jugelt and F. Pragst, *Angew. Chem.* **80** (1968) 280.

References

[370]T. Osa, A. Yildiz, and T. Kuwana, *J. Am. Chem. Soc.* **91** (1969) 3994.
[370a]J. P. Coleman, J. H. P. Utley, and B. C. L. Weedon, *Chem. Comm.* (*Chem. Soc. London*) **1971**, 438.
[371]V. D. Parker, *Chem. Comm.* **1969**, 610.
[372]A. Ronlan and V. D. Parker, *Chem. Comm.* **1970**, 1567.
[373]P. Radlick, R. Klem, S. Spurlock, J. J. Sims, E. E. van Tamelen, and T. Whitesides, *Tetrahedron Letters* **1968**, 5117.
[374]M. R. Rifi, *J. Am. Chem. Soc.* **89** (1967) 4442.
[375]T. J. Curphey, C. W. Amelotti, T. P. Layloff, R. L. McCartney, and J. A. Williams, *J. Am. Chem. Soc.* **91** (1969) 2817.
[376]J. D. Anderson and M. M. Baizer, *Tetrahedron Letters* **1966**, 511.
[377]J. D. Anderson, M. M. Baizer, and J. P. Petrovich, *J. Org. Chem.* **31** (1966) 3890.
[378]J. P. Petrovich, J. D. Anderson, and M. M. Baizer, *J. Org. Chem.* **31** (1966) 3897.
[379]R. E. Dessy and S. A. Kandil, *J. Org. Chem.* **30** (1965) 3857.
[380]V. Bellavita, N. Fedi, and N. Cagnoli, *Ric. Sci.* **25** (1955) 504.
[381]K. Gleu and K. Pfannstiel, *J. Pr. Chem.* **146** (1936) 129.
[382]B. Sakurai, *Bull. Chem. Soc. Japan* **7** (1932) 127.
[383]S. Wawzonek and J. H. Wagenknecht, *J. Electrochem. Soc.* **110** (1963) 420.
[384]S. Wawzonek and R. C. Duty, *J. Electrochem. Soc.* **108** (1961) 1135.
[385]E. K. Spicer and H. G. Gilde, *Chem. Comm.* **1967**, 373.
[386]A. F. Vellturo and G. W. Griffin, *J. Org. Chem.* **31** (1966) 2241.
[387]N. L. Weinberg and E. A. Brown, *J. Org. Chem.* **31** (1966) 4058.
[388]N. N. Melnikov and E. M. Cherkasova, *Zh. Obsch. Khim.* **14** (1944) 13.
[389]M. M. Baizer, J. L. Chruma, and P. A. Berger, *J. Org. Chem.* **35** (1970) 3569.
[390]H. Lund, *Disc. Faraday Soc.* **45** (1968) 193.
[391]N. J. Leonard, S. Swann, and J. Figueras, *J. Am. Chem. Soc.* **74** (1952) 4620.
[392]V. D. Parker, *J. Am. Chem. Soc.* **91** (1969) 5380.
[393]J. G. Traynham and J. S. Dehn, *J. Org. Chem.* **31** (1966) 2241.
[394]E. Ott and K. Kramer, *Ber.* **68** (1935) 1655.
[395]E. L. Eliel and J. P. Freeman, *J. Am. Chem. Soc.* **74** (1952) 923.
[396]R. E. Erickson and C. M. Fischer, *J. Org. Chem.* **35** (1970) 1605.
[397]R. Annino, R. E. Erickson, J. Michalovic, and B. McKay, *J. Am. Chem. Soc.* **88** (1966) 4424.
[398]J. L. Webb, C. K. Mann, and H. M. Walborsky, *J. Am. Chem. Soc.* **92** (1970) 2042.
[399]A. J. Fry and R. H. Moore, *J. Org. Chem.* **33** (1968) 1283.
[400]R. E. Erickson, G. Annino, M. D. Scanlon, and G. Zon, *J. Am. Chem. Soc.* **91** (1969) 1767.
[401]H. W. Sternberg, R. E. Markby, I. Wender, and D. M. Mohilner, *J. Am. Chem. Soc.* **89** (1967) 186.
[402]A. J. Fry and M. A. Mitnick, *J. Am. Chem. Soc.* **91** (1969) 6207.
[403]R. N. Gourley, J. Grimshaw, and P. G. Millar, *J. Chem. Soc.* (C) **1970**, 2318; *Chem. Comm.* **1967**, 1278.
[404]S. G. Mairanovskii, *J. Electroanal. Chem.* **6** (1963) 77.
[405]L. Horner and D. Degner, *Tetrahedron Letters* **1968**, 5889.
[406]J. P. Coleman, R. J. Kobylecki, and J. H. P. Utley, *Chem. Comm.* **1971**, 104.
[407]A. J. Fry and J. H. Newberg, *J. Am. Chem. Soc.* **89** (1967) 6374.
[408]R. Dietz, M. E. Peover, and B. E. Larcombe, *Disc. Faraday Soc.* **45** (1968) 161.
[409]J. A. Stocker and R. M. Jenevein, *J. Org. Chem.* **34** (1969) 2810.
[410]V. J. Puglisi, G. L. Clapper, and D. H. Evans, *Anal. Chem.* **41** (1969) 279.
[411]H. Lund, *Acta Chem. Scand.* **11** (1957) 283.
[412]J. Weimann and M. L. Bouguerra, *Compt. Rend.* **265** (1967) 751.

[413] S. Wawzonek, A. R. Zignan, and G. R. Hansen, *J. Electrochem. Soc.* **117** (1970) 1351.
[414] L. Horner and D. H. Skaletz, *Tetrahedron Letters* **1970**, 1103.
[415] S. G. Mairanovskii and L. D. Bergel'son, *Russ. J. Phys. Chem.* **34** (1960) 112.
[416] W. H. Jura and R. J. Gaul, *J. Am. Chem. Soc.* **80** (1958) 5402.
[417] L. G. Feoktistov and I. G. Markova, *J. Gen. Chem. USSR.* **1970**, 964.
[418] L. G. Feoktistov and I. G. Markova, *Elektrokhimiya* **5** (1969) 1095.
[419] J. M. Bobbitt, K. H. Weissgraber, A. S. Steinfeld, and S. G. Weiss, *J. Org. Chem.* **35** (1970) 2884.
[420] T. Shono, I. Nishiguchi, S. Yamane, and R. Oda, *Tetrahedron Letters* **1969**, 1965.
[421] T. Dickinson and P. J. Ovenden, *Chem. in Brit.* **5** (1969) 260.
[422] V. D. Bezuglyi, V. A. Ekel, M. Ya. Fioshin, E. K. Nechiporenko, and R. F. Ramakaeva, *Elektrokhimiya* **6** (1970) 1778.

4

Electrochemical Processes at Biological Interfaces

Lazaro J. Mandel

Department of Physiology, Yale University, New Haven, Connecticut

I. BIOLOGICAL OXIDATIONS

The conversion of metabolic energy into useful work is a highly efficient process in biological cells. The oxidation of glucose provides a convenient example of a common biological oxidation; the overall reaction may be represented as follows:

$$C_6H_{12}O_6 + 6O_2 \rightarrow 6CO_2 + 6H_2O + \text{energy}$$

The complete oxidation of a mole of glucose, under standard conditions, would result in a free-energy release of 686 kcal. Were this energy released in the form of heat, it would be sufficient to disrupt cellular structure. Instead, most of the energy obtained from this oxidation, as well as the oxidation of other foodstuffs (carbohydrates, lipids, or amino acids) is utilized, with few exceptions, to synthesize *adenosine triphosphate* (ATP). The free energy necessary to synthesize ATP from adenosine diphosphate under physiological conditions has been estimated somewhere between 7 and 12 kcal/mole.[1-3] This is the same amount of energy released from ATP in its transformation into useful work by biological cells. Thus, the free energy available from the oxidation of foodstuffs is transformed into smaller "packets of energy" more readily utilized by biological cells.

The efficiency of these energy conversion processes is very high: one mole of glucose provides energy for the synthesis of 38 moles of ATP, which represents an efficiency of 45% (at 8 kcal/mole per ATP).[4] Thereupon, the energy in ATP is transformed into mechanical, electrical, and osmotic work with high efficiencies, which in some cases have been calculated at close to 100% for acid secretion in the stomach,[5] active Na^+ transport in frog skin,[6] and sartorious muscle contraction.[7] These observations prove that the energy conversion process in biological cells is not limited by the Carnot efficiency factor

$$(T_b - T_e)/T_b$$

where T_b and T_e are the body and environmental absolute temperatures, respectively; in warm-blooded animals, the numerical value of this fraction rarely exceeds 0.02. The high energy-conversion efficiency, therefore, must be due to a process of direct energy conversion of chemical energy into other forms of usable energy, without passing through intermediate energy forms where losses would occur. One such possible mode of energy conversion will be discussed here: The conversion of the ATP energy into active ionic transport across a membrane by means of electrochemical "cold" combustion; this is a process in which chemical energy is transformed into electronic potential energy, much in the same way as in the operation of a fuel cell.[8,9] This would represent the most direct manner by which active ionic transport could be supplied its needed energy.

The process of active ionic transport occurs from one ionic solution to another, whereas electrochemical energy conversion requires an electrode which is electron-conductive; the conversion process would, therefore, occur at the interface between an ionic conductor and an electron conductor. Evidence for the presence of this type of interface in biological cells is discussed in the next section; Section III introduces biological models of electrochemical energy conversion, and Section IV reviews possible cellular regulatory and excitation processes proceeding via electrochemical mechanisms.

II. ARE THE ELEMENTS FOR ELECTROCHEMICAL REACTIONS PRESENT IN BIOLOGICAL CELLS?

The high efficiency of energy conversion in biological cells suggests that an electrochemical mechanism is operative. A prime requirement which cellular elements must have to be involved in electro-

II. Are the Elements for Electrochemical Reactions Present in Biological Cells?

chemical reactions is the property of electronic conduction (probably semiconduction), since these reactions occur solely on the surfaces of electronic conductors. It is important, therefore, to identify these cellular elements and describe their semiconductive properties.

1. Electronic Conduction in Biological Materials

The application of the principles of solid-state physics to biology was initiated by Szent-Györgyi in 1941.[10,11] He proposed the existence of conduction bands in proteins. Much theoretical and experimental work on the properties of biological (and organic) semiconductors followed in the intervening years, showing that semiconduction does indeed occur in biological materials. The physiological implications of this observation are far-reaching, because semiconduction in biology provides a basis for a whole new class of models of biological functions. Good reviews on the subject of semiconduction in biological materials are by Gutmann and Lyons[12] and Rosenberg and Postow.[13]

Numerous biochemical substances and cellular organelles have been found to exhibit semiconductivity, based on the criterion that their electrical conductivity varies with temperature as

$$\sigma(T) = T_0 \exp(-E/2kT) \tag{1}$$

where σ is the electrical conductivity, E is the activation energy for semiconduction, k is the Boltzmann constant, and T is the absolute temperature. Several examples of biological materials following this relationship are shown in Fig. 1. Gutmann and Lyons[12] list 116 biological substances which display semiconduction, with activation energies ranging from 1 to 3 eV. There is considerable uncertainty over the mechanism of semiconduction present in these substances, which is due to the considerable difficulty in obtaining experimental data even comparable to that available from inorganic semiconductors. The observations which have been made on biological semiconductors could be classified as follows.

(i) *Biological Systems are Naturally Impure*

The degree of purification achieved with inorganic semiconductors (which form the technological basis for the extraordinary advances in the understanding of their properties) has not been forthcoming in biological materials. Not only is purification difficult

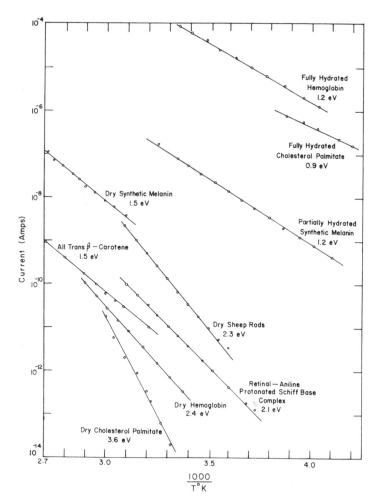

Figure 1. The log of the current plotted against $1/T(°K)$ for a variety of biological substances. The activation energies are calculated from the slopes of the lines. (From Ref. 13.)

to achieve in biological systems, but, in addition, attempts to purify biological substances often destroy their *in vivo* structure; this may cause a drastic change in their semiconductive properties or even the disappearance of semiconduction altogether. This point is illustrated by the similarity between the activation energy curves of

II. Are the Elements for Electrochemical Reactions Present in Biological Cells?

biological substances and those of heterogeneous systems of mixed inorganic crystals. In the latter, the activation energy for semiconduction changes as the proportion of each constituent in the crystal is changed. In addition, intrinsic semiconduction dominates over a large temperature range, so the break in activation energy as a function of temperature normally observed with extrinsic semiconductors is not present in these substances. The similarities between these systems led Rosenberg and Postow[13] to conclude that this intrinsic mechanism of semiconduction predominates in proteins.

(ii) *The Separation of Electronic from Ionic Currents*

In solid-state biophysics, this is an important consideration, since current is usually carried by both ions and electrons. The total conductivity for a substance may be written as

$$\sigma = \left[\sum_i Z_i C_i \mu_i + \sum_e Z_e n_e \mu_e\right] q \qquad (2)$$

where i represents the ionic species present (anions and cations) with valence Z_i, concentration C_i, and mobility μ_i; e represents the species of electronic charge carriers present (electrons or holes) with valence Z_e ($+1$ for holes and -1 for electrons), concentration C_e, and mobility μ_e; and q is the electronic charge. The contributions of ions and electrons may be separated by solid-state electrolysis; ionic currents produce electrolysis, whereas electronic currents do not. Either the increase in the evolved gas or the decrease of remanent water in a hydrated protein can be monitored. If the current were ionic, electrolysis would decrease the amount of hydration water by converting it into hydrogen and oxygen. Since conductivity decreases sharply as the state of hydration of a protein is decreased, the effect of electrolysis would be clearly observable as a function of time. Hemoglobin, with 7.5% adsorbed water, was observed by Rosenberg[14] on such an experiment to maintain a constant conductivity over a period of 140 min. A purely ionic current would have decreased by a factor of 10 during this time interval. On this basis, Rosenberg[14] concluded that more than 95% of the current was carried electronically. In a parallel experiment, Maričič *et al.*[15] could not detect any evolved gas when a potential of 150 V was impressed for seven days across a sample of crystalline hemoglobin with 9.17% adsorbed water. This experiment confirmed in a dramatic way the conclusion drawn by Rosenberg concerning electronic conduction in crystalline hemoglobin.

(iii) *The State of Hydration of the Substrate*

Figure 1 illustrates in a general manner the fact that the activation energy of a biological substrate is strongly dependent on its state of hydration. The dry substrates (on the left bottom part of the figure) have much higher activation energies than the hydrated samples (top right). Early experiments with dry proteins produced activation energies averaging around 2.3 eV; this caused Szent-Györgyi[16] to doubt that semiconduction had any physiological role in the substances investigated (there is not enough energy available in biological systems to raise an electron across a gap of 2.3 eV).

The importance of the state of hydration was demonstrated by the experiments of King and Medley,[17] Eley and Spivey,[18] and Rosenberg[19]; we shall dwell only on the experiments of the last author. A systematic investigation into the effects of hydration in proteins was made by Rosenberg on twice-crystallized hemoglobin pressed at 5000 lb/in.^{2} to form a pellet which was placed between metal electrodes. These samples were subsequently equilibrated over an appropriate saturated salt solution in a sealed chamber to study the effects of hydration. The results of this investigation and of a similar set of experiments on egg lecithin are shown in Fig. 2. The dependence of conductivity on the state of hydration is clearly demonstrated in this figure, including the saturation of this effect on the lipid and the suggestion of saturation for the protein. Studies on the effects of temperature and state of hydration on hemoglobin give the interesting result that the preexponential factor in equation (1) remains constant as a function of both variables, whereas the activation energy decreases significantly as the state of hydration is increased, as shown in Fig. 3. In the region of hydration where electronic conductivity is predominant, the hydrated activation energy E_H is observed to vary with the percentage of water adsorbed onto the protein as

$$E_H = E_D - \gamma m \qquad (3)$$

where E_D is the dry activation energy, and γ is a constant. Since

II. Are the Elements for Electrochemical Reactions Present in Biological Cells? 245

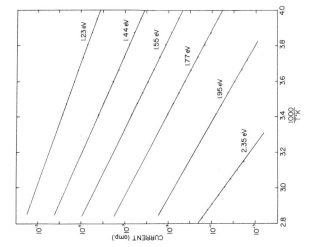

Figure 3. The log of the current plotted against $1/T$ for crystalline hemoglobin in various states of hydration. As the hydration increases, the activation energy shows a decrease, but the preexponential factor in equation (1) remains constant. (From Ref. 13.)

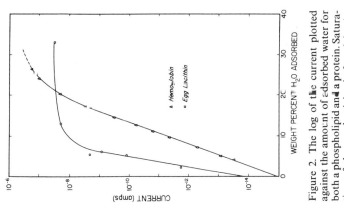

Figure 2. The log of the current plotted against the amount of adsorbed water for both a phospholipid and a protein. Saturation of the conductivity is found in the phospholipid and indicated in the protein. (From Ref. 13.)

water adsorption has no effect on the preexponential factor, equation (1) may be combined with equation (3) to yield

$$\sigma(T, m) = \sigma_0 \exp[-(E_D - \gamma m)/2kT] \qquad (4)$$

Rosenberg explains these observations by assuming that the adsorbed water has no effect on the mobility of the carriers or on the number of energy states available, rather exerting its major effect on the activation energy through a change in polarizability of the medium. The work required to move a charge from one protein molecule to its neighbor (neglecting Coulombic forces) would be

$$E = I_g - A_g - 2P \qquad (5)$$

where I_g is the gas-state ionization potential of the substance, A_g is the gas-state electron affinity, and P is the polarization of the dielectric by the free charges. The polarization is produced by relaxation of the dielectric media in a spherical region around each of the two newly created charges and may be expressed by

$$P = (q^2/2R)[1 - (1/\kappa)] \qquad (6)$$

where R is the cavity radius of the charged region and κ the effective dielectric constant of the dry protein. The dry activation energy of the protein would be given by

$$E_D = I_g - A_g - (q^2/2R)[1 - (1/\kappa)] \qquad (7)$$

A hydrated protein would have an increased effective dielectric constant κ', so its activation energy would be

$$E_H = I_g - A_g - (q^2/2R)[1 - (1/\kappa')] \qquad (8)$$

Since hydration does not change I_g or A_g, we can write

$$E_H = E_D - (q^2/2R)[(1/\kappa) - (1/\kappa')] \qquad (9)$$

which, introduced in equation (4), yields

$$\sigma(T, \kappa') = \sigma_0 \exp\left(-\frac{E_D}{2kT}\right) \exp\left[\frac{q^2}{2kRT}\left(\frac{1}{\kappa} - \frac{1}{\kappa'}\right)\right] \qquad (10)$$

This equation says that the effect of the water of hydration is to decrease the activation energy for semiconduction by increasing the effective dielectric constant of the medium. It predicts the saturation of the conductivity as κ' becomes large; this effect is observed in Fig. 2.

(iv) *The Difficulty of Hall Effect Measurements*

The literature on biological semiconductors is almost totally devoid of references to the successful measurement of Hall mobility in these substances. The few measurements quoted exhibit such variability between samples that the results are very questionable. The unavailability of this parameter makes it extremely difficult to determine which conduction mode is operating in a given biological substance. It may be argued that energy band theory is inapplicable to biological substances due to the lack of long-range order.[12] Semiconduction would therefore occur in these substances either by quantum mechanical tunneling of electrons (or holes) through energy barriers or by hopping over the barriers. Eley[20] has calculated that the hopping model is probably dominant at an intersite separation of 10 Å, whereas tunneling is prevalent at smaller distances. On the other hand, DeVault *et al.*[21] considered tunneling to be the means of charge transfer in cytochromes at low temperatures, even at a barrier height of 1 eV and width of 28 Å.

(v) *Optical Measurements*

Attempts to measure a band edge by optical means have proven futile because of the numerous other optical effects observed in biological systems, which are totally unrelated to semiconduction.

In summary, it may be stated that semiconduction has been demonstrated to occur in pressed pellets of semipurified biological substances, the properties of these substances being highly dependent on their state of hydration. The modes of semiconduction in these substances cannot be ascertained at present because of lack of experimental data. However, the methods used to study inorganic semiconductors probably still represent the best way to achieve more detailed understanding of these parameters.

One important criticism levied at these experiments on pellets has been the apparent lack of structural similarity between pellets with metallic electrode connections and biological cellular constitutents in an ionic solution. Experiments on thin lipid films have countered that criticism, since electronic conduction appears to be present in many of these. The next section is a review of these experiments.

2. Electronic Conduction in Thin Lipid Films

(i) General

There has been much work on thin lipid films following the development of techniques for their preparation in stable form in aqueous solution by Mueller et al.[22,23] The relevance of these lipid membranes to biology rests in their physicochemical similarity to cellular membranes, since they are structurally believed to consist of a bilayer of lipid molecules (for a review of this topic, see Tien and Diana[24]). Although most experiments have dealt with the passage of ions through lipid bilayers, some of them display semiconductive properties. The techniques used to show semiconductivity in films differ somewhat from those used with pellets, since electrical contacts cannot be made directly to the film, but instead must be through electrodes immersed in the solutions bathing both sides of the film. Furthermore, the passage of an electronic current through the film necessitates the occurrence of electrochemical reactions at both semiconductor interfaces to close the electrical circuit. It was argued earlier that electrochemical reactions must occur on the surface of electronic conductors, while presently the reverse argument is formulated. Both of these ideas are illustrated in Fig. 4, where a schematic drawing of a model of a semiconducting film is shown. For an electron to move from

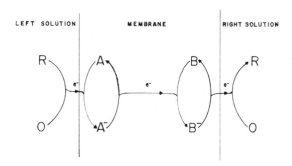

Figure 4. Lipid membrane model illustrating the passage of an electron through the membrane from left to right.

II. Are the Elements for Electrochemical Reactions Present in Biological Cells?

left to right through the membrane, a source of electrons is required in the left solution, and there must be an electrochemical reaction at this interface. In this reaction, a reduced substrate is oxidized by donating an electron (a one-electron reaction is illustrated for simplicity) to molecule A in the membrane. From there, the electron could move through a conduction band, or tunnel, or "hop" to molecule B on the right-hand side of the bilayer. An electrochemical reaction at this interface completes the passage of the electron, resulting in the reduction of substrate in the right-hand solution. The opposite scheme would, of course, be operative for conduction in the opposite direction. The same basic model could be utilized for thicker membranes, with the only modification that the electrons would be conducted over a longer span. On the basis of this model, the presence of either semiconduction or an electrochemical reaction would imply that the entire mechanism illustrated in Fig. 4 is operative. This is the approach which has been utilized by most investigators to demonstrate solid-state properties of lipid films.

(ii) Experimental

In 1967, Läuger and co-workers[25,26] reported that the normally high resistance (10^7 ohm-cm^2) of bimolecular lecithin membranes could be lowered by about six orders of magnitude with the addition of iodine to the bathing solutions. They proposed two possible mechanisms to explain the action of iodine. One mechanism was essentially ionic in nature, involving the dissociation of molecular iodine to form a fixed lecithin $\cdot I^+$ complex and free I^- which could move through the membrane. The other mechanism suggested that iodine formed a charge-transfer complex with the lecithin which conferred electron-conductive properties to the lipid film. Electrons would then be donated by iodine on one side and accepted by iodine on the other side of the membrane.

Evidence for both mechanisms has appeared in the literature since that time and, although a definitive argument favoring one of these to the exclusion of the other has not been presented, the tide seems to be turning in the direction of the second mechanism (electronic conduction). The proponents of an ionic mechanism of conduction suggest that the decreased resistance of the film in the presence of iodine may be accounted for by the movement of I^- carried by I_2 in the form of I_5^- complexes across the bilayer.[27,28]

Most of the data could, however, be interpreted with equal validity in terms of redox-mediated electronic conduction through the membrane. This latter mechanism is supported by a variety of experiments on this type of film or on similar systems. Läuger et al.[26] quoted the experiments of Kallman and Pope[29,30] in which holes were injected into anthracene crystals from aqueous redox systems such as I_2/I^-. Experiments of Rosenberg and Jendrasiak[31] show that when lipid molecules, such as egg lecithin and synthetic lecithin, are exposed to iodine vapor, they exhibit a large increase in electrical conductivity with a concomitant decrease in the activation energy from the dry state. Spectrophotometric evaluation of the lipid samples shows a new charge-transfer absorption band in the ultraviolet after exposure to the iodine vapor.[32] Similar charge-transfer complexes have been found between other donors, such as oxidized cholesterol, and acceptors, which include iodine, 2,4-dinitrophenol, picric acid, and trinitrobenzene.[33] In all of these cases, the electrical conductivity of the complex is increased and the activation energy is decreased when compared to the unreacted lipid. The effect of donor–acceptor complex formation on the preexponential factor of equation (1) was studied by Rosenberg et al.[34] These authors found that equation (1) would have to be rewritten as

$$\sigma(T) = \sigma_0' \exp(E/2kT_0) \exp(-E/2kT) \quad (11)$$

where σ_0' and T_0 (characteristic temperature) are new constants which are characteristic of the semiconducting material, independent of the electrodes used (metallic for pellets or aqueous for films). These two constants were found not to vary with state of hydration, formation of weak donor–acceptor complexes, or use of different cis–trans isomers of the substance; all changes in conductivity with temperature and the factors mentioned above could be accounted for by the variation of E in equation (11). Therefore, in these exhaustive experiments, it was shown that these biological semiconductors behave very much like compound inorganic semiconductors.[34]

Using a membrane of a slightly different composition, Jain et al.[35] recently reported an extremely low I^{131} flux across the membrane even though its resistance was lowered by seven orders of magnitude; this finding is not interpretable in terms of an ionic mechanism of conduction and suggests electronic conduction.

II. Are the Elements for Electrochemical Reactions Present in Biological Cells? 251

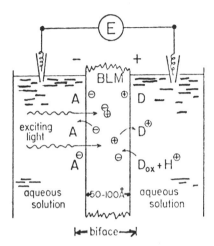

Figure 5. Schematic illustration of the basic electronic processes in photoactive BLM separating two aqueous solutions. (A) Electron acceptor; (D) electron donor; (D^+, D_{ox}) oxidized forms of D; ($\ominus$) electron; ($\oplus$) positive hole. It is postulated that electrons and holes are generated in the BLM via exciton dissociation on illumination. See text for other details. (From Ref. 36.)

In addition, Tien and Verma[36] observed open-circuit photoelectric voltages greater than 100 mV upon illumination of lipid membranes reconstituted from photosynthetic pigments. The presence of electronic reducing agents and absence of good electron acceptors greatly increased the time constant of the emf. The effects of these and other electronic modifiers suggest that the photoactive processes in these membranes are electronic. A schematic illustration of these processes is shown in Fig. 5.

The presence of electrochemical reactions in lipid bilayers (reacted with electronic acceptors) is supported by two types of observations.

(a) The voltage–current characteristics of such a film reported by Läuger et al.,[25] are shown in Fig. 6. A straight-line relationship is obtained at high voltages, illustrating the exponential dependence of current upon the applied voltage in that range. Mandel[37] pointed

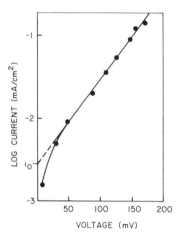

Figure 6. Logarithm of current density versus voltage characteristics of a dioleyl phosphatidyl choline membrane in 10^{-2} M KI (data from Ref. 25). The smooth curve is fitted to equation (12) with $i_0 = 2.7 \times 10^{-6}$ A/cm^2 and $\beta = 0.4$.

out that all of the experimental points are fitted extremely well by the electrochemical rate relation

$$i = i_0\{\exp[(1 - \beta)VF/RT] - \exp(-\beta VF/RT)\} \qquad (12)$$

where i is the current density, i_0 is the exchange current density at no applied potential, β is the symmetry factor, V is the applied potential, F is the Faraday, R is the gas constant, and T is the absolute temperature. The Tafel slope calculated from Fig. 6 is 0.1, which corresponds to $\beta = 0.4$, falling within the range of values observed for processes such as the present one, characterized by a single, almost symmetric, rate-determining energy barrier.[38] The rate-determining step most likely to exhibit the current–voltage characteristics shown in Fig. 6 would be the passage of electrons between one of the solutions and an electron conductor in the membrane[39] (this point is discussed in more detail in the next section). Electronic rate-determining steps of this type have been observed extensively at metal–solution[40] and semiconductor-

II. Are the Elements for Electrochemical Reactions Present in Biological Cells? 253

solution[41] interfaces, as well as across thin semiconductor films.[42]

On the other hand, Rosenberg and Pant[43] observed linear current-voltage characteristics for oxidized cholesterol membranes in iodine solutions (both sides) and rectifying behavior when only one side of the membrane was bathed by an iodine solution. In these experiments with bilayers, the conductivity followed a compensation law behavior[34] which is indicative of semiconductive properties. It appears, therefore, that these investigators were observing conditions under which the rate-determining step was the conduction of the electrons across the membrane. This step in the overall conduction process through the membrane could be either linear, or could display a power-law relationship with potential.[12] Adherence to equation (12) would only be expected from an interfacial charge-transfer rate-determining step, which would not necessarily display the exponential variation with temperature observed by Rosenberg and Pant. The differences between the shapes of the current–voltage characteristics observed by Läuger et al.[25] and Rosenberg and Pant[43] are therefore most probably due to the differences between the rate-determining steps in the different membranes used.

(b) Electroplating is possible on a lipid membrane. Pant and Rosenberg,[44] using an oxidized cholesterol membrane, added a solution of I_2 in NaI to one side and a solution of mercuric nitrate to the other side of the membrane. When a voltage of 50 mV was applied across the membrane with the mercuric side positive, a specularly reflecting metallic film was formed on the positive side. This type of film could also be formed with solutions of platinum chloride or silver nitrate replacing the mercuric nitrate. The mechanism believed to be involved in all of these cases is the reaction of the electronic charge carriers with the metallic ions, reducing them to their neutral metallic state, which could be deposited on the membrane and observed as a metallic mirror. This reaction with the metallic ions is the strongest argument in support of the idea that electronic charge carriers play an important role in the conductivity of some lipid bilayers.

3. Concluding Remarks

This section has paved the way for the next two sections in which various possible physiological roles for electrochemical reactions are presented. The theoretical considerations and experimental results discussed above show that there is considerable evidence favoring the presence of semiconduction in biological material. In addition, the work on thin lipid films demonstrates not only that semiconduction may be present in some bilayers, but also that electrochemical reactions could occur at the lipid–water interface. On the basis of these observations, it is safe to conclude that biological membranes, principally constituted by proteins and lipids, may possess the semiconducting properties which some of its constituents display.

III. ELECTROCHEMICAL MODELS OF BIOLOGICAL ENERGY CONVERSION

1. Introduction

Many of the complex physical and chemical processes occurring in biological cells depend on an organizational unit consisting of assemblies of simple molecules or macromolecules. Membranes constitute one of the most important subcellular assemblies because of their multiple vital functions. All biological cells appear to be surrounded by a thin membrane (plasma membrane, about 70 Å thick) and many cells exhibit subcellular structures either arranged on membranes or in turn surrounded by one. The plasma membrane is believed to be a permeability barrier to the movement of solutes and water into and out of cells. While many solutes and water move through the membrane following their electrochemical gradient, other solutes enter cells or are extruded from cells against an electrochemical potential gradient. The energy for this process is usually provided by ATP, and its conversion to active transport occurs at enzymes localized in the plasma membrane; this active transport process is essential for the life of a cell, since in its absence, cells are unable to function over extended periods of time. Other important biochemical reactions localized in membranes are those involved in the last few steps of foodstuff oxidation and their conversion into ATP (oxidative phosphorylation). All the enzymes for these steps are found in the inner membrane of a subcellular organelle known as the mitochondrion. In this case, the membrane may serve not only as a structure on which the enzymes are bound in the appropriate sequence for the reaction series to occur, but the membrane may

III. Electrochemical Models of Biological Energy Conversion

actually be necessary to separate two compartments held at different electrochemical potentials. This latter requirement has been the subject of considerable speculation, since it could be a universal step in biological electrochemical energy conversion. The present section considers some of the existing evidence in favor of such a generalized energy conversion process and it discusses the difficulties involved in obtaining objective experimental evidence to substantiate such a model.

2. Brief Historical Survey

The general concept that the conversion of metabolic energy into the active movement of ions may depend on oxidation–reduction (redox) reactions was based on an early suggestion by Lund[45] and subsequently used by other workers.[46–48] Redox reactions were further implicated as being responsible for active transport by the quantitative relations obtained between respiration and secretion or ionic accumulation.[49–51] The reviews of Davies[52] and Conway,[53] who coined the term "redox pump" for this general hypothesis, thoroughly cover the historical and technical aspects of this idea.

The structural relationships necessary for the coupling between this "redox pump" and directional ionic transport was discussed by Robertson.[54] He reviewed the general concept that the basic physiological process in active transport is the separation of positive (hydrogen ion) and negative (electron) charges across a proton-impermeable membrane; this author, furthermore, stresses the possibility that the mitochondrial membrane possesses this required property of redox-linked electronic conduction, which allows the separation of charge to occur across that membrane. The ideas reviewed by Robertson, in addition to the development supporting the "redox pump" theory, essentially constitute the basis of the chemiosmotic hypothesis formulated by Mitchell,[55] which links the oxidoreduction phenomena in mitochondria to ATP synthesis through electrodic processes.

3. The Chemiosmotic Hypothesis

(i) Description

The basis of the chemiosmotic hypothesis is the concept of vectorial translocation of protons across an otherwise proton-impermeable membrane. Mitchell[56] considers that such a process occurs in the cytochrome system of mitochondria as a result of the anisotropy in the oxido-reduction (o/r) through that system. This anisotropy would allow that the energy obtained from the oxidation of NADP (a reduced cofactor into which energy is channeled from the citric acid cycle—see, for example, Ref. 4) is transformed into an electrochemical gradient of protons across the mitochondrial inner membrane. This process is schematically shown in Fig. 7. Three such loops are assumed to span the mitochondrial membrane to account for the oxidation of NADP through the final reaction of citochrome oxidase with oxygen. In each loop, for each pair of electrons traversing the appropriate cytochromes, two hydrogen ions are considered to be translocated across the mitochondrial membrane; thus, through these electrodic mechanisms, six hydrogen ions would be translocated by an electron pair traversing the entire cytochrome chain. Since the mitochondrial membrane seems to be impermeable to hydrogen ions, an osmotic gradient is established by this proton translocation process. It is this osmotic gradient which, added to the preexisting membrane potential, constitutes the electrochemical force driving the anisotropic ATP synthesis reaction across the membrane.

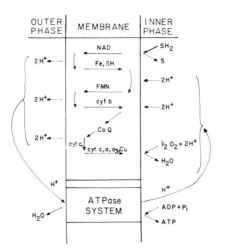

Figure 7. Chemiosmotic mechanism of phosphorylation.[37]

III. Electrochemical Models of Biological Energy Conversion

Mitchell presents the overall ATPase II reaction as

$$ATP + H_2O + 2H_R^+ \rightleftharpoons ADP + P_i + 2H_L^+ \qquad (13)$$

where the suffixes L and R stand for the aqueous phases on the left and right of the coupling membrane, respectively (the ATPase I reaction, which occurs less frequently, involves the translocation of only one proton per ATP). In this reaction, it is clearly seen that a proton gradient from left to right across the membrane would drive the ATP synthesis reaction. Therefore, the energy conversion from reduced substrate to ATP would involve, according to Mitchell, intermediate electrochemical mechanisms.

In direct contrast with this formulation stands the chemical hypothesis of oxidative phosphorylation.[58-60] Its proponents suggest that substrate oxidation leads to the formation of "high-energy" intermediate chemical substances which ferry the energy from the o/r reactions to ATP via the making and breaking of "high-energy" chemical bonds. Experimental evidence has not been able as yet to clearly establish if either hypothesis is correct by itself, although the controversy has stimulated intensive investigative efforts directed to that end. The main problem confronting the proponents of the chemical hypothesis has been the inability to isolate a high-energy intermediate, while the main difficulty of the chemiosmotic hypothesis has been the difficulty in directly measuring the electrochemical gradient across the mitochondrial inner membrane (mitochondria are characteristically less than 10 μm in their largest dimension). The chemiosmotic hypothesis is favored as a model in the present work, because many of its features satisfy the requirements established in the first section for electrochemical energy conversion, which is a feature not encountered in the alternate hypothesis. In addition, the weight of experimental evidence appears to have recently shifted in the direction of the chemiosmotic hypothesis with the successful indirect calculations of a substantive electrochemical gradient across the mitochondrial membrane.[61,62] These calculations are based on various conditions of ionic equilibration across that membrane.

The functional analog of mitochondria in plant cells, the chloroplast, has properties very similar to those of mitochondria. Chloroplasts receive their energy input from photons, and have a cytochrome system with different elements in it than mitochondria;

however, they appear to elicit the transformation of electronic potential energy into ATP via the same chemiosmotic mechanism displayed by mitochondria. Much of the experimental data favoring the chemiosmotic hypothesis has been obtained in chloroplasts, where the relative simplicity of the phosphorylation system and the larger cellular size has allowed more direct verification of this hypothesis. Thus, the arguments presented below in favor of the more general electrochemical mechanisms in cells are based on data obtained from both mitochondria and chloroplasts. The details of the voluminous literature on mitochondria and chloroplasts have been thoroughly reviewed[63-65] and are beyond the scope of the present work. Only the general concepts related to the electrochemical aspects of energy conversion will be presented here.

(ii) Some Relevant Properties of Mitochondria and Chloroplasts

Many of the elements postulated in the chemiosmotic hypothesis have been experimentally identified in the o/r system of mitochondria, such as electron and hydrogen carriers. In addition, there appear to be three "crossover points" where energy from the o/r system is tapped to be transformed into ATP. This is the observation which led Mitchell to postulate three loops in the o/r system, and they account well for the observed stoichiometries between oxygen consumption and ATP synthesis under a variety of conditions.

Another aspect involving the mechanism of ATP synthesis is the action of uncouplers in mitochondria and chloroplasts. Uncouplers are substances which halt ATP synthesis without stopping respiration; they appear, therefore, to uncouple the o/r reactions from ATP synthesis. The chemiosmotic hypothesis postulates that the action of these uncouplers is to increase the protonic permeability of the membrane. Under these circumstances, the chemiosmotic gradient would collapse and ATP synthesis would be halted. This supposition has been given strong experimental support by the finding that the degree of uncoupling produced by a substance is directly proportional to the protonic conductance which it confers to black lipid membranes.[66,67]

An important feature of mitochondria and chloroplasts is their ability to accumulate various ions by diverting energy from both the o/r system and the ATPase system. Mitchell postulated

III. Electrochemical Models of Biological Energy Conversion

that this function is performed by neutral exchange-diffusion of protons for other cations and hydroxyl ions for anions. In this manner, the electrochemical gradient of protons could be partially converted into an electrochemical gradient of other cations, such as K^+. This model of K^+ accumulation was recently confirmed through elegant experiments of Montal et al.[62] in submitochondrial particles (inside-out mitochondrial inner membranes). They first added nigericin to the particles in the presence of external K^+, which induced a K^+ for H^+ exchange with consequent accumulation of K^+. This exchange did not uncouple the particles, which continued to phosphorylate ATP until they were uncoupled with the addition of valinomycin, an antibiotic which selectively increases the membrane permeability to K^+.

(iii) Phosphorylation

The knowledge related to the ATPase system is sparse when compared to other aspects of mitochondrial and chloroplast function. Both the chemical and the chemiosmotic hypotheses converge in postulating a "high-energy" intermediate chemical compound which transmits its energy to the ATP synthesis reaction. The chemical hypothesis visualizes a direct tapping of energy from the o/r system through "high-energy" compounds to energize ATP synthesis; Mitchell, on the other hand, postulates a "high-energy" compound formed by the difference in chemiosmotic gradient across the coupling membrane.

The evidence in favor of any specific model for a mechanism of ATP synthesis is not very compelling. Some of the more important observations made about this reaction are as follows.

(a) The ATPase system is intimately tied to the o/r system since one can control the other. Phosphorylation is necessary in intact mitochondria for respiration to proceed in the o/r system, since inhibition of the ATPase with oligomycin or the absence of ADP from the experimental solution produces "respiratory control"; that is, the o/r system ceases utilizing oxygen. On the other hand, an excess of ATP could drive the respiratory chain backwards.

(b) Mitchell and Moyle[68] have shown that two protons are translocated outward for each molecule of ATP hydrolyzed by the ATPase. In chloroplasts, Schwartz[69] measured the relationship

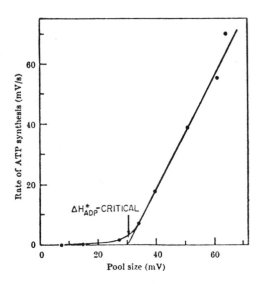

Figure 8. Correspondence between the rate of ATP synthesis and the magnitude of the light-induced steady state proton gradient. The symbol ΔH^*_{ADP}-critical represents the critical magnitude of the proton gradient required across the chloroplast membrane before ATP synthesis is initiated. The linearity between R_{ATP} and ΔH°_{+ADP} at values greater than ΔH^*_{ADP}-critical suggests that R_{ATP} is regulated by the size of the proton gradient in excess of ΔH^*_{ADP}-critical. (From Ref. 69.)

between the rate of ATP synthesis and the pH difference across the membrane. This relationship is shown in Fig. 8, where it appears to be linear beyond a critical value of the pH gradient.

(c) Cohn and Drysdale[70] demonstrated that in the reaction

$$ADP + P_i^{O^{18}} \to ATP + H_2O$$

the rate of fall in O^{18} activity is faster in ATP than in inorganic phosphate and, conversely, if the H_2O is O^{18}-labelled, the rate of appearance of the radioactive label is faster in ATP than in phosphate. These observations suggest that the phosphorylation reactions involve the presence of intermediate steps, which proponents of the chemical hypothesis have tried to identify with the elusive "high-energy" intermediates.

III. Electrochemical Models of Biological Energy Conversion

An alternative explanation for the observations described above would be a mechanism of electrodic phosphorylation[37] in which the energy in the chemiosmotic gradient is transformed into ATP synthesis via an intermediate step involving electronic potential energy. This model is discussed next.

4. Electrodic Phosphorylation

(i) Mitochondrial ATPase

In the introductory section, it was proposed that the most direct energy conversion method for ATP would involve electrochemical processes. The chemiosmotic hypothesis lays the groundwork for such a possibility, since it couples ATP synthesis to an electrochemical gradient of protons. The model of electrodic phosphorylation[37] is based on the assumption that this electrochemical gradient is first transformed into electronic potential energy by means of simple electrodic reactions occurring at both sides of the mitochondrial membrane. The reactions are of the form

$$H^+ + e^- \rightleftharpoons H \tag{14}$$

The equilibrium potential for this reaction is given by

$$\phi = \phi^0 + (RT/F)\ln(1/H^+) = \phi^0 + 2.3(RT/F)(pH) \tag{15}$$

where ϕ^0 is the standard potential, R is the gas constant, T is the absolute temperature, and F is the Faraday constant. At 300°K, the difference in electrochemical gradient between the two sides of the membrane could give rise to an electronic potential

$$\Delta\phi = \Delta\phi^0 + 0.06(pH_{out} - pH_{in}) \tag{16}$$

where $\Delta\phi^0$ is a potential difference across the membrane caused by differences in the environments or due to other electrochemical processes occurring across the membrane. Mitchell[56] calculated that $\Delta\phi$ larger than 0.210 V would poise the ATPase reaction in the direction of net ATP synthesis; this $\Delta\phi$ could be due to a combination of factors described in equation (16), rather than by any one factor alone. Thus, for example, a relatively small $\Delta\phi$ of 0.15 V and a pH differential of only one unit could provide the energy for ATP synthesis. The energetic requirements are, therefore, reasonable and achievable within the physiological realm.

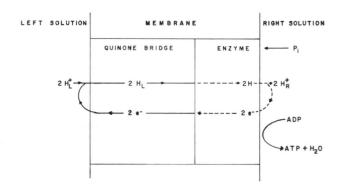

Figure 9. Possible mechanism of electrodic phosphorylation.

A schematic illustration of the electrodic phosphorylation model is shown in Fig. 9. A quinone bridge is assumed to carry hydrogen atoms and electrons from the outside to the enzyme proper, which is also an electronic conductor. The enzyme is a phosphoprotein and/or phospholipid in contact with the inside solution, which catalyzes the phosphorylation reaction. The energy transfer from the electrochemical gradient to electronic potential energy is assumed to occur between the enzyme and the inside solution in the first step of the overall phosphorylation reaction:

$$E-H_2 \rightarrow E + 2e^- + 2H_{in}^+ \qquad (17)$$

This reaction could proceed by two consecutive one-electron transfers, too; however, the end-result would be the same. The ratio of concentration of oxidized enzyme [E] to that of reduced enzyme [E—H$_2$] is given by

$$[E]/[E-H_2] = (K/[H_{in}^+]) \exp(2F \Delta\phi/RT) \qquad (18)$$

Thus, the oxidation of the enzyme would be accomplished by the electronic potential energy and the low H^+ concentration on the inside solution. The next step would be the formation of a high-energy phosphorylated radical[71,72] by the reaction of the oxidized

III. Electrochemical Models of Biological Energy Conversion

enzyme with phosphate (P_i)

$$E + 2e^- \rightarrow E^{2-} \quad (19)$$

$$E^{2-} + P_i \rightarrow P_i \sim E^{2-} \quad (20)$$

The "squiggle" denotes a high-energy compound which in turn can transfer its energy to ATP synthesis in the next step

$$2H_{out} + P_i \sim E^{2-} + ADP \rightarrow ATP + H_2O + E-H_2 + 2e^- \quad (21)$$

The phosphorylation step is a complex one as written in equation (21); it would probably occur in a series of fast, consecutive reactions resulting in ATP synthesis on the inside and the discharging of the two electrons to the outside, thereby returning the enzyme to its original state after the incorporation of two H atoms. The last step would be

$$2H_{out}^+ + 2e^- \rightarrow 2H_{out} \quad (22)$$

to bring in H atoms and remove electrons through the quinone bridge to complete the reaction sequence.

The reaction sequence postulated in the present model is consistent with experimental observations on mitochondrial ATPase. It is based on Mitchell's chemiosmotic hypothesis, and utilizes some results of Wang and co-workers[71,72] on high-energy imidazolyl phosphate radical intermediates which promote the synthesis of ATP. In addition, this model is consistent with the observations of Cohn and Drysdale[70] as well as Boyer[73] regarding the phosphorylated intermediate, since the phosphate reacting with the enzyme in step (20) need not be the same as the phosphate incorporated into ATP in step (21).

(ii) Transport ATPase

The most studied type of ATPase in biological cells is concerned with the conversion of the energy from ATP hydrolysis into active ionic transport across a membrane. Diverse forms of this ATPase (assayed by their ability to hydrolyze ATP) have been isolated within membrane fractions.[74] The membrane attachment is absolutely necessary for the enzymatic properties; the transport properties are, however, lost in the isolated ATPase. Apparently, an intact membrane separating two compartments is essential to the observation of active transport.

Many theories of active transport have been suggested, most of them involving a carrier molecule which preferentially transports certain ions in one direction while returning either with another ion or empty. The energy, in these models, enters into the cycle by an *ad hoc* assumption involving the regeneration of the active form of the carrier. The idea of carriers is helpful conceptually, and it has even been observed that certain antibiotic substances act as ionic carriers;[75] however, no actual biological carrier has been isolated from a biological membrane.

An alternative hypothesis to that of carrier-mediated active transport would involve the principles of the electrodic phosphorylation model described earlier. If ATP synthesis could involve electrochemical energy conversion, why couldn't ATP hydrolysis proceed via similar mechanisms? The mitochondrial ATPase is known to be reversible; that is, an excess of ATP could reverse electron flow through the cytochrome chain. Similarly, transport ATPase is also known to be reversible, since reversal of ionic gradients can lead to ATP synthesis.[76] In addition, both mitochondrial and transport ATPases are selectively inhibited by oligomycin,[77] whereas the cytochrome chain is unaffected by this drug.

The energy from ATP hydrolysis could be transformed into active ionic transport by a reversal of the steps involved in electrodic phosphorylation. Thus, a common mechanism for transport ATPases could be the formation of a local pH gradient across a membrane due to ATP hydrolysis.[78] The local pH gradient created in this manner could cause active transfer of ions in a specialized ion-exchange structure with pH-dependent selectivity (Fig. 10).

In the gastric mucosa, this ion-exchange structure could allow selectively for Cl^- ion movement by ion pair formation with the H^+ ions transported in the ATPase, resulting in the observed acid secretion.[52]

In frog skin, this structure would favor Na^+ ions at the high-pH side and K^+ and H^+ ions at the low-pH side. The Na^+ ions inside the cell would adsorb on the exchanger, traverse the membrane to the low-pH side, where the opposite selectivity exists, and exchange for K^+ or H^+ ions which would travel to the other side to maintain electroneutrality. The local pH gradient produced by the ATPase would thus provide the driving force for active Na^+ transport. Active transport of other ions and varying respiratory quotients

III. Electrochemical Models of Biological Energy Conversion

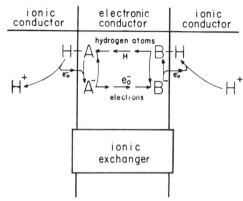

Figure 10. Proposed basic model of cellular interfaces utilizing electrodic processes.[37]

could be modeled from the basic chemiosmotic ATPases by assuming differentiated ion-exchange structures which result in the specialized active transport properties of various tissues. The resulting electrodic models of active ionic transport are found not only to be consistent with a variety of experimental observations, but also to provide novel interpretations for many observed phenomena such as the origin of the tissue potential difference, the nature of exchange diffusion, the actions of various drugs, and many other properties.[37]

Experimental evidence in favor of this model of active transport is extremely hard to obtain. The postulated local pH gradient could be localized near the double layer and, therefore, be unobservable from the solution bulk by a pH electrode. Since the two electrode reactions occur back-to-back on a membrane, it is not possible to dissect out the single electrode reactions. The situation is similar to that of a storage battery that is in a "black box" and whose properties can only be studied from the two leads emerging from the box. Under these circumstances, the only evidence in favor of this electrodic model of active transport is obtained from the voltage–current characteristics of secretory tissues. Live secretory tissues display nonlinear voltage–current characteristics, in contrast to the linear relationship of dead tissue. When these characteristics are plotted as voltage against the logarithm of the current density (Fig. 11), they display straight-line relationships at displacements

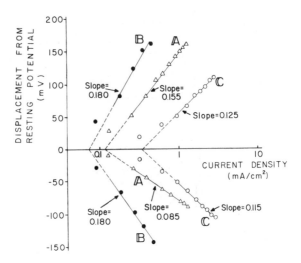

Figure 11. Displacement from resting potential versus current density for frog gastric mucosa (curves A), frog skin (curves B), perch swimbladder (curves C).[37]

from resting potential larger than 40–50 mV. The curves can be fitted extremely well by the Butler–Volmer equation (12),[39]

$$i = i_0\{\exp[(1 - \beta)V^*F/RT] - \exp(-\beta V^*F/RT)\}$$

where i_0 is the exchange current density (a constant), F is the Faraday, R is the gas constant, T is the absolute temperature, V^* is the displacement from resting potential, and β is the symmetry factor. β has been calculated, from the magnitude of the slopes of a variety of tissues, to be close to 0.5, as shown in Table 1. The range of β values encountered experimentally is 0.35–0.70, representing the normal type of asymmetry accompanying uniquely determinable processes (from the β values) such as the present one, which is characterized by a single, almost symmetric, rate-determining energy barrier.[38] The most likely process showing the voltage–current characteristics of Fig. 11 is an electrodic one in which the rate-determining step would be the passage of electrons between one of the solutions and an electron conductor in the membrane.

In these secretory tissues, an electron-transfer rate-determining step would be evidence favoring the electrodic model of active transport presented earlier. The electric current could pass through

Table 1
Values of β Calculated from Data on Secretory Tissues and Nerve Cells[78]

Animal species and membrane type	Exponential slope in mV	Calculated value of β
Frog gastric mucosa	125	0.52
Frog skin	180	0.67
Perch swimbladder	115	0.48
Rabbit oviduct	145	0.49
Loligo axon	110	0.45
Sepia axon	90	0.33
Lobster axon	155	0.61
Frog node	105	0.43

the electrodic ATPase, where the rate-determining step could be the reaction shown in equation (17).

5. Summary

A possible generalized mechanism of biological energy conversion is discussed which involves electrochemical reactions occurring on the surfaces of a membrane separating two compartments. In particular, a model of electrodic phosphorylation is proposed for both the mitochondrial and transport ATPases. The mitochondrial ATPase is assumed to utilize the protonic gradient established by the o/r system, in order to synthesize ATP via electrochemical mechanisms. Transport ATPase is modeled as a two-step process in which the first step is the electrochemical translocation of protons across a membrane, the second step being the exchange of protons for other cations. These models of electrodic phosphorylation account for a variety of observed phenomena on ATPases and, in addition, provide novel interpretations for many of them.

IV. ELECTROCHEMICAL REACTIONS AS BIOLOGICAL REGULATORS

The possibility that electrochemical reactions are at the basis of many biological regulatory functions has been proposed by numerous investigators. A novel theory proposing that intermolecular electron transfer may play a major role in biological regulation and defense as well as in cancer cell growth has been proposed by

Szent-Györgyi.[79] He claims that donor-acceptor (DA) complexes play a fundamental role in the stimulation and inhibition of cellular growth, the extent of growth being regulated by the relative proportion of donors to acceptors. One of the most active donor groups is the sulfhydryl (SH), which is known to be exceedingly important in cell division; the acceptor group could be methylglyoxal, an inhibitor of cellular growth. Normal cells could regulate their growth through the activity of glyoxalase, an enzyme which decomposes the keto-aldehyde methylglyoxal. Thus, cell division would be induced by release of glyoxalase, which reduces the concentration of acceptor groups inhibiting growth; binding of glyoxalase would restore the inhibition of growth. Cancerous cells multiply at a much faster rate than normal cells; the cause of their rapid rate of growth could be related to their inability to bind their glyoxalase and, consequently, they reproduce endlessly.

Many enzymatic reactions may be electrochemical in nature, as proposed by Cope;[80] the enzymes serving not only as a catalyst, but also as an electronic conductor between active sites. Cope shows that a reaction which is rate-limited by electron or ion transport across a particle (enzyme) obeys first-order kinetics, with a hyperbolic relationship between rate constant and the sum of substrate plus product. The reactions catalyzed by cytochrome oxidase and peroxidase, as well as the pyruvate carboxylase enzymes, are described by this hyperbolic relationship. On the other hand, a reaction which is rate-limited by electron or ion transport across an interface shows kinetics conforming to the Elovich equation

$$-dx/dt = me^{nx} \qquad (23)$$

where x is the concentration of reduced substrate, and m and n are constants related to the experimental conditions. A reaction which fits this equation is the decay of photogenerated free radicals in eye melanin particles and in photosynthetic particles of bacteria.[81]

The model of enzyme catalysis proposed by Cope is similar to the mechanism of metal corrosion[39] in which two electrochemical reactions occur in parallel at different sites on the metal, and the electrons flow through the metal (catalyst) from one site to the other.

Nerve excitation may involve electrochemical processes, as proposed by Mandel,[78] for the steady-state voltage–current characteristics of nerves (K^+ current as function of clamping voltage) are very similar in the depolarizing direction to those of secretory cells. After the series resistance inherent in the uncompensated feedback type of measurement[82] is subtracted from the original data, the voltage–current characteristics of nerves from a variety of species are described by equation (12) at V^* larger than 40–50 mV. The values of β obtained from these curves (Table 1) are seen to fall within the same range as those from secretory tissues, hinting at the possibility that electrodic processes may occur in nerve as well.

An electrodic model of nerve excitation can be proposed, based on a mechanism similar to that discussed for secretory tissue, but not metabolically dependent. The quantitative dependence of the K^+ current on applied potential[83] is modeled by a two-step process involving the primary transport of H^+ ions (electrodic current) and the secondary exchange of these H^+ ions for K^+ ions, which results in the observed K^+ current. No specific assumptions need be made regarding the process of Na^+ activation in this model; the inactivation of Na^+, however, is assumed to occur through the lowering of the local external pH by protonic transport across the membrane. The excess H^+ ions could react with the negative membrane sites to prevent the passage of the Na^+ current, as was suggested by Hille[84] for frog node of Ranvier. The reaction of H^+ ions with the membrane sites would prevent these H^+ ions from immediately exchanging for K^+ ions from the nerve cytoplasm, and thus produce the observed delay in the rise of K^+ current to its steady-state value. This model qualitatively accounts for a variety of observations made on nerve, including the results of voltage-clamp experiments, effects of varying ionic concentrations on the bathing solutions, and the action of various drugs on nerve excitation.[37]

Other biological processes which have been ascribed to electrochemical reactions are blood clotting,[85] vision and olfaction,[13] and thyroid function.[86] The diversity of biological phenomena which could have an electrochemical underlying mechanism is large; however, the difficulties involved in obtaining meaningful electrochemical data in inherently impure biological preparations has slowed down research progress in this direction. It is hoped that this situation will change with the advent of the utilization of more sophisticated electrochemical techniques in biological research.

REFERENCES

[1] T. Benzinger, C. Kitzinger, R. Hems, and K. Burton, *Biochem. J.* **71** (1959) 400.
[2] R. C. Phillips, P. George, and R. J. Rutman, *J. Biol. Chem.* **244** (1969) 3330.
[3] L. V. Eggleston and R. Hems, *Biochem. J.* **52** (1952) 156.
[4] A. White, P. Handler, and E. L. Smith, *Principles of Biochemistry*, McGraw-Hill, New York, 1964.
[5] E. E. Crane and R. E. Davies, *Biochem. J.* **49** (1951) 169.
[6] H. H. Ussing, *Ber. Bunsenges Phys. Chem.* **71** (1967) 807.
[7] M. J. Kushmerich and R. E. Davies, *Proc. Roy. Soc. B.* **174** (1969) 315.
[8] J. O'M. Bockris and S. Srinivasan, *Nature* **215** (1967) 197.
[9] J. O'M. Bockris, *Nature* **224** (1969) 775.
[10] A. Szent-Györgyi, *Science* **93** (1941) 609.
[11] A. Szent-Györgyi, *Nature* **148** (1941) 157.
[12] F. Gutmann and L. E. Lyons, *Organic Semiconductors*, Wiley, New York, 1967.
[13] B. Rosenberg and E. Postow, *Ann. N.Y. Acad. Sci.* **158** (1969) 161.
[14] B. Rosenberg, *Nature* **193** (1962) 364.
[15] S. Maričič, G. Pifat, and V. Pradvič, *Biochim. Biophys. Acta* **79** (1964) 293.
[16] A. Szent-Györgyi, *Disc. Faraday Soc.* **27**(11) (1959) 239.
[17] G. King and J. A. Medley, *J. Colloid Sci.* **4** (1949) 9.
[18] D. D. Eley and D. I. Spivey, *Nature* **188** (1960) 725.
[19] B. Rosenberg, *J. Chem. Phys.* **36** (1962) 816.
[20] D. D. Eley, *J. Polymer Sci.* **17** (1967) 73.
[21] D. DeVault, J. H. Parker, and B. Chance, *Nature* **215** (1967) 642.
[22] P. Mueller, D. O. Rudin, H. T. Tien, and W. C. Wescott, *Nature* **194** (1962) 979.
[23] P. Mueller, D. O. Rudin, H. T. Tien, and W. C. Wescott, *Circulation* **26** (1962) 1167.
[24] H. T. Tien and A. L. Diana, *Chem. Phys. Lipids* **2** (1968) 55.
[25] P. Läuger, W. Lesslauer, E. Marti, and J. Richter. *Biochim. Biophys. Acta* **135** (1967) 20.
[26] P. Läuger, J. Richter, and W. Lesslauer, *Ber. Bunsenges Phys. Chem.* **71** (1967) 906.
[27] A. Finkelstein and A. Cass, *J. Gen. Physiol.* **52** (1968) 145s.
[28] Ye. A. Liberman, V. P. Topaly, L. M. Tsofina, and A. M. Shkrob, *Biophysics* **14** (1969) 56.
[29] H. Kallman and M. Pope, *J. Chem. Phys.* **32** (1960) 300.
[30] H. Kallman and M. Pope, *Nature* **186** (1960) 31.
[31] B. Rosenberg and G. L. Jendrasiak, *Chem. Phys. Lipids* **2** (1968) 47.
[32] B. B. Bhowmik, G. L. Jendrasiak, and B. Rosenberg, *Nature* **215** (1967) 842.
[33] B. Rosenberg and B. B. Bhowmik, *Chem. Phys. Lipids* **3** (1969) 109.
[34] B. Rosenberg, B. B. Bhowmik, H. C. Harder, and E. Postow, *J. Chem. Phys.* **49** (1968) 4108.
[35] M. K. Jain, A. Strickholm, F. P. White, and E. H. Cordes, *Nature* **227** (1970) 705.
[36] H. T. Tien and S. P. Verma, *Nature* **227** (1970) 1232.
[37] L. J. Mandel, Ph.D. Dissertation, Univ. of Pennsylvania, 1969.
[38] F. H. Johnson, H. Eyring, and M. J. Polissar, *The Kinetic Basis of Molecular Biology*, Wiley, New York, 1954.
[39] J. O'M. Bockris and A. K. Reddy, *Modern Electrochemistry*, Plenum, New York, 1970.
[40] S. Srinivasan, H. Wroblowa, and J. O'M. Bockris, *Adv. Catalysis* **17** (1967) 352.
[41] M. Myamlin and Y. Pleskov, *Electrochemistry of Semiconductors*, Plenum, New York, 1967.
[42] K. L. Chopra, *Thin Film Phenomena*, McGraw-Hill, New York, 1969.
[43] B. Rosenberg and H. C. Pant, *Chem. Phys. Lipids* **4** (1970) 203.

References

[44] H. C. Pant and B. Rosenberg, *Chem. Phys. Lipids* **6** (1971) 39.
[45] E. J. Lund, *J. Exp. Zool.* **51** (1928) 265.
[46] J. S. Friedenwald and R. D. Stiehler, *Arch. Opthal. N.Y.* **20** (1938) 761.
[47] R. D. Stiehler and L. B. Flexner, *J. Biol. Chem.* **126** (1938) 603.
[48] H. Lundegardh, *Nature* **143** (1939) 203.
[49] E. J. Conway and T. C. Brady, *Nature* **162** (1948) 456.
[50] R. E. Davies and A. G. Ogston, *Biochem. J.* **46** (1950) 324.
[51] R. E. Davies and H. A. Krebs, *Biochem. Soc. Symp.* **8** (1952) 77.
[52] R. E. Davies, *Biol. Rev. Cambr. Phil. Soc.* **26** (1951) 87.
[53] E. J. Conway, *The Biochemistry of Gastric Acid Secretion*, Charles C. Thomas, Springfield, Ill. 1953.
[54] R. N. Robertson, *Biol. Rev. Cambr. Phil. Soc.* **35** (1960) 231.
[55] P. Mitchell, *Nature* **191** (1961) 144.
[56] P. Mitchell, *Chemiosmotic Coupling in Oxidative and Photosynthetic Phosphorylation*, Glynn Research, Bodmin, Cornwall, 1966.
[57] P. Mitchell, *Fed. Proc.* **26** (1967) 1370.
[58] B. Chance, C. P. Lee, and L. Mela, *Fed. Proc.* **26** (1967) 1341.
[59] B. Chance and L. Mela, *J. Biol. Chem.* **241** (1966) 4588.
[60] B. Chance and L. Mela, *J. Biol. Chem.* **242** (1967) 830.
[61] Ye. A. Liberman, V. P. Topaly, L. M. Tsofina, A. A. Jasaitis, and V. P. Skulachev, *Nature* **222** (1969) 1076.
[62] M. Montal, B. Chance, and C. P. Lee, *J. Mem. Biol.* **2** (1970) 201.
[63] A. L. Lehninger, *The Mitochondrion*, Benjamin, New York, 1964.
[64] B. Chance, *Energy Linked Functions of Mitochondria*, Academic Press, New York, 1963.
[65] E. Racker, *Membranes of Mitochondria and Chloroplasts*, Van Nostrand Reinhold Co., New York, 1970.
[66] J. Bielawski, T. E. Thompson, and A. L. Lehninger, *Biochem. Biophys. Res. Comm.* **24** (1966) 948.
[67] Ye. A. Liberman, V. P. Topaly, L. M. Tsofina, A. A. Jasaitis, and V. P. Skulachev, *Nature* **222** (1969) 1076.
[68] P. Mitchell and J. Moyle, *Nature* **208** (1965) 1205.
[69] M. Schwartz, *Nature* **219** (1968) 915.
[70] M. Cohn and G. R. Drysdale, *J. Biol. Chem.* **216** (1955) 831.
[71] J. H. Wang, *Science* **167** (1970) 25.
[72] S. I. Tu and J. H. Wang, *Biochemistry* **9** (1970) 4505.
[73] P. D. Boyer, in *Biological Oxidations*, Ed. by T. P. Singer, Wiley, New York, 1968.
[74] J. C. Skou, *Physiol. Rev.* **45** (1965) 596.
[75] B. C. Pressman, *Fed. Proc.* **27** (1968) 1283.
[76] P. J. Garrahan and I. M. Glynn, *Nature* **211** (1966) 1414.
[77] I. M. Glynn, *J. Physiol.* **169** (1963) 452.
[78] L. J. Mandel, *Nature* **225** (1970) 450.
[79] A. Szent-Györgyi, *Science* **161** (1968) 988.
[80] F. W. Cope, *Bull. Math. Biophys.* **27** (1965) 237.
[81] F. W. Cope, *Bull. Math. Biophys.* **31** (1969) 519.
[82] R. E. Taylor, J. W. Moore, and K. S. Cole, *Biophys. J.* **1** (1960) 161.
[83] K. S. Cole, *Membranes, Ions, and Impulses*, Univ. of California Press, 1968.
[84] B. Hille, *J. Gen. Physiol.* **51** (1968) 221.
[85] P. N. Sawyer and S. Srinivasan, *Am. J. Surg.* **114** (1967) 42.
[86] E. Gruenstein and J. Wynn, *J. Theoret. Biol.* **26** (1970) 343.

5

The Role of Electrochemistry in Environmental Control

A. T. Kuhn

Chemistry Department, The University of Salford
Salford, Lancs, U.K.

I. INTRODUCTION

By now, the danger in which our environment has been placed is generally appreciated. Scientists can chart mean temperatures at various points on the earth's surface, monitor the SO_2 content, or measure the amount of suspended matter in the atmosphere with tests such as those of Ringelmann. Cousteau has testified to the fact that marine life has atrophied visibly, even over the minute time span represented by his own explorations underseas. Though the consensus of scientific opinion indicates that the overall ecological balance of life on the surface crust of the earth is threatened, the experts are in many cases quite unable to agree concerning the time scale on which the threat lies, or from which quarter measurable effects will first present a major threat. An example of this might be quoted. The climate at any point on the earth's surface is largely influenced by the balance of radiant energy entering the earth's atmosphere and leaving it. There is no doubt that an increase in the amount of CO_2 would tend to raise the temperature of the surface of our globe, and this "greenhouse" effect has been the subject of several calculations. Against this factor should be placed the possibility that increasing quantities of water vapor will be deposited in the upper atmosphere by jet planes, especially the higher-flying supersonic craft. This will tend to screen off the radiation from the sun and so lower the earth's temperature.[142] There is no agreement whatsoever as to the "steady-state" concentration of these two simple species or their effect on the temperature, which will have to be viewed against a background of cyclic variation of the earth's climate, or more properly, a short-term (11 yr) cycle superimposed on a long-term (hundreds of years) cycle. It has been argued that the sea is a "buffer" which will absorb excess CO_2 and also that higher CO_2 concentrations will be counterbalanced by an increased rate of plant growth and of the photosynthetic cycle. Such arguments provide the excuse, for those who wish it, for inactivity. They find no sympathy from the author of this chapter or the editors of this book. The two forms of pollution described above are only the simplest of many. Over and above these, we can mention the thermal pollution of rivers, arising from their use as cooling waters. This leads to partial de-oxygenation, and the river becomes lifeless and foetid. Then there is the vast spectrum of chemical pollution of the atmosphere and natural waters everywhere. Lastly, and scarcely recognized in many parts of the globe, is the problem of "pollution" by noise.

Pollution of the atmosphere is mainly one of SO_2 and particulate suspended matter. Many studies have linked the incidence of bronchitic and similar diseases of the chest with SO_2 concentrations. More recently, there have been indications that the damaging effect of SO_2 is sharply accentuated by the particulate matter which invariably accompanies it when solid fuels are combusted. The damage to the fabric of civilized society, be it wear of stonework or ironwork, runs to billions of dollars annually. More recently, the Scandinavian countries have become alarmed by the increase in the acidity of their rainfall, which is causing higher leach-rates of their light soils. Apart from these SO_2 hazards, the phenomenon of "photochemical smog" is too well known to discuss here. Whether the less obvious but equally present minor combustion products of gasoline or diesel fuels such as the benzpyrenes have any serious effect cannot be stated, though these compounds are known to exhibit carcinogenic activity. The concentrations of carbon monoxide in urban conditions can also reach dangerous proportions, and the now normal practice of resuscitation of members of the Tokyo police force with oxygen is traceable to the effects of this compound.

I. Introduction

Turning from the atmosphere to the inland lakes and rivers, one sees the problems of falling biological and chemical oxygen levels, the problems of eutrophication caused partly by overhigh phosphate-based detergent concentrations, partly by leaching out from the farmlands of the heavily applied synthetic fertilizers. Lastly, there is the enormous variety of thoroughly undesirable matter, ranging from free mercury to pesticides of the DDT types, heavy metals such as copper (which poisons the simple organisms used in most sewage plants), and organic compounds such as phenols and chlorohydrocarbons. Marine life is equally threatened from these effluents, and the danger here is that greater depths of the oceans will become deoxygenated with only the oceanographers being aware of the fact.

Toxicologists have laid down acceptable levels for the human ingestion of all of these products, but here, too, complacency is to be avoided. While the level of an instantly fatal dose can be determined with fair accuracy, the longer-term effects of smaller doses and the extent to which these are cumulative are very much less certain. Second, there exist natural means by which initially small concentrations of toxic substances can be multiplied by factors of a hundred or more. Thus, certain forms of crustacea concentrate mercury compounds, as do certain fish. The metallic mercury is converted to a far more poisonous organomercurial species. The concentration of radioisotope strontium by herbivores is another example of this hazard. One may conclude this catalog of perils with two very simple statements of fact. First, the cause of pollution of every kind lies in the increasing population density (or overpopulation, as some might prefer it) of the globe. Second, the problem can be solved in a short space of time, given sufficient funds to apply initially to research and development, and then to the installation of appropriate hardware. It is utterly false to lay the blame of our present situation on the head of the technologists, who merely create what the public demands, and are forced into making it in the cheapest possible way. It is now increasingly clear that among the tools with which pollution will be fought, electrochemical methods rank powerfully. This chapter reviews those areas where electrochemical techniques are already being applied, or where their use might grow, as well as earlier, and often less successful work. Details of engineering aspects can be found in Ref. 142.

Basic Problems in the Use of Electrochemical Methods

It is worthwhile enumerating the strengths and weaknesses of the electrolytic approach to process treatment, be it chemical manufacture or effluent treatment. A plant of given initial cost has to match its competitors in terms of the capital cost per unit throughput treated or processed. That is, the initial cost has to be expressed in terms of throughput and also life of the plant. For manufacturing plants, it is common to write off over seven years or so. For effluent plants, this can be ten years or more. Then, too, the energy costs must be considered, namely the kilowatt-hours per unit volume of effluent treated. Electrochemical processes are unusual, if not unique, in that a relationship between these two terms exists, since under many conditions, a raising of the current can bring greater yields, though unit efficiency drops. Finally, there are certain effluent processes which produce a valuable byproduct and this should be credited. In consideration of plant and process costs, it should not be overlooked that dumping or lagooning costs are by no means insignificant. Thus, a 5-ton load of sludge could cost $100 to dump, through a contractor, while pumping sludge to lagoons is also far from being a negligible term in process costing. Thus any idea which avoids these problems—which almost invariably crop up in the conventional chemical "dosing" effluent treatments—generates an extra credit term. The electrochemical processes frequently score highly in this respect.

Electrolytic processes can only deal with liquid effluents. If the problem arises with gaseous materials, a scrubbing cycle has to be introduced before the electrolytic plant can be used. This can be a real disadvantage, as in the removal of SO_2 from flue gases. At other times, it is scarcely disadvantageous. Furthermore, it should be said that electrochemical methods are at their best for electrolytically conductive solutions. Fortunately, there are many effluents which fall into this category. Then, too, there are situations where ionic species can be added to render the solution conductive, and this is considered under treatment of domestic wastes. If an electrochemical method is applied to a suitably conductive solution, two further problems appear. The first relates to the problem of mass

I. Introduction

transport of the effluent species itself, be it ionic or combined in uncharged form (see the appendix). In laboratory work, high concentrations or else rotating-disc electrodes are used to overcome this problem. Until very recently, the cell designer has chosen to ignore these realities and used a cell with vertical electrodes, simply placed in an electrolyte whose flow is at best a meandering one. The result has been that the maximum effluent treatment current density i_{eff} has often been minute. Where this has been recognized, the plant has been run at low current densities, with the result that return on capital was low. In other cases, a higher current density was used and the result was gross inefficiency. It should be stated here that violent gassing at the electrodes does cause increased rates of diffusion to the electrode surface, though this is energetically an expensive way of achieving this. The foregoing question should be taken in conjunction with the problem of competitive reactions. For cathodic reactions, in aqueous solutions, the hydrogen evolution reaction (h.e.r.) is often in competition with metal deposition reactions. Sometimes, this problem can be avoided or at least mitigated by use of a high-overvoltage cathode, e.g., lead, mercury, or an amalgamated metal. But mercury as a cathode is increasingly out of favor, for its initial capital cost is high, and the inevitable metal loss during operation leads to further running costs, while generating a pollution of its own. Then too, if a metal is deposited on a high-overvoltage cathode such as lead, it behaves less and less as a lead cathode and more like an M cathode, where M is the metal plating out on its surface. One of the main problems running right through this field is ways and means by which such competition from the h.e.r. can be reduced, either by electrochemical ideas or by engineering of the system with mass transport in mind. There are many electrochemical systems in which the electrochemical cell forms a part of a closed loop. Here, one should not overlook the possibilities of adding quaternary salts to suppress the h.e.r., or complexing agents, as used in metal deposition for the purpose of giving better metal deposits. On the anodic side, too, oxygen evolution is a competitive reaction, as also is anodic dissolution of electrode material. The same arguments apply in this case.

There are two general ideas which can extend the versatility of electrolytic processes and improve their costs. First, in dealing with certain solutions, especially those dilute in the noxious species we may generate by electrochemical means a reactive species *in situ*. This can then diffuse in three dimensions, to react with the dissolved effluent and thereby mitigate mass-transport problems. An example of this, which finds application both in electrolytic destruction of cyanide and in the treatment of domestic wastes, is the addition of NaCl to generate ClO^- (hypochlorite) ions at the anode:

$$2Cl^- = Cl_2 + 2e$$

$$2H_2O + 2Na^+ + 2e = 2NaOH + H_2$$

which, on mixing, give

$$2NaOH + Cl_2 = NaCl + NaClO + H_2O$$

which is a strong oxidizing agent and a sterilant.

Another dimension of freedom open to the electrochemical process designer is to modify his counterelectrode reaction. Thus, the normal counter-cathode reaction is hydrogen evolution, though spinoff from the fuel cell research program has made available a wide variety of air cathodes, which serve to provide a counterelectrode reaction operating ~ 1 V lower in potential (the theoretical value is, of course, 1.23 V). Anodic counterelectrode reactions, normally oxygen evolution, can be replaced by anodic oxidations of hydrogen or other fuels. These ideas have been discussed in a general article by the author[2] and form the subject of at least one patent, U.S. Patent 3,103,474. This survey of the broad principles and problems will now be followed by an examination of actual applications, and also a discussion of cell designs and future prospects. Before leaving the subject, it might be apposite to point out that the treatment of effluents, being dilute solutions, bears closely on the problem of extraction of materials from the oceans. Some very thought-provoking comments on this may be found in an article by McIlhenny and Ballard in a volume edited by Levine on desalination, etc.[3]

II. SURVEY OF INDIVIDUAL APPLICATIONS

We shall now survey applications and possible future developments in electrochemical treatments of effluents, stressing the background scientific work, and the advantages and disadvantages of each application. These will be considered under the headings (i) cathodic processes; (ii) anodic processes; (iii) gaseous effluents; and (iv) flotation.

Theoretical factors governing rates of electrochemical purification processes are discussed in the appendix.

1. Cathodic Processes

(i) Copper Removal and Recovery

The electrolytic removal of copper is possibly the most important of all metal-recovery processes based on electrochemical methods. Its application can occur in two different settings. First, sulfuric acid is used as a pickling bath in the copper or brass working industry. The oxide scale on the copper articles dissolves in the sulfuric acid to give cupric sulfate, and the article is bright and scale-free when withdrawn from the bath. The basis of the electrolytic process is the reaction

$$CuSO_4 + H_2O = Cu + H_2SO_4 + \tfrac{1}{2}O_2$$

and this transformation is achieved either in a corner of the pickle bath itself, or in a separate cell, to which the pickle liquor is continuously pumped. In a normal plant, this process achieves a saving of acid, together with a regeneration of metallic copper and virtual elimination of the effluent problem. A second and less important application arises in the manufacture of printed circuit boards or in the electroforming of copper foil, and here again, the benefits lie in recovery of metallic copper and the avoidance of penalties imposed by sewerage authorities for effluents containing as little as 1 ppm of Cu^{2+}.

(a) *Basic characteristics of the copper deposition reaction.* The thermodynamics of the reaction $Cu^{2+} + 2e = Cu$ may be discussed in terms of the Pourbaix diagram. In strongly acid solutions, the reversible potentials for copper deposition and hydrogen evolution coincide. The kinetics of the reaction have been well-studied and there is little to say beyond the fact that the exchange current, as for all such reactions, is high. For details of this, as well as the theoretical aspects of metal deposition, the monograph of Bockris and Razumney[4] may be consulted. In dealing with an effluent electrolytically, one is almost invariably at or close to the region of

Table 1
Predicted Diffusion Current Maxima for Electrodeposition of Copper

Cu^{2+} (mole liter^{-1})	Maximum C.D. (A cm^{-2})	Temp. (°C)
10^{-1}	1.4×10^{-2}	25
10^{-3}	1.4×10^{-4}	25
10^{-5}	1.4×10^{-6}	25
10^{-1}	2.0×10^{-1}	85
10^{-3}	2.0×10^{-3}	85
10^{-5}	2.0×10^{-5}	85

diffusion limitation. Some idea of the magnitude of these currents may be gained from Table 1, which gives calculated values, using data given by Parsons[5] and assuming a free energy of activation of 4 kcal mole^{-1} for the diffusional process.

The appropriate picture is developed assuming the hydrodynamic model illustrated by Reddy.[6] In practice, forced flow and convective stirring will alter the situation and it is these realities

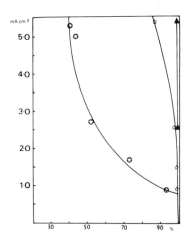

Figure 1. Electrodeposition of copper from dilute solutions (data of Meyer[7]). Current density in mA cm^{-2} as a function of process efficiency; 30-min, runs in 6% H_2SO_4: (star inside circle) 1 g Cu liter; (star) 4 g Cu liter; (triangle) 10 g Cu liter.

II. Survey of Individual Applications

which must be considered. We can now consider the results of various workers who have examined the efficiency of copper deposition. In none of the experiments is the hydrodynamic condition ideally defined. Nevertheless, the data are extremely valuable. Meyer[7] examined the efficiency of Cu^{2+} deposition in a cell which was gently air-agitated. He took solutions of 10, 4, and 1 g of Cu^{2+} per liter and electrolyzed them for 30 min at 1–5 mA cm^{-2}. His results are shown in Fig. 1. Similar data are given in Table 2, which relates to actual working measurements in an electrolytic copper refinery.[8] Soviet work by Gopius and Postnikov[9] should also be consulted, while Table 8 shows results for Hg cathodes.

Table 2
Electrorefining of Copper[a]

Stage	Temp. (°C)	C.D. (mA cm^{-2})	Current efficiency	Copper content g liter^{-1}
1	60	18	95	40–35
2	50	15	80	35–20
3	45	15	av. 60	20–1

[a]Courtesy J. S. Jacobi, James Bridge Copper Works.

Table 3
Nature of Metallic Copper Deposits

Cu (g liter^{-1})	H_2SO_4 (vol. %)	C.D. (mA cm^{-1})	Deposit	Remarks
1	6	5	Dark, spongy, poor adhesion	Strong h.e.r.
1	6	2.5	Dark, spongy, poor adhesion	Strong h.e.r.
1	6	1.7	Moderate dark, spongy	Weak h.e.r.
1	6	0.8	Smooth, compact adherent	Insignificant h.e.r.
4	6	5.4	Light, spongy poor adhesion	Weak h.e.r.
4	6	2.6–0.8	Good deposit, light red color	No (sic) h.e.r.
10	6	5.4–2.7	Very compact deposit	No h.e.r.

In some elegant work by Robinson and Gabe,[10] a rotating cylinder cathode has been employed. Though they do not explicitly quote efficiencies, the information can be deduced from their paper. Both they and Meyer[7] also consider the nature of the copper deposits. Meyer[7] characterizes his deposits by visual observation, while Robinson and Gabe[10] show microphotographs. The data of Meyer are given in Table 3.

In another series, Meyer[7] determined the limiting conditions (in terms of time, Cu^{2+} concentration and C.D.) for a satisfactory deposit is still obtainable, and these are in the region of 1–3 g liter^{-1} Cu^{2+} at 2.5 mA cm^{-2}. This result is, however, unique to the particular conditions of the experiment.

Finally, Surfleet[11] has described the recovery of copper in a fluidized bed cell. His results, recalculated by us, are shown in Fig. 2. It will be clear that no comparison between his work and that of Meyer,[7] for example, is permissible, since the concept of

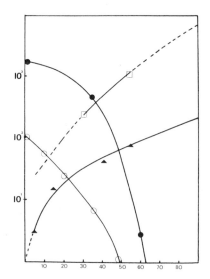

Figure 2. Electrodeposition of copper on a bed electrode (data of Surfleet[11]) and current efficiencies calculated from these results. Ordinate, Cu^{2+} concentration in ppm. Abscissa, deposition time in minutes (○, ●) and current efficiency (△, □).

II. Survey of Individual Applications

C.D. is different in the use of a three-dimensional electrode. Even attempts to make some kind of equivalence (in terms of real surface area of the bed particles) is hindered by the fact that the particles grow during the course of the experiment and their surface area thus changes. However, the results are of interest in that they show that Cu^{2+} levels below 10 ppm can be dealt with at acceptable current efficiencies. Similar work has been undertaken by Flett,[12] of Warren Spring Laboratories, by workers at the University of Southampton, and by Wilkinson of Constructors John Brown.[13]

(b) *Deposition of copper in the presence of other metals.* Only on certain occasions is pure copper pickled in the manner previously described. More often, brass or other cuprous alloys are involved; nor is the pickle acid always sulfuric acid. Though nothing approaching a scientific study of metal deposition under these conditions has been made, some data do exist. Meyer[7] examined the effect of Zn in the solution, and using conditions similar to those described above, and 30 g liter^{-1} of Zn^{2+}, concluded that no effect was apparent. Gopius and Postnikov[9] also examined the effect of zinc ions in solution and reached similar conclusions, and Gorobets and Ponomarev[14] have also considered the deposition of copper in the presence of other species. Turning to the effect of other electrolytes, Meyer has examined varying concentrations of chromic acid. At 1% sulfuric acid and 3% sodium bichromate, the copper cathode at 2.5 mA cm^{-2} actually lost weight instead of gaining it. At 2% acid and 0.1% bichromate, near-normal yields were returned.

Mitter and Dighe[15] conducted experiments aimed at recovering copper from the chromic acid used to pickle coinage in the Indian Mint. They describe a pilot plant using lead anodes, copper cathodes, and porous clay diaphragms. Using a waste stream of approximate composition

Sodium bichromate	nil
Sulfuric acid	3.1%
Available bichromate	4.5%
Copper (as sulfate)	2.0%
Nickel (as sulfate)	0.2%
Zinc (as sulfate)	0.2%

the regeneration of acid and recovery of copper proceeded with initial efficiency of 90%, falling over 8 hr to a terminal value of 40%. Initially, the copper deposited was firm, though as hydrogen evolution became more prominent, the deposit became spongy. The results of Mitter and Dighe, because of their use of a diaphragm, should not be compared with those of Meyer, who used an undivided cell. The work of Youngblood,[16] who worked with mercury cathodes, depositing from mixed solutions of Cu, Ni, and Mn, will be discussed elsewhere; other patents relating to Cu–Cr mixtures are U.S. Patent 3,271,279 and Japanese Patent 22,742 (1969).

(c) *Practical plants.* Many plants exist in copper works for the regeneration of acid and copper metal. Because these are based on a recirculatory system, there is no incentive to reduce the Cu^{2+} concentration in a given pass to a very low level, and the cells are rugged, rather than sophisticated. Typical cells are depicted in Ref. 142. Such cells have been described by Hallowes,[17] Hands,[18] Fishlock,[19] and others.[20,21] All specify a working C.D. in the range 4–16 mA cm^{-2}, with a fairly large electrode spacing of some 7.5 cm. Fishlock[19] recommends the higher C.D.'s only with stirred solutions if acceptable deposits are to be obtained. Cell voltages range from 2.5 to 3.5 V, and under these conditions, the copper content is reduced from a typical 7% to an outlet value of 2%. The power costs then work out around[18] 4.5 kW-hr kg^{-1} Cu, or[21] 890 A-hr kg^{-1}, while Boer[22] gives a range of 800–900 A-hr kg^{-1}. The latter author also prefers anodes of 2 mm thickness and copper starters 0.5 mm in thickness, and is generally in line with other authors, giving Cu^{2+} values of 25 g liter^{-1} inlet and 7 g liter^{-1} outlet and a C.D. of 6–12 mA cm^{-2}. Paparoni[23] however, specifies different conditions, which may or may not reflect special conditions obtaining in Italian industry:

C.D.	1.04–2.37 mA cm^{-2}
Temperature	48–65°C
Cu^{2+} (in)	35–52 g liter^{-1}
Free acid	125–225 g liter^{-1}

Hanna and Naylor[20] have published a note which also stands in line with the observations of other authors referred to above. Especially valuable is a survey[24] published by the British Non-Ferrous Metals Research Association in which are described a number of different ways in which a plant can be operated, citing the work of Sierp,[25] Jones,[26] and Keetman,[27] as well as a reference describing the APV-Kestner plant.[28] Other sources which may be

II. Survey of Individual Applications

consulted are two Czech patents[29] describing cells with "meandering" flow path and a Japanese patent[30] which describes a divided cell, with differential anolyte and catholyte flow rates and a regeneration of chromic acid. A similar concept is outlined in U.S. Patent 3,271,279 (to Allied Chem. Corp.).* This concludes a survey of copper recovery plants and processes.

One aspect which appears to have been little studied, especially in recent years, is the problem of copper oxide formation at somewhat higher pH values. This could certainly be a limiting factor in operation at pH 2 or higher. The suggestion has been made that a cementation process—the displacement of cupric ions by ferrous ions from scrap steel, etc.—might form the basis of a competing process, and indeed such processes do exist. But they are often unsatisfactory, for the recovered product is impure, often containing occluded iron. The future of the process would seem to be utterly assured, however, though one can foresee several improvements. Lancy Laboratories has patents pending[278] on a number of configurations in which the cathode rotates. Such a cathode can recover copper at 60% current efficiency at a level of 50 ppm. The author personally feels that such an arrangement may be more valuable than the fluidized bed, since in this latter arrangement, the copper is plated out onto glass beads coated with copper, or onto solid copper beads. Neither of these are inexpensive, and to remove the copper from them in such a manner as to restore them to their pristine condition is not easy. Flett has filed a patent[273] in which this is done in a second fluidized bed cell, acting as a stripping bath. It is true that in a fluidized bed cell (but not in an ordinary bed cell) a flow of copper beads can continuously be withdrawn. But to control the amount of copper stripped is an awkward problem. The copper, as recovered, is inevitably in a less convenient usable form than when it is stripped off as a pure sheet, from a starter cathode, as is done in refining practice. The power costs associated with operation of a fluidized bed are given in Ref. 142, and these should be compared with the data of other workers, though the latter may not always explicitly record costs.

A major challenge which remains is to plate out brass from pickle boshes in this industry. The ratio of the main metals leached out by the acid is not exactly that of the parent alloy. Nevertheless,

---

*The advantages of Ti starter cathodes, both as regards finer grain Cu deposits and freedom from pitting, are discussed by Wortley.[261]

a brass might be so recovered with the elimination of a cumbersome secondary process at present used in which zinc is chemically precipitated after electrolytic removal of the copper. It has been suggested that suitable addition agents might be found which could be recirculated with the pickle acid, and which would permit the redeposition of a brassy alloy.

(ii) Gold

It is understood that gold is recovered by electrodeposition from wash tanks used in gold-plating, though nothing may be found in the literature to confirm this. Kuti and Stamberg[31] have described an application in which waste gold was recovered from solutions containing upwards of a few tenths of a gram per liter by placing a highly basic ion exchanger between two electrodes.

(iii) Silver

The electrolytic recovery of silver from waste "hypo" fixing solutions such as are used in the photoprocessing industry is an established and successful process. Though it is sometimes stated that the "hypo" is "regenerated," this is not so, at least in the purely chemical sense. It is true in that the lifetime of the "hypo," once stripped of silver, can be extended, but not indefinitely. A good description has been given by Levenson,[32] who also referred to the competitive nonelectrochemical processes. The cathodic deposition of the Ag^+ ion is simple enough though if the potential becomes too cathodic, reduction of the thiosulfate anions commences and this can lead to a blackening of the silver deposited and even formation of free sulfur. Second, coevolution of hydrogen gives poor deposits, exactly as in the case of Cu. In hospitals and commercial movie- or still-film processing laboratories, "sets" for the electrolytic recovery of silver are commonly found on a hire basis, where the owners take, as remuneration, a set percentage of the recovered silver. The earliest cells were described by Hickman in a series of patents,[33,34] while the later work of Doffin (French Patent 783,033) made the process suitable for smaller establishments. As rented out by Pennellier Bros., the sets have a simple form of current control which reduces to a minimum the problems mentioned above. From time to time other patents and cell designs have emerged. Both U.S. Patent 3,477,926 and British Patent 730,649 are examples

II. Survey of Individual Applications

of these which, together with others which may be found in the chemical or photographic abstracting literature, share a common aim of lessening problems of mass transport as the silver becomes more dilute. A review of such cells may be found in Refs. 181 and 274.

Jean[35] has published a comparative study of the electrolytic and chemical methods which supports the contentions made above as to the advantages of the electrochemical method.

(iv) Iron

(a) *General.* Perhaps of all the metals referred to in this chapter, iron is the one with the longest and, unfortunately, the least successful history in respect of its electrolytic recovery. The story is still very much a continuing one and the vastness of the industry, together with the potential applications, means that the problem can justify research funds which would never be available in the case of less important metals.

In much the same way as for copper, at various stages in their fabrication and working processes, iron and steel are pickled to remove the "mill-scale" which forms. Sulfuric acid was the commonest pickle acid used for this purpose, and the disposal of the ferrous sulfate so formed has long been a problem. A portion of this waste liquor can be treated to give hydrated ferrous sulfate, and up to the level of the market—which is not large—this is the best solution to the problem. Beyond this, the only solutions are to treat the liquor chemically or electrochemically, or to pay for its dumping. A number of fairly successful processes exist in which the ferrous salt (sulfate or chloride) is thermally treated to give iron oxide, which may be used or fed back into the main steel-making process. The electrolytic processes which have been proposed are, of course, unique in that they give elemental iron as a product. Before describing these processes, it is well to recall to mind some kindred electrochemical technologies. Electroforming of iron pipes and sheets has been employed for over half a century. The electrochemical manufacture of iron powders is a process carried out today, though it suffers severe competition from chemically formed iron powder. Though these processes have used all manner of ferrous feedstocks, there was throughout the idea that pickle liquor might be used, not only as a suitable source of ferrous metal, but also because regeneration of the sulfuric acid provided a further cash credit term. Electroforming of iron pipes and similar objects in Europe has been largely described by Engelhardt,[36] while Steele has reviewed[37] the production of iron powders. Finlayson et al.[38] gave a description of the Woodall–Duckham Peace River Process, which provides strong competition for the electrochemical routes. In addition, Silman has reviewed, in a series of interesting papers,[39] the electrolytic plants which make iron in the U.S. The use of electroforming processes in British iron industry has not apparently been documented,[173] though techniques are known to have been developed in Wales between the world wars. A very thorough survey of the history of this field has been made (but apparently not published) by Bridgwater.[184]

(b) *Fundamental data on iron electrodeposition.* In contrast to copper or silver, the reversible potential for the equilibrium $Fe^{2+} + 2e = Fe$ lies below that for the h.e.r. The competition from the latter reaction is thus much harder to contend with. The data in Table 4 are derived from the work of Lee[40] and show, perhaps surprisingly, that even at relatively low pH values, iron deposition can occur fairly efficiently, especially at higher temperatures. However, all serious attempts to translate iron deposition into an operational process have made use of a diaphragm or ion-exchange membrane in order to create a catholyte of higher pH (~ 7). The solitary exception to this is the work of Hohn and co-workers,[42] who have used a high-overvoltage (Hg) cathode, as described below.

Table 4
Current Efficiency for Iron Deposition[a]

Temp. (°C)	0.01 N Acid	0.05 N	0.1 N
18	86.8%	61.6%	21.4%
37	89.0%	52.0%	19.6%
55	90.8%	60.0%	26.0%
75	94.8%	71.8%	48.6%
95	—	—	78.8%

[a] At 320 A m^{-2}.

(c) *Technological aspects of iron recovery. Cells using a porous diaphragm.* Such cells emulate closely practice in the chlor-alkali industry, and the best paper on the subject is that of Pienaar.[41]

II. Survey of Individual Applications

Though many of his results are relevant only to this system, the author, using only the most primitive diaphragm, concludes that the system is viable. His data qualitatively support those of Lee[40] in that the efficiency of iron deposition is related to pH and he suggests that the most favorable method of cell operation would not seek to strip the solution of all the iron (and thus restore the initial acidity), but rather to strip iron between the levels 8% Fe (input) and perhaps 4% Fe (return feed to pickling tanks). There is no clue in this paper as to the reasons why the work did not lead any further, though with modern diaphragms and a better understanding of the factors affecting metal deposition, the process costs might well have been transformed. Several patents describe this type of process, including the two-compartment cell in German D.A.S. 1,054,803 in which the spent liquor is fully neutralized in a pretreatment tank, and D.A.S. 1,183,698 (to Ionics) which describes a three-compartment cell, and which quotes D.A.S. 1,155,610 (Japanese Patent 8910/60 also discloses such a system). The latter incorporates a diaphragm and an ion-exchange membrane. The former comprises a cell with three compartments and two diaphragms. A hard coating of iron is deposited on the cathode, while acid is drawn off from the central compartment, water being fed into the anode compartment, spent pickle liquor going into the cathode compartment. Hodge,[43] in an overall review article on pickling, mentions some of the very first attempts at electrolysis with diaphragm-type cells, some of which incorporate SO_2 purges. This idea might find favor once more in combination with flue-gas scrubbing. He refers to the patents of Ramage (U.S. Patents 788,064; 984,703; and 1,007,388) and the work of Farnham (U.S. Patent 1,006,836). His other reference to the "Gaver process" is, unfortunately, not traceable. Swindin[44] also mentions attempts of this time, and the same author updates this in Ref. 45.

Cells using ion-exchange membranes. Electrolytic cells using ion-exchange membranes to separate anode and cathode compartments (otherwise known as electrodialysis cells) can clearly be applied in the same way as porous diaphragms. They are more versatile, and permit other cell reactions and configurations to be employed. These are set out by Quitmann.[46] In its simplest form, the cell has two compartments divided by an anion-exchange membrane. Iron is deposited at the cathode and acid is regenerated in the anode compartment. In a second variant, a three-compartment cell is built using two anion-exchange membranes. Spent pickle liquor is fed into the central compartment, and sulfate migrates through to regenerate acid in the anode compartment. The cathode compartment is filled with caustic soda, and from this, hydrogen is evolved at the cathode while hydroxyl ions migrate through to the central compartment, where ferrous hydroxide is precipitated. Periodically, the cathode compartment must be filled with water, while acid is recovered from the anode section and hydroxide sludge is removed from the central compartment. In a final variant described by Quitmann, a two-compartment cell is again used, but with a cation-exchange membrane. Spent pickle liquor is fed into the anode compartment; the cathode compartment is filled with sodium sulfate. Acid is regenerated in the anode compartment, while ferrous hydroxide is precipitated in the cathode compartment. The sodium sulfate is recycled from this filtration operation back to the cathode compartment again. The working of this cell depends on the rather surprising fact that iron passes through the membrane to the virtual exclusion of H^+ species; Quitmann indicates that this is indeed the case.

Quitmann does not emphasize, as does U.S. Patent 3,072,545, one special advantage of three-compartment cells. This lies in the fact that they may be used to treat acids other than sulfuric, without liberation of chlorine from, say HCl, when this is the pickling acid. Thus, in such a case, iron would be recovered from the cathode compartment, regenerated HCl would flow from the center compartment, while the anode compartment would be fed with water and would contain sulfuric acid.

Of these various ways of operating an electrodialysis cell, the first has received the closest attention. All workers have recognized that use of ion-exchange membranes must entail higher operating costs, both on account of higher cell voltages and also because replacement costs of the membranes are an important factor. In compensation was the fact that pH could be precisely controlled. The pioneering investigations in this field were due to Bramer and Coull[47] and Horner et al.,[48] with later evaluations by Lewis and Tye[49] and Mantell and Grenni.[50] The latter authors also costed the electrodialytic process and reached two important conclusions: First, that the size of the iron powder market would never approach

II. Survey of Individual Applications

the amount of iron released as dissolved $FeSO_4$. Second, that the process costs were not viable. Both these conclusions bear a second examination. The dramatic expansion of the use of iron powder is common knowledge, and figures as well as predictions for its future growth (500,000 tons in 1973) can be found in an article by Finlayson and Morrell.[38] There is no doubt that in this respect Mantell's analysis has been overtaken by change. In his costing of the process, Mantell charges electric power at 1 c $(kW-hr)^{-1}$, which is a high figure, especially for the U.S. and Canada. There is no doubt that a revised costing would provide a different set of figures, and perhaps for this reason there are reports that a number of companies are renewing their interest in the field. Sybron Corp., a Division of the Pfaudler Group, has published recent work[51] and U.S. Patent 3,394,068 also relates to their findings. In other publications, they estimate their operating costs at 3.55 c gal^{-1} on a 200,000-ton steel plant handling 1.65 million gallons annually. These figures include credit given for recovered acid, though not for the electrolytic iron made. Other figures in their papers are also worth quoting. Dumping costs, where the process is not used, are set at 2.75 c gal^{-1}. Membrane life is given at "better than 32 days" with no observed fouling from iron oxides. The regenerated acid, at 28.5% strength, contained less than 200 ppm Fe. The metallic product was a fast-settling iron oxide. This was precipitated in the alkaline catholyte, and no loss of this alkali or of a recirculating metal sequestrant was reported. It is known that similar work is in progress at Daimler Benz, West Germany, and at least one Dutch company, Water Wetkapoer, is known to be closely following developments. A large U.S. company in the field, which holds several important patents in the area, is Ionics Inc.[52] Their position is not clear at this time. Tye,[49] in an extended paper on this subject, suggested that the work at Permutit reached the conclusion that the electrodialysis process was not viable in its own right, but felt that social pressures might, in due course, force its acceptance. Ionac Chemical Company were also active in the field, at least until recently.[53] Other work includes that of Parker,[54] who describes a two-compartment anion-exchange cell, while references to three-compartment cells, including descriptions of a fairly large semi-technical plant, are Ref. 55 and the work of Berger and Ströhlein,[56] while the operation of the process in conjunction with the monohydrate or heptahydrate crystallization process is described by Vogel.[57] Some of the key patents covering electrodialytic recovery of pickle acid are: U.S. Patent 2,810,686 (to Rohm and Haas), a two-compartment anion exchange membrane cell, and British Patent 654,474 (to National Carbon Co.). Soviet Work is described in Refs. 58 and 59, while a review of the work of Ionics Inc. is to be found in Ref. 179. A recent Hungarian paper is Ref. 180.

(d) *Electrodialysis process costs.* Some authors quote process or power costs. Parker[54] quotes 2–12 U.S. cents per 100 sq. ft. of iron pickled. Quitmann[46] sets up costs for a 1000 tons per month throughput pickling plant, assuming 0.8 % loss of weight on pickling, and concludes that the plant will cost 6400 D.M. monthly to run, inclusive of 10-yr amortization. His power consumption is 4.5–5.0 kW-hr per kg of recovered sulfuric acid. Lewis and Tye quote 2.5–8.4 kW-hr per lb of $FeSO_4$ treated, while Horner *et al.*, who show extensive calculations, reach a figure of 1 cent per gallon treated. Mantell and Grenni suggest a loss of $1477 per day on a plant producing 10 tons of iron per day. Against this must be set the advances in electrochemical engineering and the knowledge that desalination of seawater is now a viable process under many circumstances. Clearly, in this field, the experts in electrodialysis and pickle treatment should join forces. The literature survey of Bridgwater[184] should certainly be consulted in this context.

(e) *The use of high-overvoltage cathodes.* It has been seen that even in acid solutions iron may be deposited with high efficiency on the "high-overvoltage" metals such as Pb or Hg. The problem then arises that after some time in operation, the metal becomes coated with iron and behaves like an iron electrode. It was this problem that Hohn and co-workers[42] in Vienna set out to overcome, using a mercury cathode. In their initial work,[42] the iron was deposited on the mercury, which was recycled to a thermal plant where it was distilled off. This process was later modified[60] to one where two electrolysis cells operate in tandem, one a mercury cathode cell, the other a simple diaphragm cell. The same workers[61] have published a valuable paper describing the powder-metallurgical properties of electrodeposited iron. In their revised process, the spent pickle liquor is circulated through a mercury cathode cell, there being a diaphragm above the surface of the mercury. The amalgam is treated in a separate reactor with weakly acidic Fe(III) sulfate solution and the iron-free mercury is returned to the mercury cell. The Fe(III) sulfate is converted to Fe(II) sulfate, still weakly acidic. This is the solution from which the iron powder is recovered in a secondary cell. The weakly acidic Fe(III) sulfate, with which the mercury is denuded, comes from the anode compartment of the secondary electrolysis cell. The plant has been scaled up to three mercury cells of 2 m^2 area each; they are fitted with Pb–Ag anodes. The secondary cells are of conventional construction. The authors state that trials are being conducted with industry, and the project is obviously a serious one, in contrast to some of the processes reported in a number of the papers referred to in this review, which are literally beaker experiments. The quoted energy costs (including ancillaries) are stated to be 12.5–13.5 kW-hr per kg of iron, together with the associated amount of sulfuric acid. This would be $120 per ton of iron, with under two tons of sulfuric acid (at 100%). The authors do not state what capital or operating costs other than power would be involved, though an estimate of these can be obtained by analogy with the costs of the chlor-alkali mercury and diaphragm cells. If one were to seek a straw from which to predict the commercial success of the venture, it could well be the fact that Duisburgerkupferhütte abandoned their electrolytic mercury amalgam process for zinc recovery, after years of operation, on the grounds of over-heavy capital costs of bearing the mercury inventory, as well as making good the mercury losses. The economics of recovering zinc on the one hand, and iron and sulfuric acid on the other, are not very different.

(f) *Other electrochemical processes for treatment of pickle liquors.* Certain other processes have been disclosed which do not fit into the above categories. The U.S. Patent 2,865,823 (U.S. Steel Corp.) covers the regeneration of HCl pickle liquors by bubbling HCl gas into the spent pickle liquor and so precipitating FeCl$_2$ crystals. These crystals can then be electrolyzed in an ion-exchange cell to give chlorine and iron. The chlorine is burned in hydrogen (which has to be supplied) to form HCl, which is recycled. British Patent 796,213 discloses a method for electrochemically oxidizing ferrous to ferric hydroxide, using alternating current, though the philosophy behind the invention is not altogether clear. An unusual patent is

U.S. 3,111,468, in which "active lead" is used as an exchange material. Thus $FeSO_4 + Pb = Fe + PbSO_4$, and the lead sulfate is then regenerated in a separate cell or in another part of the first cell. In an ingenious attempt to harness the otherwise wasted[52] evolution of oxygen at the anode, Aravamuthan *et al.* set up a cell in which a cathodic reaction—the deposition of iron from pickle liquor—was coupled with anodic production of MnO_2 (itself a successful industrial process). They state their intention of scaling up the process in which 0.83 kW-hr of powder produce 1 lb of MnO_2 and 0.8 lb of iron. Such processes involving discrete reactions carried out at the anode and cathode are notoriously difficult to operate. However, theoretically, they offer large possible energy savings, and if, in an appropriate manufacturing complex, a need arose for performing both anodic and cathodic reactions, then such a process, whatever the nature of the anodic reaction, should be considered. Eckhardt,[63] in a short communication, described electrolytic recovery of iron from pickle liquors in a manner which is insufficiently precise to enable it to be classified.

Finally, a straightforward electrolysis in which ferrous chloride is converted to iron powder and chlorine is described in British Patent 574,818 (to British Titan Co.).

At this point, one might claim to have completed a thorough survey of the field, but in the last year or so, work has been resumed on this concept. All the processes described in this survey can claim a certain economic merit, by virtue of making iron powder, or iron oxide, together with regeneration of the pickle acid. The iron powder can, in certain circumstances, command a premium price by virtue of its magnetic and other physical properties. In most cases, a further process credit is the saving in alkali or other costs associated with neutralization or dumping. The removal of iron from the cathode has been, till now, a laborious operation. Whether the use of endless cathode belts of a steel to which iron powder adheres only weakly, with mechanical or magnetic removal of the product, is feasible, remains to be seen.[188] Taking up an idea conceived some years ago in the BISRA laboratories, work is now proceeding at the Electricity Council's Research Centre in which pickle acid yields not iron powder, but foil.[64] This is a completely new product. Its intrinsic merits and consumer acceptance are largely unknown quantities, but if all goes well, it could make

II. Survey of Individual Applications

significant changes in the flow pattern of pickling. The basic electrochemistry of the process is scarcely different from that previously described here. One special situation in which such a plant has to be considered closely is in a smaller economy, perhaps an insular one, which is of insufficient size to support a full-sized steel-making plant. Nevertheless, the installation of a plant such as this would enable scrap steel to be converted directly into foil for manufacture of cans, which might then be used to export the primary produce of that particular country.

(v) Arsenic

Arsenic occurs in many waste streams, notably in the electrowinning or electrorefining of copper and other metals. Its removal is highly desirable, not only from the viewpoint of public health, but also because it would enable the acid to be reused in many more circuits of the plant. The removal of this arsenic by electrochemical reduction to arsine has several times been proposed or examined. Thus, Engelhardt describes work using lead or stainless steel cathodes. Jangg et al.[65] appear to have been the first workers to use mercury cathodes, on which they claim that efficiency is much higher. Their results are reproduced in Tables 5–7. (See also Fig. 3, page 327.)

(vi) Antimony

Data relevant to the recovery of antimony are given in Table 8, where it can be seen that, at least in comparison with other metals, arsenic has been studied in some detail; what is now awaited is an engineering approach.

(vii) Nickel and Stainless Steels

Little appears to have been published on recovery of these metals, though the economic desirability of such an operation equals or surpasses that of copper recovery. The Pourbaix diagram for Ni implies that the competition from h.e.r. is more severe than on other metals previously discussed. A series of reports from the U.S. Defense Agencies describe work on the electrodeposition of alloys of nickel and copper and iron, using a mercury cathode. The origin of the work was the treatment of highly radioactive waste materials from reactors. Youngblood,[16] in one of the latter reports, described an intermediate-sized cell, with which

Table 5
Current Efficiencies for Recovery of Arsenic from As(V) State

C.D. (A m^{-2})

g As(V)liter^{-1}	H$_2$SO$_4$[a]	750			1000			1250			1500		
		As	AsH$_3$	Total	As	AsH$_3$	Total	As	AsH$_3$	Total	As	AsH$_3$	Total
20°C													
25	500	50.1	26.7	76.8	56.7	17.2	73.9	56.0	18.8	74.8	50.3	23.2	73.5
25	750	52.9	27.9	80.8	54.6	22.8	77.4	66.5	10.4	76.9	62.8	14.8	87.6
25	1000	56.4	22.8	79.2	54.6	22.9	77.5	61.2	10.1	71.3	62.8	12.0	74.8
5	500	44.9	25.9	70.8	52.9	13.2	66.1	53.6	16.2	69.8	42.4	25.2	67.6
10	500	48.9	26.3	75.2	56.5	15.1	71.6	54.3	16.7	71.0	46.7	24.1	70.8
25	500	50.1	26.7	76.8	56.7	17.2	73.9	56.0	18.8	74.8	50.3	23.2	73.5
10	1000	58.8	21.2	80.0	57.9	16.6	76.5	56.5	16.6	73.1	43.2	26.8	70.0
25	1000	56.4	22.8	79.2	54.6	22.9	77.5	61.2	16.1	77.3	62.8	12.0	74.8
50	1000	69.6	20.0	89.6	59.4	25.8	85.2	60.4	22.6	83.0	70.8	11.8	82.6
50°C													
25	500	43.4	31.2	74.6	69.1	12.0	81.1	43.9	31.0	74.9	38.6	35.4	74.0
25	750	79.6	11.2	90.8	65.5	22.8	88.3	70.5	16.0	86.5	46.8	26.7	73.5
25	1000	35.8	9.0	44.8	47.4	4.4	51.8	29.0	20.8	49.8	43.5	13.3	56.8
80°C													
25	500	78.0	15.6	93.6	91.9	1.2	93.1	83.7	11.4	95.1	84.1	12.9	97.0
25	750	86.5	13.4	99.9	95.2	4.7	99.9	93.5	5.8	99.3	67.6	23.2	90.8
25	1000	62.6	2.6	65.2	64.9	4.4	69.3	47.6	5.9	53.5	40.6	8.2	48.8
5	500	41.0	15.4	56.4	51.2	14.9	66.1	31.3	14.7	46.0	19.2	21.6	40.8
10	500	53.9	6.9	60.8	54.8	14.0	68.8	58.6	7.1	65.7	42.4	48.4	50.8
25	500	78.0	15.6	93.6	91.9	1.2	93.1	83.7	11.4	95.1	84.1	12.9	97.0

[a] In g/liter

Table 6
Current Efficiencies for Recovery of Arsenic from As(III) State

g As liter^{-1}	H_2SO_4[a]	C.D. (A m^{-2})										
		250			500			1000			1500		
		As	AsH$_3$	Total	As	AsH$_3$	Total	As	AsH$_3$	Total	As	AsH$_3$	Total
1	750	55.6	6.1	61.7	29.8	3.7	33.5	25.8	5.7	31.5	17.8	7.6	25.4
5	750	99.0	0.7	99.7	88.8	2.9	91.7	61.6	1.4	62.5	43.6	3.5	47.1
10	750	99.2	0.6	99.8	83.5	7.3	90.8	65.0	1.5	66.5	47.2	2.0	49.2

[a] In g/liter.

Table 7
Current Efficiency for Recovery of Arsenic[a]

C.D. (A m^{-2})	18/8 Steel			Smooth Pt		
	As	AsH$_3$	Total	As	AsH$_3$	Total
250	14.7	1.1	15.8	16.0	2.7	18.7
500	1.8	4.6	6.4	12.5	3.1	15.6
750	0.3	1.6	1.9	3.7	7.4	11.1
1000	0.4	0.5	0.9	5.6	3.7	9.3
1250	2.8	1.1	3.9	6.4	3.0	9.4
1500	5.6	6.4	12.0	15.3	3.1	18.4

[a]25 g As(V) per liter, 500 g H$_2$SO$_4$, 80°C.

Table 8
Efficiency of Cu, Ni, and Sb Deposition[a][65]

C.D. (A m^{-2})	5 g Cu per liter	5 g Ni per liter	1 g Sb per liter		
			Sb	SbH$_3$	Total
500	28.0	15.6	7.2	11.8	19.0
750	25.3	9.6	7.0	1.8	8.8
1000	19.8	6.4	12.3	1.2	13.5
1250	9.7	4.2	7.3	2.3	9.6
1500	6.2	6.2	2.6	3.2	5.8
2500	8.5	8.4	0.7	1.0	1.7

[a]750 g H$_2$SO$_4$ per liter, 80°C, Hg cathode.

considerable operating knowledge had been gained, while Rhodes et al.,[66] in the earlier reports, also studied the effects of Cr, U, and Mn on the process. Much of the work was done in deuterium oxide. Problems arose with the regeneration of the mercury, which contained many dissolved alkali metals. Though these could be anodically dissolved out, much of the mercury was also lost at this stage.

Much data are contained in these reports, which are worth further reading for those interested. The basic approach, which resembles that of Jangg and co-workers[61,65] in Vienna, would seem to be too expensive for an iron recovery process, though by a small margin. Were the efficiencies to be raised, it might form a

II. Survey of Individual Applications

viable process for stainless steel wastes, and the problem clearly merits further study. One of the best known (nonelectrochemical) methods of treating nickelliferrous solutions is the "integrated Lancy process,"[189] which produces $NiSO_4$ or similar salts. However, a modification of this process exists whereby the Ni may be recovered in metallic form by subsequent electrolysis of a neutral solution, and this is believed to be in use in, for example, the U.S. (See also Table 8 for work on Hg cathodes, from Ref. 65.)

(viii) Tin

The electrolytic recovery of this metal from waste plating solutions is referred to in Ref. 189, though no other references to the process have been located.

(ix) Chromium

Chromic acid and the chromates are widely used for metal deposition and pickling baths. A number of papers describe electrolytic methods for purifying, reducing, and oxidizing spent baths, though the reason for so doing is not at first sight apparent. Schulze[67] describes the conventional methods for reduction of chromic acid wastes, including the use of sulfites and ferrous sulfate. He describes a straightforward electrolytic reduction quoting work of Ibl,[68] but it is clear that such a method is not practicable under normal working conditions. The main method described by Schulze was developed by VEB Wasserversorgungs und Abwasserbehandlung, Dresden, and uses ferrous sulfate for reduction:

$$2H_2CrO_4 + 6FeSO_4 + 6H_2SO_4$$
$$= Cr_2(SO_4)_3 + 3Fe_2(SO_4)_3 + 8H_2O$$

The ferric sulfate is then electrolytically reduced back to the ferrous state, with regeneration of sulfuric acid. Lead anodes and cathodes, and porous diaphragms are used in the cell. Iron (consumable) anodes are also discussed. The process is stated to be in use with a large plating firm. A Japanese paper by Imai[69] also describes the electroreduction of chromic acid waste solutions. He describes work with Pt, carbon, graphite, Fe–Pb, Pb–Ag, and Pb–Sb electrodes, and the starting solution contained 50 g liter^{-1} of $Na_2Cr_2O_7$ and 15 g liter^{-1} of sulfuric acid. This was diluted 1:200 with water and electrolyzed at 25 and 40°C at 0.001–0.1 A cm^{-2}, with an optimum found at 0.02 A cm^{-2}. Other work on reductions of chromates is cited in Refs. 190–194.

A series of patents also describe reoxidation of chromic acid solutions. British Patent 1,102,899 (to Lancy Laboratories) describes the use of electrodialysis in removing foreign ions such as Cu, Zn, Fe, Ni, and Cd, and reoxidizing the chromium(III) to chromic acid. A similar process, using a diaphragm, is cited in British Patent 1,109,624 (to Pittsburgh Plate Glass). The Lancy process is understood to have operated on a pilot plant scale at least, and a detailed memorandum is available from the Laboratories. Indian work describes the regeneration of chromic acid pickles used in the Mint there. The technology of such reoxidation processes is close to that used in the chemical industry for the oxidation of anthracene to anthraquinone using recycled chromic acid, as described by Mantell.[71] Both U.S. Patent 3,271,279 and Japanese Patent 22,742 (1969) are also relevant.

(x) Radioactive Materials

The economics of recovery of radioactive materials from their solutions are not to be compared with those for stable isotopes of the same elements. The radioactive isotopes are often required in a particular form for storage or disposal or even subsequent use. A number of patents or papers discuss this. A series of U.S. Government reports[16,66] discuss the treatment of stainless steel wastes from radioactive plants and the treatment of HRT fuel solutions, in all cases using a mercury cathode. Culler and Blanco[73] also describe treatment of waste products, while French Patent 1,369,431 describes a technique for decontamination of radioactive waters by an apparatus based on electrochemical and conventional filtration techniques.

(xi) Convenience

Where convenience is a factor, economic considerations are clouded. In the management of the domestic environment, this is especially true. Work referred to in Ref. 74 would suggest that studies are in progress to effect electrochemical removal of tin cans from household refuse. These are relatively bulky objects which cannot be disposed of by incineration and so attract the attention of local authorities wishing to reduce the volume of refuse collected.

II. Survey of Individual Applications

Approximate calculations show that the power costs for a week's supply would be 1 ¢ per 20 cans, concentrated to bulk iron. Whether the housewife is prepared to enter the field of electrorefining in order to assist the local authorities remains to be seen!

2. Anodic Processes

Under the heading of anodic processes, we may list first, the true anodic oxidation reactions of effluents. Second, we may list the *in situ* generation of oxidizing species at the electrode, such as the formation of chlorine, hypochlorite, or oxygen itself, all of which have been used for treatment of domestic sewage wastes, etc.

(i) *Anodic Decomposition of Cyanides*

Of the many toxic effluents generated by industrial society, few are more lethal and persistent than the cyanides, which find widespread applications in the metal-finishing and metal-treatment industries. Many methods have been proposed and adopted for their treatment. Of these, the electrolytic destruction technique has had much success. At the same time, the theory and basic electrochemical principles underlying it have never been fully discussed.

The efficacy of the electrolytic method can only be judged in the context of other techniques for disposal of cyanides. These include methods of dilution and acidification to give HCN with liberation of this gas from a stack. The most widely practiced method is treatment with chlorine, hypochlorite, or similar species. All these and other methods are well reviewed in Refs. 70 and 72.

Electrochemical treatment of cyanide consists of two entirely different techniques. The first of these is based on the *in situ* liberation of chlorine-containing species by electrolysis of a solution to which brine has been added. The second method is the straightforward anodic oxidation of the cyanide itself. These techniques will both be examined in depth, though the former can only be fully discussed by reference to the nonelectrochemical technique of adding free chlorine or similar species,[75,76] Drabek and Komendova[77] review both methods, and provide some further data of their own.

(ii) Anodic Oxidation of Cyanides

The CN radical can be anodically oxidized just like any other carbonaceous species, and this observation goes back to the last century. Much work has been done using electrodes which dissolve anodically in the electrolyte. This is referred to in Ref. 78, and will not concern us here. Clearly, under these circumstances, complex cyanides can be formed and the reaction is largely nonelectrochemical. Although many of the cyanide complexes are less toxic than their uncomplexed equivalents, there is only one metal—iron— which might be used in such a way with any hope of being economical.[75]

Schlagdenhauffen[79] reported CO_2 and NH_3 formation, with generation of a brownish precipitate, when KCN was electrolyzed at a Pt anode. He failed to detect CNO^- ions. Luckow[80] found cyanogen when an HCN solution was electrolyzed, but this broke down further to give CO_2 and N_2. In KCN, however, the same author observed potassium and ammonium carbonate formation. Bartoli and Papasogli[81] also used KCN solutions and reported formation of a dark red substance. They added HCl to their electrolyte, and so their work might be better considered in the next section. The first scientific study, as it might now be understood, was that of Hittorf,[82] who used Pt electrodes with a KCN solution and a U-type cell. He detected no gas evolution at the anode, but the formation of the dark red/brown species (ppt) together with a liquid substance whose color changed from yellow to brown with time. Brochet and Petit[83] also used KCN, with Pt electrodes, but failed to detect KCNO, which they felt was unstable. They found CO_2 and NH_3. Perhaps surprisingly, they reported that iron was unattacked as an anode. Paterno and Panain[84] report cyanate formation, as does a German patent.[85] An industrial process for the manufacture of cyanates is discussed in Ref. 86, though it is worth noting, in the light of what follows, that the much less soluble sodium salt was specified, and the cyanate formed was precipitated. Clevenger and Hall[87] reported carbonates and cyanates in their solutions. Tsukamoto,[88] too, reports cyanates in a cell using Pb anodes.

Schmidt and Meinert[78] repeated the work of Hittorf, and confirmed his results, in particular that cyanide was oxidized on Pt without oxygen evolution being visible at C.D.'s of 55 mA cm^{-2}.

II. Survey of Individual Applications

The dark brown substance formed at the anode was partly solid, partly colloidal. They point out that polymeric forms of HCN (known as azulmic acid) have several times been reported in the literature. There appear to be at least two reaction mechanisms for its formation, one being alkali catalyzed polymerization, the other being a reaction of cyanogen with alkali or water; thus,

$$(CN)_2 + 2KOH = KCN + KCNO + H_2O \qquad (I)$$

or

$$(CN)_2 + H_2O = HCN + HCNO \qquad (II)$$

respectively, with subsequent polymerization of the HCN to azulmic acid. These conclusions are broadly supported by Fitzgerald.[89] Schmidt and Meinert[78] discuss the two alternatives for formation of a polymeric cyanide type of species. That originally proposed by Hittorf[82] was that cyanogen simply formed paracyanogen, which is also a brownish polymeric form of cyanogen. Such a mechanism differs from that described above, in which the cyanogen undergoes hydrolysis. Schmidt and Meinert resolved this by measurement of CNO^- concentration before and after the electrolyses. They observed a significant increase in the concentration of this species, which, in their opinion, supports their mechanism and contradicts that of Hittorf. Further proof was supplied when these authors repeated their work in strong alkali. The formation of HCN was now prevented, and thus mechanism (II) was not to be expected, and nor indeed was any brown coloration observed. The ratio of anode to cathode gas volumes was measured. Minimal anode gassing was observed. This, again, lends support to mechanism (I). The same authors continued their study by using acidified solutions of KCN, with a view to seeing whether free cyanogen was formed. There are indications—from vapor pressure measurements—that acidic aqueous solutions of cyanogen are stabilized by complex formation.[90,91] Schmidt and Meinert,[78] using a weakly acidic solution ($pH = 6.5$), found no trace of brown coloration during electrolysis, though subsequent mixing of anolyte and catholyte—to give a weakly alkaline solution—did produce coloration. Clearly, a "reserve" of cyanogen, stored as a complex, was available for formation of the polymer, once alkalinity (pH was 7.5 after mixing) enabled mechanism (II) to proceed. At the slightly higher pH of 5.0, similar results were obtained, and in both cases, failure to observe a colored product, together with the fact that the ratio of anode to cathode gas volume was 10:1 or 20:1, appears to confirm that cyanide ions are being discharged, presumably to form cyanogen.

To summarize the very important contribution of Schmidt and Meinert, one may simply quote their statement that at all pH values from 5 to 13 (the limits of their own work), the sole electrolysis product is cyanogen, and that any other compounds detected result from the homogeneous reactions of this compound with the solution.

A more rigorous electrochemical study was conducted by Sawyer and Day,[92] whose findings support those of Schmidt and Meinert.[78] Using voltammetric and chronopotentiometric techniques, they find that only CN^- is electroactive, and that the undissociated acid is inert. They further conclude that the rate of anodic oxidation is faster, at the potentials studied, than the rate of dissociation of the acid, for which data are available.[93] In the range $CN^- \leq 0.025\,M$, the reaction was diffusion-controlled, and n, the stoichiometric number, was found to be 0.83, that is to say, 1. Having established these facts, the authors proceed with a parametric study of the reaction using galvanostatic pulses, with various strengths of CN^- up to the figure mentioned above, at $pH = 9.5$ and 10.5. Their findings are complex. The slopes of $\log i$ versus V (Tafel slopes) vary systematically as a function of cyanide concentration, from 170 mV at low concentrations ($M = 1 \cdot 10^{-3}\,CN^-$), rising to 330 mV at the upper concentrations. Calculations show that k_0, the electrochemical rate constant, does not change. However, the result of this change of slope is that increase of cyanide concentration reduces oxidation currents at constant E. This is explained in terms of a poisoning mechanism, where the processes

$$Pt(OH)_2 + CN^- = Pt(OH)_2 CN^- \qquad \text{fast}$$
$$Pt(OH)_2 CN^- = Pt(OH)_2 + CN^{\cdot} + e^- \qquad \text{slow}$$
$$2CN^{\cdot} = (CN)_2 \qquad \text{fast}$$

are increasingly hindered, at higher CN^- concentrations, by the substitution of species such as $Pt(CN)_2$ for the oxygen-containing

II. Survey of Individual Applications

surface. Sawyer and Day[92] report NH_3 as an oxidation product, again in terms of reactions of the cyanogen with the solution. The studies of Meinert and Schmidt[78] on the one hand, and of Sawyer and Day[92] on the other, present a picture which, if not complete, is at least coherent. Many other papers, of a largely technological nature, suggest mechanisms which differ from that discussed above, while an unfortunate dichotomy exists in that the few mechanistic studies relate to Pt, while the many technological papers report results with graphite and other electrodes. The limited data of Dart et al.[94] indicate no difference between results obtained on graphite and platinized titanium. Thus, Sperry and Caldwell[95] discuss an anodic reaction:

$$CN^- + OH^- = CNO^- + H + e^- \qquad (III)$$

while Lure and Genkin[96] also state that this reaction occurs as well as

$$2CNO^- + 4OH^- = 2CO_2 + N_2 + 2H_2O + 6e^-$$

Drogon and Pasek[97] suggest reaction (III) as well, while Dart et al.[94] again state that a reaction

$$CN^- + O^{\cdot} = CNO^-$$

occurs. Easton[98] states that both

$$CN^- + 2OH^- = CNO^- + H_2O$$

and

$$2CNO^- + 4OH^- = 2CO_2 + N_2 + 2H_2O$$

take place, in addition to the hydrolyses which are possible, though the rationale behind these mechanisms is not stated. Meyer et al.[136] appear to be the only workers to have considered the thermodynamics of the reactions. For the reaction

$$2OH^- + CN^- = CNO^- + H_2O + 2e^-$$

they quote $E^\circ = 0.96$ V (this is presumably positive although the paper is an American one). Then, taking reactions

$$2OH^- = \tfrac{1}{2}O_2 + H_2O + 2e^- \qquad E^\circ = -0.40 \text{ V}$$

and the sum of the previous two,

$$\tfrac{1}{2}O_2 + CN^- = CNO^-$$

$E^\circ = 1.36$ V. If, however, the mechanism is the coupling of CN^- to form cyanogen as proposed by Sawyer and Day,[92] with all subsequent reactions in the homogeneous phase, the reversible potential is different, and is approximately 0.05 V at $pH = 0$. This implies that at pH 13 or 14, one would not expect any currents below the oxygen reversible potential, and that the process would require study by analysis of CN^- disappearance rates, while the main Faradaic process would be oxygen evolution.

Ruml and Topinka[99] appear to be alone in suggesting that the cathode reaction also plays a role in the decomposition:

$$CN^- + 2OH^- = CNO^- + H_2O + 2e$$
$$CNO^- + 2H_2O = NH_4^+ + CO_3^-$$

In view of the fact that the first reaction is in fact an oxidation, with a reversible potential referred to above, and the second reaction is not an electrode reaction at all, their suggestion is hard to accept.

(iii) Discussion of Mechanisms

One may start from the hypothesis of Schmidt and Meinert,[78] supported by Sawyer and Day,[92] that (on Pt at least) the mechanism is electrolytic discharge of cyanide ions to form cyanogen, with all subsequent reactions being homogeneous. The latter assert that under their conditions of study (25°C and 0.025 N KCN), the reaction was diffusion-controlled. On the other hand, many other workers, notably Dart et al.,[94] provide evidence that on graphite electrodes, in ~ 1 M solutions at 75°C, the process was reaction-controlled, and the current efficiency was 0.53 mole Faraday^{-1} at all C.D.'s up to 90 mA cm^{-2}. The solutions used by Lure and Genkin[96] were of the order of 100 mg liter^{-1} more dilute than those of Sawyer and Day. The evidence from their paper is that the reaction order in CN^- is zero until levels of less than 40 mg liter^{-1} are reached, at which point diffusion control sets in. This is deduced from the fact that their Fig. 1 (Concentration of CN^- as a function of time) gives a rectilinear plot. Their Table 2 also shows that the effect of stirring enhances the current by only 7% at a 65 mg liter^{-1} level. This evidence is unfortunately in direct contradiction with that of Sawyer and Day, for the nature of the electrode material is quite irrelevant under these circumstances. Lure and Genkin[96]

II. Survey of Individual Applications

must have had considerable anodic evolution of oxygen in their system, for they quote current efficiencies of only 30% overall, though this will have varied through the course of the reaction. The overall shape of their curves, i.e., zeroth order ending in diffusion control (i.e., first order) agrees with the data of Sperry and Caldwell.[95]

Not only is it impossible to ascertain on what grounds mechanisms of the type (III) are advanced, but the very fact that it has an $E°$ of 0.96 V on the one hand, while on the other hand, Sawyer and Day show that at potentials cathodic to this, the reaction is already sufficiently far from reversibility to give a rectilinear Tafel plot, argues that the hypothesis is a mistaken one, though one must not preclude the assumption of type (III) mechanism (p. 305) at higher potentials. Such discrepancies make a clear case for the carrying out of further studies, which should be of three different types: (a) steady-state parametric analyses of the reaction; (b) studies under defined hydrodynamic conditions, i.e., with the ring-disc electrode; and (c) surface studies, using techniques such as ellipsometry.

Such studies would serve to prove or disprove the thesis of Sawyer and Day—poisoning of the electrode surface—as well as the hypothesis of Schmidt and Meinert that the only significant electrode reaction is the formation of cyanogen. The thermodynamics quoted by Serfass et al.[136] are of limited interest, though nonetheless of value, since the reaction would rarely proceed in the standard state for which the $E°$ values are quoted. Thus CN^- would often be 50 times lower, tending to raise the E_{rev} value by some 90 mV, while the CNO^- could be very low indeed; in fact, this appears to be the case. Thus the reversible potential might be around 1.27 V assuming 10^{-3} M CNO^-. The $E°$ value for the reaction

$$CN^- = CN^· + e^- \qquad \text{(IVa)}$$

$$2CN^· = (CN)_2 \qquad \text{(IVb)}$$

is not quoted in the papers referred to here, though, in fact, if the mechanistic assumptions of Schmidt and Meinert or Sawyer and Day are correct, this is the vital figure. Using data given by Latimer,[27] an $E°$ value of 0.195 V (at $pH = 0$) is as correct as one can estimate it for reaction (IV), though the observations of Naumann[91] on the reduced vapor pressure of cyanogen suggest that the real value may be lower than this. Enns and co-workers[137] at the University of Waterloo have also reported on the kinetics of the reaction, but a direct approach to them has failed to elicit further information.

It is also believed that fundamental work is in progress at the Battelle Institute.

(iv) Technical Information on Cyanide Oxidation

The earliest papers on the application of electrolysis to the destruction of cyanides appear to be the (fortuitously) juxtaposed contributions of Sperry and Caldwell[95] and Oyler.[101] The latter recommends temperatures close to 100°C and high current densities, though it is recognized that this implies loss of efficiency. He also makes the important points that where the cyanide is a Cu or Ni salt, the process allows recovery of the metal without further modification and CN-complexes are also removed by the method.

Sperry and Caldwell,[95] in their detailed study, conclude that the temperature should be raised as high as possible, and in this they are supported by Dart et al.[94] However, when it comes to the question of agitation of solutions, they oppose this. The author is unable to understand why this should be, on the basis of the facts shown, for although their quiescent "run No. 3" is shown to reach zero CN^- concentration after 13 hr instead of 15 hr for the agitated run, its initial total CN^- content is given in Table II as some 20% greater. The work of Lure and Genkin[96] suggests quite definitely that agitation, especially in the later stages of the reaction, can only be beneficial, and they give tables illustrating their findings. It is worth noting that purely thermal decomposition of cyanide (by boiling) takes place to a significant extent, and Ref. 95 records this. At C.D.'s of 30 mA cm^{-2}, the rate of destruction of a solution containing 4.5 oz. gal^{-1} was less efficient than at half the current density. In general, some 20 hr were required to decompose the normal cyanide effluent. A similar picture is reported by Connard and Beardsley,[103] who report good results using carbon or graphite anodes, though there is a slow decomposition of anode material. Attempts to use ac were unsuccessful—in contrast to the findings of Dodge and Reams—with iron anodes, though in the experience of the author, the corrosion problems induced by ac are very rarely overcome. In contrast with Lure and Genkin,[96] Connard

II. Survey of Individual Applications

and Beardsley[103] achieved little success with Pb anodes, while Lure and Genkin[96] report the disappointing use of magnetite anodes, and Dart et al.[94] evaluate stainless steel similarly. In both cases, the materials are resistant, but produce less decomposition of cyanide than do lead or graphite. Dart et al.[94] also report good results, though with heavy Pt loss, from anodes of platinized titanium.

Using effluents of 20,000–60,000 ppm total cyanide, Connard and Beardsley[103] report decomposition times of 24–96 hr and conclude that process costs are well beneath those for boiling hypochlorite destruction. More detailed costs are reported by Easton,[98] who quotes 4.3 U.S. cents per pound cyanide decomposed. He reports decomposition times of up to 18 days and includes details of an analytical method for cyanide. The same author describes this work in Ref. 104. Work reported in Ref. 105 also confirms the picture, and quotes a figure of 67 g cyanide per kW-hr decomposed, while Lure and Genkin[96,113] obtain of 0.012 kW-hr per g of CN^{-1}, a figure very similar to that of Dart et al.[94] Easton[98] gives ~ 5 kW-hr per kg of $Cu(CN)_2$. A cell with unusually close interelectrode spacing was studied by a large engineering firm in Manchester. Graphite anodes $11 \times 4 \times \frac{1}{2}$ in., were spaced $\frac{1}{8}$ in. apart from mild steel cathodes. This cell, fed with a solution 78 ppm in Cd^{2+} and 100 ppm in CN^- ions, reduced the concentrations of these species to 18 and 12 ppm, respectively, with current efficiencies of 7% with respect to Cd^{2+} and 22% to CN^-. Problems were experienced with cell voltage slowly increasing, and this was attributed to gas-locking. It was believed that application of suction or pressure to the cell, or increasing the flat rate of electrolyte, might overcome this, but the project was abandoned for other reasons unrelated to the electrochemical problems. A recent patent (DOS 1,931,123) to Nordnero is very similar to those described above.

The B.N.V. Process handled by Kampschulte and Cie consists of passing back the wash liquor from dip tanks, through a series of four tanks. In some fashion that the authors[134] do not explain, this process obviates all need for cyanide treatment in any form whatever. The author suggests that this process may in fact be an electrolytic decomposition, though occurring at the anode of the plating vat. Where the latter is an inert, and not a sacrificial anode, there seems no reason why this should not be an entirely acceptable solution.

(v) Destruction of Cyanides by Electrochlorination

The first report of this process appears to be that of Byrne et al.[106] The concept is a simple one, namely that addition of NaCl to the electrolyte gives rise to active chlorine species either at the electrode or in solution. These Cl species in turn react with cyanides to form cyanates in the conventional manner, and from this point on, the chemistry does not appear to differ from normal chlorination procedures for cyanide decomposition. The method has several advantages over the straight anodic oxidation previously described. First, the addition of NaCl improves the conductivity of the solution and lowers the cell voltage. More important is the introduction of a new electrode reaction. The discharge of the chloride ion is an inherently fast reaction, with an i_0 of 10^{-3} A cm^{-2} on the noble metals. Whether, at the discharge potential of the Cl$^-$ ion, the cyanide oxidation reaction is diffusion-limited or is activation-controlled, but with a lower i_0 value, the addition of chloride ions will accelerate the reaction. At low CN$^-$ concentrations, the chlorine or hypochlorite can diffuse and react with cyanide in the bulk of solution. Having done so, the Cl$^-$ ion is reformed, and in this way becomes available once more for charge transfer. A side reaction which must be minimized is the formation of chlorate, rather than hypochlorite, and to control this, Byrne et al.[106] specify an operating temperature of 40–50°C. Nagendra et al.[107] quote results for various NaCl concentrations and current densities. In their virtually sealed cell, time for decomposition of cyanide is directly proportional to current up to 120 mA cm^{-2}. They recommend at least 70 g liter^{-1} of NaCl. At these levels, diffusion control does not appear to set in till the 30-ppm point. Energy consumption (500–5 ppm CN$^-$) can be as low as 10 W-hr per g of CN$^-$. Lure and Genkin claim that addition of Cl$^-$ ions increases current efficiency, as indeed it should, for if the escape of chlorine is prevented, there seems no way in which 100% efficiency should not be reached. They, too, claim 7–10 W-hr per g of CN$^-$, with best results at 0.1 A dm^{-2}. Their data indicate that Cl$^-$ concentration should be 3–5 times greater than that of the CN$^-$.

II. Survey of Individual Applications

Two papers give details and results for operating plants. Drogon and Pasek[108] describe a flowing system. They state that current efficiency is not high, but then, somewhat confusingly, quote that 90 g of CN^- requires 4–4.5 kW-hr theoretically, whereas their plant gives a consumption of 20 kW-hr. On what premises their "theoretical" consumption is based is not clear. In a similar paper, Kurz and Weber[109] describe the CYNOX[111] process. They specify 3% brine solution, pH 10–11, with electrolysis to achieve a chlorine level of 80–200 mg liter^{-1}. Their paper gives a fuller description of cell design and hardware (pumps, etc.) than Ref. 108, together with a description of operational problems arising from sludge handling, etc. As regards efficiency, Kurz and Weber[109] quote the requirements of 1 g CN per 7 g chlorine, which appears to imply ~50% efficiency based on the stoichiometry of the decomposition equation. They estimate 5.5 kW-hr per kg Cl_2. This is a figure which can be compared with modern chlorine cell production, which exceeds it by some 90%. In a sense, this is the crux of the process economics—that, as a method of chlorine generation, it is inefficient, though against this has to be set the avoidance of chlorine handling and transport. Taking into account power and alkali costs, Kurz and Weber[109] conclude that the running costs of the process are some 50% less than those of the straightforward chlorination. From this saving, a contribution toward plant writeoff has to be set aside. The fullest description of process operation is given by Ruml and Topinka.[99] These authors show the effect on process efficiency of virtually every parameter. Among their findings are the relative unimportance of brine concentration in the range 15–60 g liter^{-1}. It appears that current efficiency starts to fall off above 4.4 A dm^{-2}, though no exact data are given. Not surprisingly, flow through the cell is critical; the faster the flow, the lower the efficiency. Anode–cathode distance is not very critical; over the range 2–5 cm, there is perhaps a 10% change in amount of CN^- destroyed. Surprisingly, the anode to cathode ratio appears to be more important, the optimum being 0.9:1. The authors also state that the microporous PVC diaphragms installed in the laboratory cell had no effect. Similar Polish work is quoted in Ref. 114, and other work is given in Refs. 195–203 and 277.

(vi) Phenolic Effluents

(a) *The anodic destruction of phenolic effluents.* The principles underlying this operation resemble those described earlier for the destruction of cyanides in that the phenol may be anodically oxidized or reacted with *in situ* generated hypochlorite derived from anodic discharge of added chloride ions. The earliest workers to examine the reaction with effluent treatment in mind appear to have been Pasynkiewicz[115] and Sakharnov,[116] while in the U.K. several patents have also been taken out.[117] Lure and Genkin,[118] reference to whose work has already been made, have studied this topic as well, while a Polish patent[119] in the name of Trache also relates to this field. Other work is cited in Refs. 204–208.

(b) *Electrochemistry of the phenols.* It can be said at the outset that little is known in this area and that what is reported cannot be said to be reliable. Allen[120] describes a series of reactions which are based on the work of Renard[121] and Fichter.[122] In this scheme, the phenol may be oxidized to the dihydric catechol or alternatively, via hydroquinone, and quinone to maleic acid. Polymerization may also occur anodically. Not too much emphasis should be placed on these schemes, for in the case of an effluent stream which is both dilute and flowing, there is a tendency for oxidation to be arrested at a point where the reactant has left the system. In those plants where sodium chloride is added and hypochlorite ions are formed, the chemistry is different, and chlorinated phenols seem to be the initial product, with subsequent oxidation. Such a hypothesis is borne out by the work of Mulhofer,[123] who found asym-chlorxylol after chlorination of *m*-xylol. See also Ref. 275.

(c) *Electrode kinetics of phenol oxidation reactions.* There are a number of cyclic voltammetric studies in the literature, though not much mechanistic information can be extracted from them. Hedenburg and Freiser[124] studied the effect of *p*H on the half-wave potential, and found the latter decreases with increasing height up to a value of *p*H 9. From this, it might be deduced that the oxidation of the phenolate ion is favored in comparison with that of the un-ionized phenol and indeed the former would be expected to adsorb more strongly on the electrode. Other studies of the anodic oxidation are given in Refs. 125–130. Surfleet[131] examines the

II. Survey of Individual Applications

relative oxidation rates of the phenols and their chlorinated analogs and concludes that the latter are more difficult to oxidize. His study also suggests that the anodic discharge of the Cl^- ion in 0.1 N brine at 25°C is inhibited by the presence of phenol, though the origin of this conclusion is not clear. This is a surprising result, suggesting that there is preferential adsorption of the phenol. The same work indicates that Cl^- discharge can occur at potentials as low as 400 mV w.r.t. SCE and this, in itself, is a rather unusual result.

(d) *Technology of the system.* The cells used differ little from those employed for cyanide decomposition. Anodes are of platinized titanium, graphite, or stainless steel, with mild steel cathodes. Cell voltages vary from 2 to 10 V, and various workers have used different concentrations of salt, from Raine[117] (nil) to Sakharnov[116] (200,000 ppm), as well as a range of temperatures, without any clear picture emerging.

Surfleet[131] quotes electricity costs within a factor of three of costs incurred by dumping or other methods of disposal, and though he does not take into account capital costs, he is also less generous than he might be in allowing any benefits from better cell design. Last of all, he in no way considers effluent processes based on anodic oxidation at low potentials. He raises an interesting point in mentioning ozone treatment as a viable competitive method. Ozone, however, has a very high power generation cost per unit of oxidation capability. It is clear that its advantages stem from the minimal mass-transport problems associated with use of a soluble gaseous species. It should also be noted that tastewise, chlorophenols are more objectionable than their unsubstituted parent molecules.

3. Other Applications of Anodic Oxidation

(i) General

Apart from cyanides and phenols, certain other effluent streams have been discussed as lending themselves to treatment by anodic oxidation. Surfleet[132] has studied the decomposition of acetates in perfumery effluents by means of a Kolbe reaction on a massive Pt electrode. He reports optimal results at anode potentials of

2.7 V (R.H.E.) with C.D.'s of 50–100 mA cm^{-2} and pH 4–5, with the temperature at 30°C or less. Though the results were not unpromising at high effluent concentrations (65,000 ppm), the power consumption per unit decrement in C.O.D. rose sharply as the dilution increased. This is hardly surprising, for in the potential range considered, the Kolbe coupling of acetate ions has to compete with oxygen evolution. Once again, it is regrettable that this problem was not attacked by operation at low potentials, where acetic acid is known to be oxidized. Whatever the rates, it is clear that absence of any other Faradaic reaction must have led to higher current efficiencies and lower cell voltages, and the only disadvantage might be one of low plant throughput.

Several authors who consider cyanides also mention, in passing, cyanates, thiocyanates, and sulfides.[204] Thus, Lure and Genkin[96] refer to phenol, cresols, and thiocyanates as do Raine[117] and the Coal Industry (U.K.) patents,[117] and the latter include treatment of thiosulfates among their claims. The South Eastern Gas Board (U.K.) has filed patents[133] with a slightly different approach, in which a three-compartment cell using porous membranes or semipermeable membranes is employed. The liquors are fed to the center compartment. Impurities are removed by formation of elemental sulfur from thiosulfates, and precipitation of Prussian Blue from ferrocyanides. A reservoir of scrap iron is kept in the bottom of one compartment and sulfur dioxide gas is vented out through the top. Each compartment is automatically kept at a predetermined pH by addition of acid or alkali. It is not known whether this operation is currently being pursued. Another application of anodic oxidation, which appears to be reasonably well engineered, is the oxidation of Malathion waste waters, carried out by Albright and Wilson and mentioned by Hillis.[185] Dimethyldithiophosphoric acid is oxidized at a platinized titanium anode to a disulfide compound, which has a much lower solubility than the reduced species, and which can thus be mechanically removed.

(ii) *The Treatment of Domestic Wastes by "in situ" Chlorination and Other Methods*

The electrolytic treatment of domestic sewage goes back to the last century, and work done in the 1920's in the U.S. is described

II. Survey of Individual Applications

by Mantell,[138] while earlier work has been reviewed by Miller and Knipe,[187] as well as in Refs. 209–211 and 213. In these early plants, the oxidation was done in moderately alkaline media (by adding lime) with oxygen evolution. The design of the cells was rudimentary, and a large interelectrode gap was required to accommodate the wooden paddles which scooped out the solids and also agitated the electrolyte. Not surprisingly, the process passed into abeyance, though the concept of water sterilization by electrolytic oxygenation has been the subject of several patents[146] and the use of Ag^+, electrochemically injected, has also been described in patent form, and discussed in a paper by Lutsch.[139]

The idea of using seawater, which can be electrolyzed to give a sterilizing solution of sodium hypochlorite, goes back to the work of Mendia[140] and Føyn.[138,141] In the earliest development of the method, the sewage was mixed with seawater and passed through an electrolytic cell. It was later recognized that pre-electrolyzed seawater, subsequently mixed with the effluent, was preferable, since in this way, no solids were introduced into the cell. The use of seawater conferred an additional benefit, since the magnesium hydroxide formed at the cathode acted as a flocculant. A plant based on this system has been operational on the island of Guernsey for some years now, and its recent extension implies that the method is successful in every way. However, for more conventional sites, bottled chlorine is a competitive oxidant, while the conventional bacteriological process is still the preferred one, where site availability is no problem. A fuller description of the Guernsey plant can be found in Refs. 142 and 212. A number of Soviet workers have also experimented with similar processes and they include Sergunina,[143] who used a bed electrode of graphite or magnetite particles, and Maslennikov.[144] Also see Refs. 253–256. In addition, Kunina has examined the effect of cathodically formed $Mg(OH)_2$,[145] while another view is given in Ref. 147 and Japanese work is described in Ref. 148.

4. The Treatment of Gaseous Effluents

(i) General

Under this heading fall mainly the gases SO_2 and CO_2, though one can conceive of many other gases which might lend themselves well to electrochemical treatment. Among these would be CO, which has been used in fuel cells quite successfully, and which is a major pollutant from the internal combustion engine. Oxides of nitrogen are equally undesirable, and these, too, are electrochemically oxidizable and reducible. However, no report of work aimed at this goal is known.

(ii) Electrochemical Scrubbing of SO_2 from Flue Gases

Until very recently, all papers and patents[151] in this area described processes in which the flue gases were treated with a more or less aqueous wash which absorbed the SO_2 and was then passed to another stage of the process, where it was electrochemically treated, to provide some byproduct such as sulfuric acid, followed by regeneration of the wash liquor. Such processes suffered the universal drawback that the maximum temperature of the wash liquor could never greatly exceed 100°C, with the result that the flue gases were cooled, and required booster fans to propel them skyward. Very recently, work has been reported[149,150] with molten salt scrubbing, and this opens the way to a high-temperature electrochemical process. This apart, however, we shall consider here the aqueous processes, many of which have become quite highly developed. All electrochemical wash processes have the attractive feature that the regeneration of the wash liquor can largely be done at off-peak times, with a resulting diminution of energy costs, only partly offset by storage expense.

An individual description of processes will now be made.

(a) *Stone Webster/Ionics method.* This process is the most fully developed of all the electrolytic methods, and according to Ref. 152, in which the process is described, a pilot plant has operated during the latter half of 1967 at Gannon Power Station, one of the Tampa Electricity Group, in Florida. The process operates as follows.

Sulfur dioxide is scrubbed with caustic wash liquor, NaOH + SO_2 = $NaHSO_3$; wash liquor and the bisulfite are then passed to a second tower, where, mixed with sodium bisulfate, they are heated by steam-jacketed tubes to give Na_2SO_4:

$$NaHSO_3 + NaHSO_4 = Na_2SO_4 + H_2O + SO_2$$

The heart of the process is the electrodialysis cell, which converts sodium sulfate solution to sulfuric acid, hydrogen, sodium bisulfate,

II. Survey of Individual Applications 317

and caustic soda and oxygen. The acid stream goes to a neighboring plant where it is enriched by the SO_2 piped in from the steam-heated tower, having first been oxidized to SO_3 in a contact vessel. The caustic recirculates to the scrubbing tower, the bisulfate to the steam-heated tower. The gases hydrogen and oxygen are presumably vendable. From an electrochemical point of view, the cell is of interest since it appears to be the first time that sodium sulfate solutions have been electrolyzed to give three liquid streams. During the war, a series of plants were operated to recover caustic soda and sulfuric acid from waste liquors in the rayon viscose industries, and these plants, using mercury cathodes, Pb/Ag anodes ("Tainton anodes"), and diaphragms of rubberized canvas or similar materials are described in Ref. 153. Presumably, by suitable mixing of the acid and alkali streams, they, too, could be engineered to give caustic, acid, and bisulfate as well. However, the current trend is not favorable to mercury cells, for reasons of contamination by that metal. Process economics are given in Ref. 152 by the operators of the process. It is clear that, as quoted, the process would not be viable under open commercial conditions. In the event of a subsidy being allowable for the antipollution plant, the process is in the same cost bracket as that of other nonelectrochemical methods. However, the fact that no further information has been released by the manufacturers, in spite of direct approaches, must imply its own conclusions, although the process is still referred to in the literature.[182]

(b) *Pintsch-Bamag process*. British Patent 950,204 (to Pintsch–Bamag) discloses a method very similar to the Stone–Webster/Ionics one in that it employs caustic soda wash liquor and a cell equipped with an ion-exchange membrane. In this process, it is clear that not only the SO_2, but also the CO_2 is scrubbed out and the latter gas is expelled as the wash liquor becomes more acid during the electrolysis stage. What happens to the CO_2 in the Stone–Webster process is not stated, but it is presumably released when the bisulfite is formed. The patent describes the Pintsch–Bamag cell, which is unusual in that it is fitted with two cationic membranes. In one form, it also has a mercury cathode. The spent wash liquor is fed into the central compartment, and sodium and hydrogen ions migrate through one of the cation membranes. Hydrogen is discharged at the mercury cathode, while sodium amalgamates with it. The mercury amalgam is circulated in the same way as the normal chlorine cell, being treated with water in a separate "denuder" vessel to give caustic soda (which is returned to the scrubbing tower) and clean mercury, which returns to the cell. Loss of sodium ions from the central compartment of the cell lowers the pH, so that SO_2, CO_2, etc. are given off. Hydrogen ions are replaced by migration from the anode chamber, which is fed with water, and where oxygen is evolved. The patent, which also shows a two-compartment cell, claims that addition agents, such as amino acids or arsenious acid, can thus be continuously recirculated round the process without being electrochemically oxidized. By comparison with the Stone–Webster process, the economics of operation must be similar, though perhaps rather more expensive in accounting for capital charges in the mercury inventory. There is no evidence that the process was ever operated on a significant scale, and an approach to the manufacturers has not elicited any further information.

(c) *The Simon Carves process.* This process is disclosed in British Patent 930,584 (to Simon Carves, Ltd.) and differs completely from the previously described approaches. The flue gases are first scrubbed with sulfuric acid, and, after electrolysis, the product is persulfuric acid, which leads to hydrogen peroxide or the per-salts. The problems associated with the corrosive nature of the 40–80% sulfuric acid which is mentioned in the patent must be formidable, and the actual kinetics of the process are not fully discussed. The sulfuric acid also contains hydrogen peroxide and persulfuric acid, the latter compounds causing the SO_2 to be oxidized to SO_3 which dissolves in the acid. The latter, now stronger in H_2SO_3 and leaner in H_2O_2 and persulfuric acid, is passed to an electrolysis cell of the type used to manufacture per-acids electrochemically (such cells are described in Ref. 153). Yet another problem must arise from the heat-exchange equipment. The entire "moving burden" of acid and per-acid has to be alternately heated up to flue-gas temperature (in the scrubbing tower) and cooled (in separate coolers) before the electrolysis stage can commence. Commercial practice shows that the electrolysis operates best at 15°C or even below. The cost of such heat-exchange equipment in acid-resisting metals is not

II. Survey of Individual Applications

negligible. There is no indication of the efficiency of the scrubbing process. A spontaneous thermal decomposition of the hydrogen peroxide to form water and oxygen (either in the gas phase or catalytically on solid surfaces) as well as the per-acids would be expected, and this would lower the efficiency of the system. Once again, no evidence exists that the process was ever scaled up.

(d) *The Lockheed process.* A press release[154] revealed that Lockheed was working on an electrochemical method of flue gas scrubbing using a bed electrode. No further information was released, and a recent approach to the company brought the answer that the project had gone into abeyance because of shortage of funds.

(iii) *Discussion of Electrochemical Scrubbing*

This survey of published methods for electrochemical flue-gas washing may not seem encouraging, yet it should be read in the overall context of the problem where no one solution has emerged as being preferable. It may well be that on a *factory* scale, which is the only one where something approaching economic recovery could be practiced, electrochemical methods do not prove the best. However, there is also the problem of emission from domestic chimneys. In this case, a different set of economic factors operates, and convenience factors may override purely financial ones. The small height of the domestic chimney makes it a far more important source of pollution than the high factory chimney. In addition, the geometry of the system is different, with a much lower ratio of gas flow to orifice area than that of the industrial chimney. An electrochemical device for removal of SO_2 here, provided it was convenient to install, maintain, and operate, would certainly be of the greatest value. Rough calculations show that it would not be impossible to fit, within the domestic chimney itself, a cell to oxidize SO_2 to SO_3 and to wash this down into a separate drain to earth. The oxidation might be accomplished using any of the principles outlined above, or others, such as the electrochemical generation of H_2O_2 by reduction of air as described in Ref. 153. In another configuration, SO_2 could be oxidized to SO_3 and air could be reduced at the cathode. The poor solubility of SO_2 in acid media could be partially overcome by use of a matrix type of cell. Much data already exist regarding the kinetics of SO_2 electrooxidation, and these can be found in Ref. 155, with additional work described in Refs. 156 and 157.

The suggestion has also been made that electrochemical methods be used to remove sulfur from fuel oil. Clearly, electrochemical processes conducted in the hydrocarbon itself are not possible. The suggestion was for sodium metal to be used to extract sulfur with the sodium being subsequently regenerated. This idea presents at least one difficulty common to reactions in liquid/solid systems, namely that the sodium sulfide forming on the metal itself will "passivate" it and prevent further reaction. Second, from an energetic standpoint, the idea is similar to the Moss (nonelectrochemical) process,[158] with the drawback that the energy required for decomposition of sodium sulfide is far greater than the (only thermal) energy required to convert the $CaSO_4$ or CaS back to CaO in the Moss process. It is always possible that a solvent extraction process could be developed in which the sulfur is extracted from the fuel oil, and electrochemically removed from the extraction solvent. This certainly appears to be the direction in which the laboratory electrochemist must apply his thoughts. It might be pointed out here that both the petroleum and the gas industries have made some use of electrochemical techniques for removal of sulfur-containing compounds from their process streams. Thus Fiske and Miller[183] describe the removal of mercaptans by means of an alkali wash with electrochemical regeneration, and this is supported by Jackson,[153] who states that several small such installations are operating in Europe and the Far East. In the gas industry, the so-called "Stretford Process" uses a cell in which electrochemically generated oxygen is used to oxidize sulfur compounds (British Patent 1,046,973), while British Patent 1,044,518 is similar. The former uses a filtering diaphragm, the latter an ion-exchanger. See also p. 316 and Refs. 220 and 228.

(iv) The Electrochemical Removal of CO_2

Carbon dioxide can be removed from closed atmospheres by electrochemical reduction. Kuhn[159] gives calculations of free energy of reduction, based on constructed entropy and enthalpy values, and assesses the commercial prospects for the process as a synthetic route to carboxylic acids. The paper concludes that recent reports of highly efficient (800 mole % per Faraday) reduction

II. Survey of Individual Applications

must be ill-founded. It can further be said that the use of impure CO_2 (derived from fermentation processes) cannot account for these unusual results. It should be clear that microorganisms, in a closed system, are bound to follow the same laws of thermodynamics, including the conservation of energy, as do all other reactant species, and the superefficient reduction results of Bewick must remain enigmatic.

Without plagiarizing the paper referred to above, one might cite the papers which refer specifically to CO_2 reduction as a means of removal of the species from closed environments, especially since they occur in nonelectrochemical literature. The work of Andersen et al.[161] describes a detailed mechanistic analysis of the electrode process, while two U.S. Government reports are of a technological nature.[162,163] It is clear that both in space and in undersea exploration (the latter probably being very much more important in the near future) electrochemistry offers a means both of removal of CO_2 and also of providing oxygen by water electrolysis. Nothing is known, however, if any attempt has been made to combine these two functions in a single cell, though the amount of CO_2 formed is in direct proportion to the quantity of oxygen called for, and its reduction would obviate hydrogen evolution, at least in part. One of the major aerospace contractors has also examined CO_2 removal[164] by electrochemical means. Indian workers[160,271] have used ac in a study of this reaction.

5. Electroflotation

(i) General

In the same way that sewage treatment is of limited interest to the electrochemist, since it is based on the thoroughly well known hypochlorite formation reaction, electroflotation is based on the regeneration of tiny bubbles by water electrolysis. The genesis of the idea goes back to work on ore beneficiation and mineral dressing. In an unpublished technical memorandum on the subject[165] privately circulated by Saint Gobain, Techniques Nouvelles, the idea is traced back to 1888,[166] while other early patents to Elmore[167] and to Lockwood[168] are also cited. The first unequivocal application to waste-water treatment appears to be a series of patents to Metallgesellschaft,[169] and this has since been followed by many others,[170,215] including a modified technique disclosed by the Fairbanks Morse Engineering Corp. in which brine is added to the waste, thus reducing the resistivity of the solution and at the same time making possible a sterilizing action in the manner previously discussed on p. 310. The basic engineering has been described in Refs. 171 and 172. The stream containing the colloid or slurry flows into the cell above the electrodes and so meets the cloud of minute bubbles rising from the electrodes. These bubbles attach themselves to the particles and bear them upward to the point where a conveyor or scoop system can skim off the second (solid) phase. The clear water is extracted below the electrodes. It has been suggested that in addition to the gas-lift effect of the bubbles, an additional role is the discharge at the electrodes of at least a part of the charge which stabilizes the colloid. It is clear that in order to achieve this, the water has to flow across the electrodes, and because of this, some of the electrogenerated gas bubbles may rise above the slurry, without being able to gas-lift anything to the surface. At the present, it is not known how important the first effect is. If it were found to be significant, one might envisage a redesigned cell, perhaps incorporating two sets of electrodes at different potentials, where the first set acted to neutralize the charge on the colloid, while the second set provides the mass of bubbles for the gas-lift action. Alternatively, a single set of electrodes placed in a fast-flowing horizontal stream might combine both actions.

In the U.K., the method has been applied successfully on a pilot plant scale by the Simon Engineering Group, and by the Denco Engineering Group to a variety of problems which include domestic sewage, oily waste waters from a steelworks, and paint-containing waste waters, and at least one full-scale plant is now operating in the U.K. with more scheduled in the next year or two. In addition to the French reference already cited, Lutsch[139] cites French Patent 1,230,728. In practice, it is not always possible to predict which problems will respond well to this technique. The author has experience of at least one two-phase system which could not be separated, and certain cutting-oil emulsions are also notoriously difficult to separate. In their memorandum,[165] St. Gobain lists the following industries as being areas of applicability: oil industry; metal-working industry; slaughter house industry; vegetable oil refineries; food canning plants; dairy industry; paper

II. Survey of Individual Applications

mills; fiber-board plants; glass-fiber industry; asbestos and textile fiber producers; Lucerne wastes; steelworks. In most of these applications, it is claimed that dissolved solids can be reduced below 30 mg liter^{-1}. Two detailed descriptions are also shown in Tables 9 and 10, (from Ref. 165).

Table 9
Application of Electroflotation to Steel Rolling Mill Waste[a]

Inlet waters:	Suspended solids (mainly Fe powder)	150–350 mg liter^{-1}
	Palm oil content	300–600 mg liter^{-1}
	Flow rate	75 m^3 hr^{-1}
Treated water	Solids	30 mg liter^{-1}
	Oils	40 mg liter^{-1}

[a]Plant description: 25 m^2 electrode area, cell volume 25 m^3. Electrodes of Pt/Ti 100 A m^{-2} c.d. 8 V dc. Overall power consumption 275 W/m^3, 275 Wh m^{-3}, no previous conditioning. Bladed concentrate recovery.

Table 10
Application of Electroflotation to Paper-Mill Wastes[a]

Inlet waters:	Suspended solids (fibers, kaolin, etc.)	1 g liter (dry weight, 110°C)
	Flow rate	100 m^3 hr^{-1}
Treated water:	Solids[b]	30 mg liter
Sludge water content		90–95%

[a]Plant description: Plant size as above. Electrodes 18/10 low carbon steel, 80 A m^{-2}, 10 V. 200 Wh m^{-1}.
[b]Includes use of 30 mg liter^{-1} of Al$_2$(SO$_4$)$_3$ with kaolin present

(*ii*) *Electrodes for Electroflocculation Process*[174]

Descriptions of earlier plants have specified a wide range of electrode materials, including carbon, graphite, stainless steel, silicon iron, and platinized titanium and lead with Pt inserts.[177] Modern plants appear to be moving toward a situation where cell polarity is periodically reversed; this prevents electrode fouling due to mineral (Ca salt) deposition. In this event, platinized titanium electrodes such as those supplied by Imperial Metal Industries or

Marston Excelsior are preferable. Graphite and carbon are not acceptable anode materials, nor is mild steel. The use of Al under some conditions is of interest since its slow dissolution provides a continuing injection of Al salts into the system, which thus aid flocculation. However, the metal is the most expensive source of Al salts, not least when installation costs are involved. Lead dioxide and magnetite are two electrodes which could show much promise in this application, especially now that methods have been developed for their deposition on light-weight Ti grids and sheets.[174] Apart from a technical advertisement and an accompanying article,[175] few other open references to the method exist. German Patent 1,177,081 refers to this technique, while some Soviet work is described by Maslennikov and Zhadanova, who discuss its application in the thickening of active slimes. According to them, the skimmed solids contained approximately 97% water, while 90% of suspended matter was removed. It is interesting to note that they found digested solids unsuitable for treatment by this technique. Typical power consumptions are quoted as 400 W-hr m^{-3}—a figure very much in line with the results quoted by St. Gobain.[165] See also Refs. 257 and 258.

(iii) *Miscellaneous*

At some point in a treatment such as this, there must be an enumeration of techniques which, because they employ more than one principle or no known or understood principle, are difficult to describe. There is a great body of literature in which waste solutions were quite simply "electrolyzed" without too much thought as to the *modus operandi* of the process. In many cases also, it was appreciated that attack on the electrodes by anodic dissolution created a concentration of Fe or Al salts which aided flocculation.[240–251] Thus, U.S. Patent 3,340,175 falls in this class, while U.S. Patent 3,523,891 combines this with ozonization. Some of the extensive Soviet work in this field is described in Refs. 221–227; e.g., the work of Matov[221] describes an 80 m^3 hr^{-1} plant of (apparently) a production design which also uses the electrophoretic effect of dc fields to agglomerate the solids; a similar description is given by Romanov.[225] The U.S. Patent 2,852,455 describes the use of ac to produce a "resonance" effect. The ac was superimposed on dc in the electrolysis of sulfite and textile waste waters. Another process for the decolorization of

II. Survey of Individual Applications

waste waters is described in U.S. Patent 3,485,729 and uses electrogenerated hypochlorite, with current reversal. Onstott[229] describes a process for Cl^- ion removal with large, porous C electrodes, by what appears to be specific adsorption of these ions. British Patent 1,173,258 is not readily classifiable, but uses a diaphragm cell. The widespread use of electrodialysis is treated in Refs. 228, 229. The use of electrochemical processes for manufacture of the acids required to operate electrodialysis plants is described by Schaffer and Knight.[230] A cell with one gas-depolarized electrode and one gas-evolving electrode (half-fuel cell) can be used to this end. The oxidation of SO_2 in the presence of H_2S is described by Vitkov et al.,[231] while its reduction is treated by Kravtsov.[232] Simple electrochlorination cells are described by Yanko,[233] and the works of Hartkorn,[234] Fainstein and Mamakov,[235] and Wilk[236] all describe some form of electrolytic treatment.

Negreev and Eligulashvili[237] describe a method for softening temporary hardness of water, while an interesting idea from Beer, the inventor of the RuO anode now widely used in the chlor–alkali industry, is found in German Offen Patent 2,003,426, which describes the use of silicone coatings on electrodes to prevent deposition of Ca salts. The electrochemical metering in of Ag^+ ions to sterilize water (U.S. Patent 3,528,905) is similar to the work in Ref. 139 (p. 315). Kucharski[238] again describes work in which ions are removed by adsorption, followed by change of electrode potential, while the waste stream is removed. In the work of Smith et al.,[239] large-area carbon electrodes are used for the same purpose, though a degree of ion-specificity is claimed here. A method for water treatment which uses no electric current, but which is electrochemical in its operation, is described in U.S. Patent 3,392,102. The water passes through a bed of Mg granules separated by plates. The idea is a type of "bed" electrode, battery powered.

6. The Role of Electrochemistry in Corrosion Protection

The application of electrochemistry to the cathodic or anodic protection of metal structures is too well known to require further elaboration here. It should, however, be pointed out that all types of effluent treatment plant are especially prone to metallic corrosion, and that their protection is specially desirable on this account.

The application of electrochemical techniques in this particular area is well described by Berkeley.[186]

CONCLUSIONS

In this chapter, we have discussed applications of electrochemical processes to the treatment of effluents. It will be agreed that in many cases, the electrode kinetics are straightforward, and that what is lacking is better cell design, and in particular better mass-transport conditions. In this context, the fluidized bed electrode has already been referred to. It is believed, however, that this device is more usable as a cathode than as an anode, and whether it could be used, for instance, in the oxidation of cyanides is uncertain. It is also known that this type of three-dimensional electrode suffers from reactions in which gases are used or evolved. All of these conditions tend to increase the contact resistance between adjacent beads of the bed, with obvious deleterious results. The work of Lancy using rotating electrodes has also been referred to, and Jackson[178] discusses other cell designs in which problems of mass transport can be mitigated. An exciting new idea has been disclosed by RCI (Resources Control Inc., of West Haven, Conn.).[216-219] Here, the space between two planar electrodes is filled with particles of a "semiconducting bed," and this matrix is filled with electrolyte. The concept appears to be that each particle is of such high resistance that a sufficiently great range of metal–solution p.d.'s exists for both anodic and cathodic reactions to take place. The idea is illustrated in Ref. 142, and the company brochures (issued in conjunction with Stauffer Chemical) gave impressive performance data, though it has been learned that certain operation problems exist at the present stage. Other types of cell design may prove equally interesting. The use of forced flow systems through cells of small interelectrode gaps creates a highly turbulent situation, which promotes mass transfer of reacting species. While additional energy is required (for pumping of electrolyte) in these cases, the same is also true for all other three-dimensional electrode cells, and those in which moving parts promote mass transfer. Problems of gas-locking also occur, and much thought has to be devoted to this. See also Refs. 142, 276.

A practical example of this effect was described in Ref. 259, it is seen in Table 11 that the efficiency of the reaction increases

Conclusions

Table 11

Interelectrode spacing	0.64 cm	
Feed—Cd 72 ppm	Outflow—Cd 18 ppm	Current eff. 1/2%
—CN 80 ppm	—CN 30 ppm	
Interelectrode spacing	0.32 cm	
Feed—Cd 78 ppm	Outflow—Cd 18 ppm	Current eff. 7% (Cd)
—CN 100 ppm	—CN 12 ppm	22% (CN)

markedly with decreasing gap. The work has been confirmed by unpublished[276] and continuing work in the author's laboratory. An idea which is as yet little developed is the electrochemical generation of H_2O_2 by reduction of air, with its subsequent use as a sterilant. If the technology described by Jackson[153] proves workable or capable of being improved, it should prove of real value in this context.

The outlook for electrochemical effluent treatment is therefore most encouraging. There is, however, another context in which industrial electrochemical processes should be mentioned here, even though they are purely manufacturing operations and in no sense effluent treatment processes. It is often the case that a given function—for example, the machining of a piece of stainless steel,

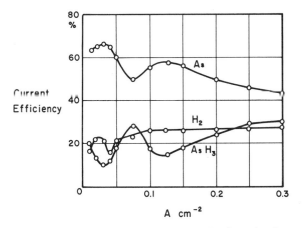

Figure 3. Current efficiency for arsenic/arsine formation from aq. H_2SO_4 solution of As(V) (20°C).

or the oxidation of a complex hydrocarbon—can be achieved by both electrochemical and nonelectrochemical methods. Until now, competitive methods have been judged simply on their economic merits. It is becoming increasingly clear that the economics of such processes will be considerably modified when the cost of effluent treatment is added in, as it will surely have to be. In the main, electrochemical processes are far cleaner than their nonelectrochemical counterparts. This fact, seen against a background of cheaper electric power which nuclear energy promises, and the growing scarcity of natural resources, including fossil fuels, adds up to the fact that electrochemistry has not one, but two roles to play in the improvement of our environmental conditions.

APPENDIX

1. Factors Influencing the Rate of Electrochemical Effluent Processes

The rate at which electrochemical effluent processes proceed is governed by the same laws that describe all electrode reactions, though the physicochemical situations in the present case are rarely as simple and well-defined as those obtaining in well-designed laboratory experiments. However, since the rate is so crucial to the economic viability of this type of process, it is as well to consider the factors at least in outline.

It is customary to consider electrode reactions as being *either* activation-controlled *or* mass-transport-controlled. In the first case, the reaction will respond to changes of electrode potential, while in the second, it will not. Such a picture is an oversimplification in two important respects. First, there is a region for most reactions where activation control gives way to mass-transport control. In this intermediate region, however, the reaction will still respond to changes of potential, though no longer in the classical fashion where

$$i_{\text{eff}} = k(\text{effluent species})e^{\alpha \eta F/RT}$$

(α is here merely an exponential coefficient $0 < \alpha < 1$ and is not intended to represent the transfer coefficient. This equation only holds for overvoltages η greater than $\alpha F/RT$). The reaction will respond to changes in concentration in either of the two cases described above, though the response will almost certainly differ in each case. The second important qualification is that increase in current will in most cases accelerate the rate of a concurrent and undesirable reaction—often hydrogen or oxygen evolution. The rate of gassing in these reactions can change the factors affecting transport of the effluent species to the electrode so that increase in gassing will usually bring slight increases in effluent reaction rate. A quantitative treatment due to Ibl and Venczel[272] which treats this situation leads to the relationship

$$K_L = DV^{0.5}/(1.50 \times 10^{-3})$$

where K_L is the mass transfer coefficient (cm/sec), D is the diffusion coefficient (cm^2/sec), and V is the gas evolution rate (cm^3/cm^2 min).

Of the reactions described in this chapter, it can be said with confidence that cyanide decomposition reactions are activation controlled in the case where the effluent concentration is ≥ 1 M. A special case arises in solutions of metal ions (such as Fe^{2+} or Cu^{2+}) where the rate of reaction is practically limited by crystallization overvoltage or the practicability of obtaining a coherent deposit. Such factors are well-treated in Ref. 4 and will not be elaborated here.

2. Conditions of Mass Transport Control

We consider here the forces and means by which electroactive species are brought to the electrode–electrolyte interfacial region. We shall distinguish between *diffusion* and *convection*, defining these as follows.

Diffusion: The motion of an electroactive species to the electrode, bearing with it its hydration sheath(s). The driving forces for this motion are the gradients of concentration—chemical potentials and also, in the case of charged species, a gradient of electrical potential. Fick's laws govern this type of behavior.

Convection: The transport of electroactive species to the electrode, being borne along with motion of the bulk liquid. Such motion may arise from purely thermal gradients, or from any form of stirring, pumping, or forced flow, as well as so-called "natural convection."

In reality, it is clear that there is no such thing as a truly quiescent, i.e., convection-free, solution, though the use of horizontal electrodes in laboratory cells mitigates this effect. What is more, we assume that whatever the motion of the bulk liquid, there is a layer next to the electrode (the diffusion layer) which, though it may vary in thickness, can be treated as a static layer where convection is unimportant.

3. Mass Transport under Diffusion Control

The rate of mass transport under pure diffusion control has been treated by many writers, assuming various models, and often using heat-transfer analogs. In this latter context, the work of Carslaw and Jaeger[262] is often referred to. The simplest models assume diffusion down a pipe, where concentration gradients exist only along the x axis. As a special case of this, an instantaneous reduction of the concentration at one end of the tube to zero sets up a concentration gradient, and the mass flux, which is directly proportional to the concentration gradient, is given by

$$\partial c/\partial t = D\, \partial^2 c/\partial x^2 \qquad (A.1)$$

(in the generalized form) or

$$\partial c/\partial x = [c°/(\pi Dt)^{1/2}] \exp(-x^2/4Dt) \qquad (A.2)$$

where $c°$ is the concentration in the bulk of solution, and t is the time after the setting up of the gradient. x is the distance from the plane of zero concentration in the tube. For the special case of the concentration gradient at this plane ($x = 0$), we obtain

$$(\partial c/\partial x)_{x=0} = c°(\pi Dt)^{-1/2} \qquad (A.3)$$

Equations such as those above may be solved mathematically using real values of x, t, and c and the results plotted on c versus x graphs to show concentration profiles at different times. Graphically, one may obtain the concentration gradient at $x = 0$ from such curves for various values of t, and it will be seen that (the diffusion layer thickness) $\delta = (\pi Dt)^{-1/2}$, where δ is defined as a layer of electrolyte in which substantial exhaustion of the electroactive species has occurred. In the steady state, similar calculations may also be made. Conway et al.[263] made first calculations of this kind for the times of exhaustion of a solution of its active species in preelectrolysis purification of solutions. Setting up the conditions that

$$\partial c/\partial t = i_L/-V \qquad (A.4)$$

where V is the volume of solution to be exhausted and other symbols have the usual meanings, and expressing $i_L = DAc/\delta$, the solution, treated as first-order reaction kinetics, is

$$c = c°e^{-DAt/V\delta} \tag{A.5}$$

which gives the reduction in impurity levels during a time t. The model of semiinfinite linear diffusion, implicit here, is the most widely used. The solution for nonconvective diffusion to a cylindrical electrode has been approximated by Cairns and Breitenstein,[264] who used a wire electrode; a corrected and more precise solution due to Hodgkins is contained in an appendix to a paper by Kuhn and Sunderland[265]; it has the form

$$i = \frac{4nFDc°}{\pi^2 a} \int_0^\infty \frac{\exp(-Dt/a^2 z^2)}{z[J_0^2(z) + Y_0^2(z)]} \tag{A.6}$$

where a is the radius of the cylinder, and the other symbols, as well as a treatment of integrated forms of this equation, and solutions, are to be found on pp. 336 et seq. of Carslaw and Jaeger's book.[262]

4. Mass Transport under Convective Control

The classic monograph in this field is that of Levich,[266] which should be consulted. It is important at the outset to establish two extreme conditions which may exist, namely those of *laminar* flow across the electrode and *turbulent* flow. The use of the rotating-disc electrode gives the laboratory worker a system in which laminar flow conditions hold, and the solution to the hydrodynamics of this case, and also for the important case of the mercury drop electrode which expands outward, are given in Refs. 260 and 266. These problems are not of immediate relevance here. The first important conclusion is that, outside the Nernst layer (thickness δ), the forces of convection completely outweigh those due to diffusion, as defined above. A measure of this fact is given by the Schmidt number:

$$Sc = Pe/Re \tag{A.7}$$

where Pe and Re are the Peclet and Reynolds numbers, respectively. These are defined as $Pe = V_0 l/D$, where V_0 is the flow rate in the bulk and l is the length of flow, and $Re = V_0 l/\nu$, where ν is the kinematic viscosity (η/ρ). For water, under normal conditions, $Sc = 10^3$.

Further details may be found in Ref, 260. Given these facts, the problem of calculating limiting currents resolves itself into one of knowing δ under a set of conditions. Levich[266] quotes for flow across a parallel plate (laminar)

$$\delta_0 = (v_1/V_0)^{1/2} = 1/(\text{Re})^{1/2} \tag{A.8}$$

where δ_0 is the Prandtl layer thickness, and in the same case,

$$\delta = 3D^{0.33}v^{0.166}l^{0.5}V_0^{-0.5}$$
$$\delta = 0.1\delta_0 \tag{A.9}$$

Thus we can calculate, using Fick's laws, the limiting diffusion current across the Nernst layer of thickness δ, under given conditions, and compare this with the transport by convective flow.

The foregoing treatment applies purely to laminar flow. Turbulent flow is far more complex, but probably represents a closer approximation to conditions in a well-designed cell. The subject has been considered by Vielstich,[267] who derives the equation

$$\delta = l^{0.1}V_0^{-0.9}v^{17/30}D^{0.33} = l\text{Re}^{-0.9}\text{Pr}^{-0.33} \tag{A.10}$$

and this has been confirmed by various workers as cited in Ref. 286.

The maximum current obtainable from a planar electrode of width W and length L can therefore be obtained. Considering a lamina of length dL, the limiting current is

$$i_L = \eta DWc° \, dL/\delta \tag{A.11}$$

and considering in turn, the case for laminar flow, we have

$$i_L = 0.1 DWc° \eta (V_0/v)^{1/2} \int_{L=0}^{L=L} (l/L) \, dL \tag{A.12}$$
$$= 0.2 DWc° \eta (V_0 L/v)^{1/2} \tag{A.13}$$

while for turbulent flow, using the preceding equations, we obtain

$$i_L = \eta D^{1.33} Wc° V_0^{0.9} v^{17/30} \int_{L=0}^{L=L} L^{-0.1} \, dL \tag{A.14}$$
$$= \eta 1.1 D^{1.33} Wc° V_0^{0.9} v^{17/30} L^{0.9} \tag{A.15}$$

One interesting consequence of these equations is that when laminar flow gives way to turbulent flow, the effect of further increase in V_0, the flow rate, increases proportionately. In the laminar flow model, the limiting current is proportional to $V_0^{1/2}$, while in the turbulent flow model, the proportionality constant is 0.9, that is, unity. It is clearly vital for cell designers to ensure that turbulent flow has set in under operational conditions in their cells. For water flowing along a smooth plate, the critical Reynolds number Re is

$$\text{Re} = V_0 l/\nu = 1.5 \times 10^3 \qquad (A.16)$$

though this can be considerably reduced by roughening of the surface, introduction of baffles, or other obstructionary protusions or indentations.

5. Natural Convection

This condition, which, as its name implies, again involves motion of bulk liquid, arises from changes in electrolyte density as a result of the electrode reaction taking place at the phase boundary. The explanation goes back to 1906, when it was discussed by Karaoglanoff.[269] Theoretical treatments by several workers, including Levich, Agar, Wagner, and Tobias, are all discussed in Ref. 268; important confirmation, not only theoretical but also experimental, was reported by Ibl et al.[268,270] In the deposition of Cu from sulfate solutions of 0.2–1.0 mole liter^{-1}, they showed that δ was almost completely independent of solution concentration. There was, however, a marked change of δ with current density. At 4.4 mA cm^{-2}, $\delta = 0.33$ mm; at 8.9 mA cm^{-2}, $\delta = 0.25$ mm; while at 40 mA cm^{-2}, $\delta = 0.17$ mm.

6. Conclusion

In the calculation of maximum diffusion-limiting currents, the key to the problem is the prediction or measurement of the diffusion layer thickness δ. The main causes in the variation of this parameter, and the quantitative relationships governing these changes, have been quoted or cited. The conclusion to be drawn is that electrodes of too great a length are not advantageous, and second that the benefits of fast, that is, turbulent, flow would appear to increase as the flow rate increases in an acceleratory fashion, over the range where transition from laminar to turbulent flow occurs.

ACKNOWLEDGMENT

The author would like to express his thanks to the many workers in industry and research associations whose correspondence and advice have been of the greatest value.

REFERENCES

[1] J. O'M. Bockris and A. K. N. Reddy, *Modern Electrochemistry*, Plenum, New York, 1970.
[2] A. T. Kuhn, *New Scientist* **1967** (5 October) 21.
[3] W. F. McIlhenny and D. A. Ballard, in *Selected Papers on Desalination*, Ed. by S. N. Levine, Dover, New York, 1968.
[4] J. O'M. Bockris and H. Razumney, *Electrocrystallization*, Plenum, New York, 1967.
[5] R. Parsons, *Handbook of Electrochemical Constants*, Butterworths, London, 1959.
[6] A. K. N. Reddy, in *Electrosorption*, Ed. by E. Gileadi, Plenum, New York, 1967.
[7] A. Meyer, *Galvanotechnik u. Oberflaechenschütz* **2** (1961) 91.
[8] J. Jacobi, private communication.
[9] A. E. Gopius and G. S. Postnikov, *Tr. Gos. Nauchn. Issled i Proektn. Inst. Splavov i Obrabotki Tsvetn. Metall.* **21** (1963) 134–6; *Pickling of Brass, Methods for Regeneration of Solutions*, Izd. Gos. Kom. Sov. Min. po Chem. i Tsvetn, Metall., Moscow, 1962.
[10] D. J. Robinson and D. R. Gabe, *Trans. Inst. Metal Finishing* **48** (1970) 35; paper presented at Ann. Mtg. of Inst. Metal Finishing, Torquay, 1970.
[11] B. R. Surfleet, Electricity Council Res. Rept. ECRC/R251, Feb. 1970.
[12] D. S. Flett, *Chem. and Ind.* **1971** (13 March) 300.
[13] J. Wilkinson, paper at Ann. Mtg. of Inst. Metal. Fin., Torquay, 1970.
[14] G. P. Gorobets and V. D. Ponomarev, *Chem. Abstr.* **69** (1968) 48634m.
[15] G. C. Mitter and S. G. Dighe, *J. Sci. Indus. Res. (India)* **2** (1943) 11.
[16] E. L. Youngblood, ORNL 2923.
[17] A. P. C. Hallowes, *Copper* **5** (1959) 18.
[18] S. Hands, *Bull. Inst. Metals* **2** (1953) 48.
[19] D. J. Fishlock, *Product Finishing* **9** (1956) 72.
[20] K. R. Hanna and C. E. Naylor, *Plating Notes (Australia)* **1950** (August) 124.
[21] Anon., *Metal Industry* **1952** (November) 370.
[22] P. Boer, *Metallwissenschaft u. Technik* **14** (1960) 1107.
[23] F. Paparoni, *Il Rame (Milano)* **3** (1965) 3.
[24] *Effluent Treatment in the Copper and Copper Alloys Industries*, BNFMRA, Vicarage Road, Edgbaston, Birmingham, U.K.
[25] F. Sierp, in *Handbuch der Metallbeizerei (Nichteisenmetalle)*, Verlag Chemie, 1961.
[26] J. L. Jones, *Trans. Am. Electrochem. Soc.* **32** (1917) 229.
[27] E. Keetman, *Wire and Wire Products* **34** (1959) 1085, 1151.
[28] Anon., *Electroplating* **5** (1952) 227.
[29] Czech. Pats. 116,366 and 119,592.
[30] Jap. Pat. 22,742 (1969).
[31] J. Kuti and J. Stamberg, *Ionenaustauscher Einzelldarstell* **1** (1961) 365.
[32] G. I. P. Levenson, *Brit. J. Photogr. Almanac* (1951).
[33] U.S. Pats. 1,954,316; 1,959,531; 1,905,467; 1,900,893.
[34] K. C. D. Hickman, W. J. Weyerts, and O. E. Goehler, *Ind. Eng. Chem.* **25** (1933) 202.

[35] A. Jean, *Kemija und Industriji* **14**(5) (1965) 321.
[36] V. Engelhardt, *Handbuch der Technischen Elektrochemie*, Akad. Verlag, Leipzig, 1933.
[37] J. N. Steele, in *Industrial Electrochemical Processes*, Elsevier, 1971.
[38] P. C. Finlayson and A. P. Morrell, paper at 9th Commonwealth Mining and Metallurgical Cong. (1969).
[39] H. Silman, *Metal Finishing* **1969** (December) 36; *Design Engineer* **1970** (November) 88; E. Konrad, *Chemical Engineering* **1953** (April) 189.
[40] C. E. Lee, quoted in Ref. 36, Vol. I, Pt. (i), p. 233.
[41] N. Pienaar, *South African Industrial Chemist* **12** (1958) 51–8.
[42] H. Hohn, E. Fitzer and G. Jangg, *Stahl u. Eisen* **78**(21) (1958) 1462.
[43] W. W. Hodge, *Ind. Eng. Chem.* **31** (1939) 1364.
[44] N. Swindin, Paper presented to Inst. Chem. Engrs., 25 April 1944.
[45] N. Swindin, *Iron and Coal Trades Rev.* **159** (1949) 941.
[46] H. Quitmann, *Draht* **10**(5) (1959) 225.
[47] H. C. Bramer and J. Coull, *Ind. Eng. Chem.* **47** (1955) 67.
[48] C. Horner, A. G. Winger, G. W. Bodamer, and R. Kunin, *Ind. Eng. Chem.* **45** (1955) 1121.
[49] D. J. Lewis and F. Tye, *J. Appl. Chem.* **1959** (9 May) 279.
[50] C. L. Mantell and E. Grenni, *J. Water Poll. Control Fed.* **34**(9) (1962) 951.
[51] C. Calmon, *Industrial Water Engng.* **1970** (March) 38.
[52] U.S. Pat. 3,103,474; Brit. Pat. 911,386; U.S. Pat. 3,072,545 (and see citations therein).
[53] F. Heit and D. L. Prober, Electrochem. Soc. Mtg., Boston, Mass. (1968).
[54] W. H. Parker, *Metal Progress* **1966** (May) 133–4.
[55] F. Meinick, *Industrie Abwässer*, G. Fischer Verlag, Stuttgart, 1968, p. 200.
[56] K. C. Berger and J. L. Ströhlein, *Proc. 17th Indus. Waste Conf. Purdue Univ. Engng. Ext. Series* 112, 1962, pp. 1–7.
[57] O. Vogel, *Handbuch der Metallbeizerei*, 2nd rev. Ed., Verlag Chemie, Weinheim, 1951.
[58] V. E. Genkin and A. V. Evlanova, *Vestn. Ekon. Issled Gos. Kom. Min. SSSR po Khim* **1961** (6–7) 44–5.
[59] A. V. Evlanova *et al.*, *Vodosnabizhenie i Sanit. Techn.* **1959**(5) 15–19.
[60] F. Aigner and G. Jangg, *Berg u Hüttenmännische Monat.* **114**(1) (1969) 12–18; also Anon., *Chem. Eng.* **1960** (May 16) 82; G. Czisi, *Chem. Eng. Tech.* **36** (1964) 686.
[61] G. Jangg, F. Aigner, and G. Ibounig, *Planseebericht f. Pulvermetallurgie* **15 1967**(2) 103 115.
[62] V. Aravamuthan *et al.*, *Bull. India Sect. Electrochem. Soc.* **11** (1962) (3) 69–71.
[63] W. Eckhardt, *Oberflaechentechnik* **15/16** (1942) 74–5.
[64] Anon., *Metals and Materials* **3**(12) (1969).
[65] G. Jangg, G. Czisi, and L. Hausleitner, *Metall* **11** (1963) 1099.
[66] D. R. Anderson and D. W. Rhodes, IDO-14570, Feb. 12 (1962); M. E. McLain and D. W. Rhodes, IDO-14533, Sept. 30 (1960); K. T. Faler and D. R. Anderson, IDO-14536, March 2 (1961).
[67] G. Schulze, *Galvanotechnik (Saulgau)* **58** (1967) 475.
[68] N. Ibl, *Galvanotechnik (Saulgau)* **55** (1964) 499.
[69] Y. Imai, *J. Metal Finish Soc. Japan* **16**(7) (1965) 284.
[70] *Methods for Treating Metal Finishing Wastes*, Ohio River Valley Water Sanitation Commission, 1953.
[71] C. L. Mantell, *Electrochemical Engineering*, 4th ed., McGraw-Hill, New York, 1960.

[72] Rept. on control of CN in plating shops, *Plating* **1969** (October) 1107.
[73] F. L. Culler and R. E. Blanco, in *Peaceful Uses of Atomic Energy, Proc. 3rd Int. Conf.*, Vol. 10, United Nations, 1964.
[74] Anon., *Ideal Home Magazine* **97** (1969) 140.
[75] B. F. Dodge and W. Zabban, *Plating* **1951** (June) 561.
[76] Alkali chlorination of CN wastes, DuPont Technical Information Sheet SP 33-264.
[77] B. Drabek and V. Komendova, *Fortschritte der Wasserchemie* **3** (1965) 103.
[78] H. Schmidt and H. Meinert, *Z. Anorg. allg. Chemie* **293** (1957) 214.
[79] F. Schlagdenhauffen, *J. Pharm.* **44**(3) (1863) 100.
[80] C. Luckow, *Analyt. Chem.* **19** (1880).
[81] A. Bartoli and C. Papasogli, *Gazz. chim. Ital.* **13** (1883) 42.
[82] W. Hittorf, *Z. Phys. Chem.* **10** (1892) 616.
[83] M. A. Brochet and J. Petit, *Bull. Soc. Chim. France* **31** (1904) 742.
[84] E. Paterno and E. Panain, *Gaz. chim. Ital.* **34**(II) (1904) 152.
[85] DRP 368, 520 (1919).
[86] C. Marie, *Manuel de Manipulations d'Electrochimie*. Dunod et Pinat, Paris, 1906.
[87] G. H. Clevenger and M. L. Hall, *Trans. Electrochem. Soc.* **24** (1914) 271.
[88] T. Tsukamoto, *Mitsui Chemical Industries* **29** (1952) (11 January) 1950.
[89] J. S. Fitzgerald, *Chem. and Ind.* **74** (1955) 17.
[90] H. E. Williams, *Cyanogen Compounds*, Arnold Press, London, 1948, p. 121.
[91] R. Naumann, *Z. Elektrochem. angew. Phys. Chem.* **16** (1910) 191.
[92] B. T. Sawyer and R. J. Day, *J. Electroanal. Chem.* **5** (1963) 195.
[93] N. Tanaka and T. Murayama, *Z. Phys. Chem. (Frankfurt)* **15** (1959) 146.
[94] M. C. Dart, J. D. Gentles, and G. D. Renton, *J. Appl. Chem.* **13** (1963) 55; also Brit. Pat. 1,025,282.
[95] L. B. Sperry and M. R. Caldwell, *Plating* **1949** (April) 343.
[96] Yu. Yu. Lure and V. E. Genkin, *Zh. Prikl. Khim.* **33**(2) (1960) 384.
[97] J. Drogon and L. Pasek, *Electroplating and Metal Finishing* **1965** (September) 310.
[98] J. K. Easton, *Plating* **1966** (October) 1341; *J. Water Poll. Contr. Fed.* **39** (1967) 1621.
[99] V. Ruml and M. Topinka, *Metalloberflaeche* **1969** (August) 225.
[100] W. M. Latimer, *Oxidation Potentials*, 2nd ed., Prentice Hall, Englewood Cliffs, N.J., 1952.
[101] R. W. Oyler, *Plating* **1949** (April) 341-2.
[102] B. F. Dodge and D. C. Reams, *Disposal of Plating Room Wastes II*, Jenkintown, Pa., A.E.S., 1949.
[103] J. M. Connard and G. P. Beardsley, *Metal Finishing* **1961** (May) 54.
[104] J. K. Easton, *Journal WPCF* **1967** (October) 1621.
[105] Anon., *Water and Waste Treatment J.* **1961** (July/August) 412.
[106] J. T. Byrne, W. S. Turnley, and A. K. Williams, *J. Electrochem. Soc.* **105**(10) (1958) 607.
[107] R. Nagendra, N. V. Parthasaradhy, and K. S. G. Doss, *Plating* **1967** (February) 179.
[108] J. Drogon and L. Pasek, *Electroplating and Metal Finishing* **1965** (September) 310.
[109] H. Kurz and W. Weber, *Galvanotechnik und Oberflaechenschutz* **4** (1962) 92.
[110] Alkali chlorination of cyanide wastes, DuPont Technical Information SP 33-264.
[111] Belg. Pat. 613,851; German Pat. 1,193,920; German Pat. 1,188,568.
[112] Polish Pat. 49,679.
[113] Yu. Yu. Lure and V. E. Genkin, *Ochista Stochrykh Vod. Vses. Nauchn.-Issled. Inst. Vodosnabzh Kanaliz. Gidrotetchn. Sooruzhenii i Inzh. Gidrogeol.* **1962**, 50.
[114] P. Sztafrowski and B. Kotulski, *Przemysl Chem.* **40** (1961) 339.

References

[115] J. Pasynkiewicz, *Gas Woda i Techn. Sanitarna* **41** (1967) 331.
[116] A. V. Sakharnov, *Lakokrasochnye mat. i ikh. primenenie* **2** (1962) 27; *Zh. Vses. Khim. Obschch. im D.I. Menedeleeva* **6** (1961) 162.
[117] Brit. Pat. 888,654; 922,148 and 901,204; 922,148.
[118] Yu. Yu. Lure and V. E. Genkin, *Purification of Industr. Water Ref.*, Moscow, 1968.
[119] Pol. Pat. 39652 (1957).
[120] M. J. Allen, *Organic Electrode Processes*, Chapman Hall, 1958, 125.
[121] A. Renard, *Compt. Rend.* **91** (1880) 125.
[122] F. Fichter, *Organische Elektrochemie*, Steinkopf Verlag, 1942; reprinted Salford University Bookshop, 1970.
[123] H. Mulhofer, Diss. Munich T.H. (1905).
[124] J. F. Hedenburg and H. Freiser, *Anal. Chem.* **25** (1953) 1355.
[125] V. I. Ginsburg, *Zh. Fiz. Khim.* **33** (1959) 1504.
[126] R. N. Adams, J. H. McClure, and J. B. Morris, *Anal. Chem.* **30** (1958) 471.
[127] J. D. Voorhies, Diss. Princeton Univ. (1958).
[128] D. Hawley and R. N. Adams, *J. Electroanal. Chem.* **8** (1964) 163.
[129] G. E. Penketh, *J. Appl. Chem.* **1** (1957) 512.
[130] V. F. Gaylor, P. J. Elving, and A. L. Conrad, *Anal. Chem.* **25** (1953) 1078.
[131] B. Surfleet, Electricity Council Research Centre Rept. No. 204 (Oct. 1969).
[132] B. Surfleet, Electricity Council Research Report ECRC R/165 (July 1968).
[133] Brit. Pat. 1,180,464.
[134] L. Blanke, W. Nohse, and R. Wollman, *Galvanotechnik (Saulgau)* **53** 1962 (May) 220.
[135] B. Drabek and V. Komendova, *Forschritte der Wasserchemie* **3** (1965) 103.
[136] W. R. Meyer, R. F. Muraca, and E. J. Serfass, *Plating* **1953** (October) 1104.
[137] K. Enns, J. J. Byerley, and R. B. Valencia, paper pres. at AIME, Washington, 18 Feb. 1970.
[138] C. L. Mantell, *Electro-Organic Processing*, Noyes Dev't. Corp., 1968.
[139] P. Lutsch, *Wasser, Luft u. Betrieb* **12** (1968) 702.
[140] L. Mendia and E. Buonincontro, *Ingeniere Sanit.* (*Ital.*) **1958**, No. 4.
[141] E. Føyn, *Int. Ver. Angew. Limnol. Verhandl.* **15**(ii) (1962) 569–579; German Pat. 1,189,025.
[142] J. O'M. Bockris (ed.), *Electrochemistry of Cleaner Environments*, Plenum Press, New York, 1971.
[143] L. A. Sergunina, *Gigenia i. Sanit.* **33** (1968) 16–27.
[144] N. A. Maslennikov, *Nauchn. Tr. Akad. Kommun. Khoz.* **1963**, 97–113.
[145] L. A. Kunina, *Nauchn. Tr. Akad. Kommun. Khoz* **1963**, 81–5.
[146] French Pats. 1,321,895 and addn. to French Pat. 82,434.
[147] K. W. Marson, *Mod. Water Treatment J.* **7**(2) (1967) 71–3, 75–7.
[148] Japan. Pat. 7082 (1959).
[149] T. R. Kozlowski, R. F. Bartholemew, and H. M. Garfinkel, *J. Inorg. Nucl. Chem.* **32** (1970) 401–9.
[150] Development of a molten carbonate process for removal of sulfur dioxide from power plant stack gases, Parts I–VII, Atomics Int'l., Canoga Park, Calif. (PB-191 957, 8, 9, 960, 1, 2, 3); U.S. Gov't. Clearinghouse, Sect. 68 of 10 Aug. 1970.
[151] U.S. Pat. 3,475,122 and 3,515,513; Can. Pat. 843,763.
[152] Anon., *Sulphur* **82** (May/June 1969) 31; Anon., *Chem. Week.* **1968** (10 August) 51.
[153] C. Jackson, in *Industrial Electrochemical Processes*, Elsevier, 1971.
[154] Press Release, Lockheed Aircraft Corp., 14 May 1969.
[155] E. T. Seo and D. T. Sawyer, *Electrochim. Acta* **10** (1965) 239–252.

[156] M. de Kay Thompson and N. J. Thompson, *Met. Chem. Eng.* **1916** (15 December) 677; M. de Kay Thompson and A. P. Sullivan, *Met. Chem. Eng.* **1918** (15 February) 177.
[157] A. E. L. Kumm, SO_2–SO_3 regenerative fuel cell, Tech. Doc. Rept. ASD-TDR-62-162, Contract AF 33(616)-7975 to Air Research Manuf. Co., Phoenix, Ariz.
[158] Press Release, Esso Petroleum Co., Ltd., NAPCA contract award for Esso research; also handout presented by Dr. G. Moss at 1st Int. Conf. on Fluid Bed Combustion, Hueston Woods, Ohio, Nov. 1968.
[159] A. T. Kuhn, *British Chem. Engr.* **17** (Jan. 1971) 64–8.
[160] S. C. Srivasta and S. N. Shukla, *Electrochim. Acta* **15** (1970) 2021.
[161] T. N. Anderson *et al. Studies of Tropical Oceanogr.* **5** (1965) 22.
[162] R. W. Treharne and C. M. Cox, Tech. Rept. AMRL-TR 68-36 AD 679597, Aug. 1968.
[163] F. H. Meller, AD 678427, April 1968; Contract Nonr. N00014-66-CO 139.
[164] *Chem. Abstr.* **69** (1968) 92374a.
[165] Technical memorandum on electroflotation, St. Gobain, Techniques Nouvelles, Courbevoie Aug. 1967.
[166] Anon., *Annee Scientifique et Industrielle* **1888**, 363.
[167] Brit. Pat. 13,578 (1904); U.S. Pat. 826,411.
[168] U.S. Pat. 1,329,127.
[169] Fr. Pat. 1,031,614; Brit. Pat. 676,854.
[170] Brit. Pat. 1,149,362; Brit. Pat. 1,067,746; German Pat. 1,203,702; Fr. Pat. 1,440,999; Fr. Pat. PV 65,450; Fr. Pat. PV 107,318; Brit. Pat. 1,120,312.
[171] W. E. Mayo, presented at Water Pollution Control Reduction Congress, Atlantic City, N.J., Oct. 1965.
[172] H. S. Thomlinson and M. G. Fleming, *Revue de l'industrie minerale* (numero special A) **1964**, pp. 677–698.
[173] A. T. Kuhn, *Iron and Steel* **44**(3) (1971) 173.
[174] P. M. Wright, in *Industrial Electrochemical Processes*, Elsevier, 1971.
[175] Anon., *Chem. Engng.* **1967** (4 December) 840; **1968** (29 July) 82; *Water Waste Treatment* **12**(4) (1968) 140.
[176] N. A. Maslennikov and T. M. Zhadanova, *Sb. Nauchn. Rabot Akad. Kommun. Khoz.* **1961**, 230–253.
[177] E. L. Littauer and L. L. Shreir, in *Proc. 1st Int. Cong. Metallic Corrosion. London 1961*, 1962.
[178] C. Jackson, in *Industrial Electrochemical Processes*, Elsevier, 1971.
[179] W. Juda, T. A. Kirkham, and E. J. Parsi, *Purdue Univ. Eng. Bull. Ext. Ser.* **1960**(106) 556.
[180] I. Orszag, F. Ovari, *et al.*, *Veszpremi Vegyip Egyet Kozlem* **11**(2) (1968) 181–96.
[181] M. L. Schreiber, *J. Soc. Motion Picture Television Engrs.* **74** (1965) 505.
[182] Anon., *Eur. Chem. News* **1971** (12 February) 30.
[183] C. E. Fiske and R. Miller, *Petrol. Engr.* 1954 (November) C.48.
[184] A. V. Bridgwater, M.Sc. thesis, Univ. of Aston, Dec. 1966; *The Electrolysis of Iron—A literature Survey*, Dept. of Chem. Eng., Univ. of Aston, 1968.
[185] M. R. Hillis, *Effluent and Water Treatment J.* **1970** (January) 35.
[186] K. G. C. Berkeley, *Chem. and Ind.* **1971** (13 March) 287.
[187] H. C. Miller and W. Knipe, AWTR-13 U.S. Dept. H.E.W., March 1965.
[188] Ger. Patent 1,018,286.
[189] R. Pinner, *Metal Finishing J.* **1967** (October) 3–8.
[190] N. Ibl and A. M. Frei, *Galvanotechnik u Oberflächenschutz* **5**(6) (1964) 117.
[191] H. Teichi, M. Shigeru, and T. Schushi, *Nagoya Kogyo Gijutso Shikensho Hokoku* **10** (1961) 554 (*C.A.* **56**, 7066i).

[192] Yu. Yu. Matulix and A. Yu. Mitskene, *Lietuvos TSR. Mokslu Akad. Darbai, Ser. B* **1959**, 45 (*C.A.* **53** 12882h).
[193] S. Uneri, *Kim. Muhendisligi* **3**(27) (1968) 10 (*C.A.* **69**, 64016s).
[194] Polish Pat. 49679 (1961) (*C.A.* **64**, 17246f).
[195] Czech. Pat. 124107 (1967) (*C.A.* **68**, 117053).
[196] V. Ruml and M. Topinka, *Vod. Hospod* **18**(5) (1968) 205.
[197] H. Okuda, *Kagaku To, Kogyo (Osaka)* **43**(1) (1969) 23.
[198] D. Koniecka, *Gaz Woda. Tech. Sanit.* **41**(6) (1967) 201.
[199] G. V. Ivanov, I. G. Gol'berg, and I. M. Bakrak, *Sanit. Tech.* **1967**, 113.
[200] S. Stemparski and J. Drogon, *Gaz Woda Tech. Sanit.* **36**(4) (1962) 158.
[201] V. E. Genkin, *Vest. Tekn. i Ekon. Inform. Nauchn.-Issled. Inst. Tekhn. Gos. Kom. Sov. Min. po SSR Khim.* **1961** (6–7) 51 (*C.A.* **58** , 3195a).
[202] G. Allais and M. G. Noisette, *Eau* **44** (1957) 131.
[203] K. N. Kollau and M. J. Riedt, *TNO Nieuws* **23** (1968) 186.
[204] I. Yu. Zhalnerius, *Tr. Molodykh Spets. Vses. Nauchn.-Issled. Inst. Vod. Kanaliz. Gidretechn. Sooruzhinzh Gidrogeol, Ochista Stochn. Vod.* **1967**, 73.
[205] N. I. Dedusenko, P. A. Egorov, L. V. Korchangin, *et al., Ukr. Khim. Zh.* **34**(10) (1968) 1069.
[206] A. I. Gladysheva and V. I. Lavrenchuk, *Uch. Zap. Tsent. Nauchn.-Issled. Inst. Olovyan Prom.* **1966**(1) 68.
[207] A. V. Sakharnov, *Lakokrasochnye Mater i ikh. Primen.* **2** (1961) 26.
[208] Yu. Yu. Lure and V. E. Genkin, *Ochista Prom. Stochnykh Vod. Akad. Stroit* **1958**, 61 (*C.A.* **57**, 8371a).
[209] G. W. Fuller, *Engineering News Record* **89** (1922) 658.
[210] R. Eliassen and G. Tchobanoglous, *Environ. Sci. Technol.* **3** (1969) 536.
[211] U.S. Pat. 3,035,992.
[212] Anon., *Chem. Eng.* **72** (4 January 1965); **73** (9 May 1966); **75** (17 June 1968).
[213] Brit. Pats. 15760 (1887), 15374 (1890), 27270, all quoted in Ref. 214.
[214] J. C. Raine, *Water Waste Treat.* **10** (1966) 633.
[215] S. Afr. Pat. 6,801,071 (1968); Brit. Appl. (1967); Ger. Offen. 1,803,229.
[216] Ger. Offen. 1,949,129 (1970).
[217] Anon., *Chem. Week* **107** (1970) 54.
[218] Anon., *Chem. Engng.* **77** (1970) 80.
[219] Anon., *Environ. Sci. Technol.* **4** (1970) 201.
[220] Brit. Pat. 687,381; L. Gerasimov and D. N. Karabanov, Oil Institute, Uffor, quoted by Ref. 214.
[221] B. M. Matov, *Izv. Vssh. Ucheb. Zaved. Pisch. Tekhnol.* **1967**, 84.
[222] G. A. Selitskii, USSR Pat. 242,764.
[223] V. A. Borisov, G. V. Golub, and I. F. Chuprin, *Vod. Sanit. Tech.* **1969**, 4.
[224] A. P. Martynova, *Tr. Khar'kovsk. Inst. Inzh. Zheleznodov. Transp.* **75** (1965) 42.
[225] N. I. Romanov, *Vses. Nauchn. Tekh. Konf. po Gidreol, Dohyche Uglya Mosk.* **1959**, 705.
[226] G. A. Archakova, *Vodotvedenie Ochista Vod* **1969**, 82.
[227] Y. Nakagawa, S. Honda, Y. Ozasa, and G. Kondo, *Osaka Kogyo Gijutsu Shikensho Kiho* **17**(2) (1966) 113.
[228] J. Zhalnerius, Yu. Yu. Lure, and V. S. Belevtseva, *Liet. TSR. Makslo. Akad. Darb., Ser. B* **2** (1969) 103.
[229] E. I. Onstott, *J. Electrochem. Soc.* **111** (1964) 966.
[230] L. H. Schaffer and R. A. Knight, *J. Electrochem. Soc.* **116** (1969) 1595.
[231] Ts. D. Vitkov, D. G. Ivanov, and L. G. Nikolova, *Khim. Ind. Sofia* **3** (1969) 115.
[232] E. E. Kravstov, *Nauchn. Zap. Lugansk Sel'skokhoz Inst.* **7** (1960) 187.
[233] V. I. Yanko, *Nauka Tekh. Gor. Khoz.* **9** (1967) 86.

[234] K. H. Hartkorn, *Städtehygenie* **20** (1969) 263.
[235] L. B. Fainshtein and A. A. Mamakov, *Elektron. Okhrab. Mater.* **1** (1970) 50.
[236] I. J. Wilk, Am. Chem. Soc. Mtg., Div. Water Air Waste Chem., Gen. Paper 99 (1969).
[237] V. F. Negreev and Z. S. Eligulashvili, *Tr. Inst. Khim. Akad. Nauk. Azerb. SSR* **20** (1964) 150.
[238] J. Kucharski, Polish Pat. 59630 (1970).
[239] F. Smith, J. Blair, B. B. Arnold, *et al.*, Dechema Mono. 47 No. 805-834 (1962) 639.
[240] V. L. Losev, *Vodosnabzh. i Sanit. Tekhn.* **3** (1965) 18.
[241] Ya. M. Pashenkov, E. A. Silin, *et al.*, *Vestn. Sel'skokhoz Nauki Vses. Akad. Sel'skokhoz Nauk.* **9** (1964) 91.
[242] E. A. Silin and Z. Ya. Yaroslavskii, *Tr. Vses Nauchn.-Issled. Inst. Gidrotekhn. i Melior* **44** (1964) 219.
[243] S. Shimada, Japan. Pat. 4886 (1959).
[244] U.S. Pat. 3,210,262.
[245] G. L. Aslan, *Rev. Aluminium* **35** (1958) 1265.
[246] P. De. Arieta Araunabena Ruiz, Spanish Pat. 369,660 (1970).
[247] C. F. Albright, R. Nachum, and M. D. Lechtman, Sci. Techn. Aerosp. Rept. NASA-CR-65738.
[248] T. Yamamura, H. Koyama, and T. Teraoka, Jap. Pat. 4236 (1957).
[249] S. Akiya, Jap. Pat. 4237 (1957).
[250] G. Okamoto, T. Morozumi, *et al.*, *J. Electrochem. Soc. Japan* **24** (1956) 269.
[251] S. Asano, *Netsu Kouri* **8** (1956) 46.
[252] Jap. Pat. 4888 (1959).
[253] E. Føyn, *Tek. Ukeblad* **19** (1956) 433; in *Proc. 1st Int'l. Conf. Waste Disposal in Marine Environment* **1959**, 279; *Tek. Ukeblad* **25** (1963) 583.
[254] L. Mendia and E. Buonincontro, *Ing. Sanit.* **6** (1958) 223; *Aqua Industriale* **2** (1960) 164.
[255] H. W. Marson, *Water Poll. Contr.* **66** (1967) 109; *Effluent and Water Treatment J.* (7 July) 71.
[256] J. P. Axell, *Inst. Public Health Engr. J.* **64** (1965) 218.
[257] Brit. Pat. 1,120,312 (1965).
[258] Anon., *Water Waste Treat.* **12** (1968) 140; *Chem. Eng.* **1968** (29 July) 82.
[259] A. T. Kuhn and D. Lawson, *New Scientist* **1969** (26 June) 705.
[260] J. Koryta, J. Dvorak, and Bohackova, *Electrochemistry*, Methuen, 1970.
[261] J. P. A. Wortley, Paper Pres. at AIME Meeting New York City, 2 March 1970.
[262] H. S. Carslaw and J. C. Jaeger, *Conduction of Heat in Solids*, O.U.P., 1959.
[263] A. M. Azzam, J. O'M. Bockris, *et al.*, *Trans. Farad. Soc.* **46** (1950) 918.
[264] E. J. Cairns and A. M. Breitenstein, *J. Electrochem. Soc.* **114** (1967) 764.
[265] A. R. Blake, A. T. Kuhn, and G. Sunderland, *J. Electrochem. Soc.*, in press.
[266] B. Levich, *Chemical Hydrodynamics*, Prentice Hall, 1962.
[267] W. Vielstich, Z. *Elektrochemie* **57** (1953) 646.
[268] K. J. Vetter, *Electrochemical Kinetics*, Academic Press, New York, 1967.
[269] Z. Karaoglanoff, Z. *Elektrochemie* **12** (1906) 5.
[270] N. Ibl, Y. Barrada, and G. Trumpler, *Helv. Chim. Acta* **37** (1954) 583.
[271] K. S. Udupa, G. S. Subramanian, *et al.*, *Electro. Chem. Acta* **16** (1971) 1593.
[272] N. Ibl and D. Landolt, *J. Electrochem. Soc.* **115** (1968) 713.
[273] Brit. Pat. Appl'n. 28701/70.
[274] M. L. Schreiber, *J. of SMPTE*, **74** (1965), 505.
[275] N. L. Weinberg, *Chem. Rev.* **68** (1968), 449.
[276] A. T. Kuhn and B. Marquis, *J. Appl. Electrochem.*, in press.
[277] M. R. Hillis, Electricity Council Rep. ECRC/R/481, Feb. 1972.
[278] Brit. Pat. Appl'n. 57509/69 (to Lancy Labs. Ltd.).

Index

A

Activity, optical, 213
Additions, anodic, 195
Addition reactions, 214
Adenosine triphosphate (ATP), 239
Adsorbate,
 electrophilic, 27
 in quantum-mechanical terms, 8
Adsorption,
 and organic electrochemistry, 184
 ionic, 25
 of carbon dioxide at surfaces, 36
 of ethylene, 89
 of ions, and oscillatory effects, 100
Aircraft, supersonic, and screening of radiation, 273
Anion radicals, 168
Anodic oxidation, 313
Anodic processes, 301
Antimony, 295
Applications, survey, to environmental problems, 278
Arsenic, 295
 recovery of, 296, 297
Atoms, chemisorbed interaction of, 22
ATP,
 and chemiosmotic mechanisms, 258
 and chloroplast function, 259
 and energy efficiency, 240
 and oxidation–reduction reactions, 256
 hydrolysis, 264
Autocatalytic effect, 80

B

Baizer, and intramolecular hydrodimerization, 201
Behavior, oscillatory, 47
Benzaldehyde, 221
Binding energy, at surface, 30
Biological cells, and electrochemical reactions, 240
Biological regulators, and electrochemical reactions, 267
Biological systems, and impurities, 241
Bonhoeffer, first mathematical models for oscillations, 79
Born–von Karman boundary, 12
Buffer solutions, processes and organic electrode, 141
Butler–Volmer equation, 266

C

Cancer, and smog, 274
Carbon dioxide,
 electrochemical removal, 320
 importance to environment, 273
Carbonium,
 and reduction of alkyl halides, 165
 intermediates, 205
 reactions, and Grignard reagents, 170
Cell interface, in electrodic terms, 265
Charge density, at surface, 17

341

Charge-transfer complex, 31
Chemical approach to electrode processes, 127
Chemical reactions, and charge transfer, 143, 144
Chemiosmotic hypothesis, 255
Chloroplast, 257
and energy input from photon, 257
Chromium, 299
Chronoamperometry, 148
Chronopotentiometry, 155
Circuits, with electrochemical cells, 97
Conduction,
and temperature, 250
electric in biological materials, 241
electronic, in the body, 240
Conduction band, and surface, 18
Controlled concentration, 156
Conway, treatment of adsorption isotherms, 139
Cope,
and electrochemical nature of enzyme reactions, 268
and enzyme catalysis excitation, 268
Copper,
deposition, 279
in the presence of other metals, 282
on bed electrode, 282
electrorefining, 281
metallic deposits of, 281
removal by cathodic processes, 279
Coulomb energies, in surface calculations, 33
Coulombic efficiency, in electrochemical kinetics, 124
Coulometry, 147
Criteria, for close trajectory, 72
Crystal plane, surface index of, 27
Crystals,
band structure at, 9
surface energy states, 8
Current and temperature, in biological reactions, 245
Current–time relations in oscillations, 111
Currents, electronic and ionic, separation of, 243
Cyanide,
anodic oxidation, 302
discussion of mechanisms, 306

Cyanide *(con'd)*
electrochlorination, 310
mechanisms of removal, 305
oxidation and poisoning, 308
Cycles, limits of, 89
Cyclic voltammetry, 151
Cyclization, 225
anodic, 204
Cytochromes, 257

D

Degradation, of concentration polarization, 105
Dehydrogenation, 199
De Levie, and oscillations theory, 108
Delocalization, and EPR spectroscopy, 121
Deposition, of copper, 279
DeVault, and tunneling, 247
Diffusion,
and inhibition, 104
of ethylene, 90
Diffusion-coupled oscillations, 86
Dimer peak-height measurement, 175
Dimers, 215
Domestic wastes, *in situ* destruction, 314
Double layer, structure of, 3

E

Effect of inductance, 115
Effects of blocking, 89
Effluents,
and electrochemical clean-up, 309
and electrolytic processes, 276
treatment by electrochemical means, 315
Electrochemical approach to electroorganic reactions, 128
Electrochemical effluent processes, factors controlling them, 328
Electrochemical mechanisms, their general applicability to biological research, 269
Electrochemical methods,
for investigating organic processes, 137
their problems, 276

Electrochemical processes, in blood clotting, 269
Electrochemical reactions, types, 2
Electrochemical theory of vision, 269
Electrochemistry,
  and corrosion protection, 324
  in environmental control, 326
  thin-layer, 158
Electrode processes,
  investigatory mechanisms, 135
  mixed, 53
  organic, and methods of investigation, 137
  under diffusion control, 140
Electrodes
  for electroflocculation, 323
  graphite, 180
  passive potential in, 57
  quantum effects at, 7
Electrodialysis, 292
Electroflotation, 321
  application to steel rolling mill waste, 323
Electroforming, of iron pipes, 287
Electrolysis, at controlled potential, 131
Electrolyte, conductivity and membrane oscillations, 95
Electron transfer,
  adiabatic, 5
  theory of, 4
Electrons, donated by iodine and lecithin membranes, 249
Electro-organic chemistry, 121
  and difficulties, 123
  and intermediates, 163
  and synthetic applications, 122
Electroplating, at a lipid membrane, 253
Electrosorption, and oscillations, 102
Elovich equation, in reactions involving cytochrome, 268
Energy bands, at surface, 12
Energy conversion, biological, 254
Energy levels,
  of surface complex, 44
  species in, 5
Environment and electrochemistry, 273, 294
Environmental control, by electrochemistry, 326
Enzyme catalysis, and Cope, 268
Epelboin and real part of impedance, 113
EPR measurements, and spectra of β-deutero anions, 166
EPR spectroscopy and radical detection, 161
Equations, for wave functions at surface, 16
Equilibrium,
  of autonomous systems and parameters, 69
  point, 68
Equivalent circuits, with and without diffusion, 109

F

Faradaic rectification, 157
Feedback, positive and autocatalytic effects, 60
Fermi distribution law, 8
Flade potential, and oscillations, 81
Fluctuations, irregular, 64
Flux in steady-state oscillations, 91
Formaldehyde and oxidation, 62
Formalization, less approximate, 27
Free energy, and linear correlation, 127
Fuel cells, and electrochemical reactions in the body, 240

G

Galvanostatic curves, 153
Gold, 268

H

Hall measurements, and conductivity in biological membranes, 247
Hamiltonian,
  for dilute alloys, 43
  for surface states, 40
  for surfaces, 28
Hemoglobin, Rosenberg, and electronic conductivity, 244
High-overvoltage cathodes, uses of, 292
Hofer–Moest oxidation, 179
Holes, and mechanism of oscillation, 88
Homogeneous distribution, 57

Hückel,
  calculations, all-valency-electron, 37
  calculation for 6 × 6 array, 36
  LCAO calculations, 34
  π-electron model, 35
Hydration, and electronic conductivity, 245
Hydrocarbons and adsorption, 185
Hydrogen,
  adsorbed, 23
  at surface, 41
Hydrogen atoms,
  adsorption of, 37
  recombination, 4
Hydrogenation, 194

I

Impedance, at different frequencies, 112
Inductance, and generation of oscillations, 115
Instability during adsorption, 103
Interaction, of adsorbed atoms, treatment of, 42
Interface of cell, in electrodic terms, 265
Interfaces, biological, electrochemical processes at, 239
Intermediates,
  adsorbed, and electrochemistry, 162
  and electro-organic chemistry, 163
  detected by transients, 154
  high-energy, 260
Ion-exchange membranes, 289
Iron, 287
  electrochemical recovery of, in industry, 291
  polarization of, 80
  recovery, technology aspects, 288

K

Katchalsky and Spangler model, 94
Kinetics, oscillatory, 87
Kolbe reaction, 167, 183
Kolbe synthesis, 121
Kronig–Penney model, 10

L

LCAO
  approach, 15
  extension to adsorbed species, 21
  treatment of surface effects, 23
Lecithin membranes, 249
Levels, acceptable, of pollution, 275
Lipid films and electric conduction, 248
Llewellyn criterion, 112
Lockheed process, 319

M

Madelung constant, 26
Mandel,
  and nerve excitation, electrochemical mechanism for, 268
  and voltage–current characteristics in biological systems, 251
Mass transport
  control, 329
  and diffusion control, 330
  and organic reactions, 134
  under convection control, 331
Materials, radioactive, 300
Measurements, optical, 247
Mechanisms,
  during copper dissolution, 54
  in electro-organic chemistry, 126
Metals,
  anodic processes in, 49
  incorporation into organic compounds, 182
Methanol, oxidation of, and platinum, 60
Methods, electrochemical, 136
Mitchell,
  and chemiosmotic hypothesis, 255
  and electrochemical theory of biological reactions, 261
Mitochrondria
  and chloroplasts, 258
  functional analogue in plant cells, 257
MO calculations, at surface, 28
Models,
  chemical or kinetic, 77
  for oscillators, electrical, 96
Mucosa, gastric, 264

Index

N

Natural convection, 333
Nearly-free-electron approach, 13
Negative resistance, 107
Nerve excitation, and electric model, 269
Nerves and electrochemistry, 268
Nickel, 295
Noisiness, 87

O

Olfaction, and electrochemical mechanism, 269
Orbital, surface orientations, 38
Organic substances and oscillatory behavior, 60
Oscillating systems, models for, 78
Oscillations,
  and Flade potential, 51
  and oxidation of hydrogen, 99
  and pitting corrosion, 56
  and PN junction, 58
  and rest potential, 64
  at passive electrodes, 82
  cathodic, 63
  diagrammatic, 50
  during anodic oxidation, 60
  during copper dissolution, 54
  during methanol oxidation, 60
  electrochemical basis, 49
  electro-osmotic, 65
  in biological systems, 48
  in cathodic processes, 63
  in Franck–FitzHugh model, 84
  kinetic, 73
  of potential, 56
    during oxidation, 61
  relaxed, 73, 115
  relaxation of, 74
  relaxation, discontinuous theory, 74
  relaxation, elastic rebound, 75
  semiquantitative model, 51
  under galvanostatic conditions, 101
  with nickel anodes, 52
Oscillators,
  and Warburg impedance, 107
  electrochemical, models for, 65
Oscillatory theory,
  appendix of term in, 116
  summary of, 115

Oxidation,
  anodic, of nonmetals, 59
  biological, 239
  of alkyl halides, 181
  of formaldehyde and oscillations, 61
  potentiostatic, 63
Oxygen, chemisorbed, involvement for oscillations, 62

P

Passivation, and instability, 97
Passivity, cathodic, 63
Periodicity,
  during electropolishing, 53
  in electrochemical systems, 48
Perturbation approaches, 39
Phase-plane configuration, with limit cycles, 70
Phase-plane portrait of Franck–Fitz-Hugh, 83
Phase-plane representation, 66
Phase portrait, 67
Phenol destruction by electrochemistry, 312
Phosphorylation, 259
  electrodic, 261
  electrodic mechanism, 262
Photosynthesis, 288
Pickle liquors, electrochemical treatment of, 293
Pintsch–Bamag process, 317
Poincaré–Bendixson theorem, 71
Poincaré's theorem, 76
Poisoning of electrode surfaces by cyanides, 307
Poisson's equation, at surface, 20
Polarization curve and instability of active–passive transition, 98
Polarogram, 110
Polarography
  advantages, 145
  and constant potential processes, 139
Pollution, by SO_2, 274
Polymerization, 178
Potentiostatic currents, and mechanism of organic reactions, 138
Potentiostatic curves, 153
Potential, at surface, 11
Potential–time behavior, 149
Potential sweep methods, difficulties of, 151

Product identification, and organic electrochemistry, 160
Processes, regenerative, 94
Proton donors, 169
Protonation, 142, 164
Protons in metals for ATP hydrolysis, 259

Q

Quantum aspects of adsorption, 1

R

Radicals, and cations, 172
Radioactive materials, 300
Reaction order, 160
and isotherms, 139
Reaction schemes, 133
and polarographic chemistry, 130
commonly observed, 129
Reactions,
addition, cathodic, 193
anodic, 191
cyclization, 200
elimination, 197, 224
organic, as surface reactions, 186
substitution, 187
stereospecificity, 209
Reagents, ring-forming, 203
Rearrangements, Stevens, 190
Reducibility, polarographic of bonds, summarized, 195
Reduction, electrochemical, 202
Relaxation, 58
cycle, phase portrait of, 78
techniques, 157
Reorganization, 6
Resting potential, displacement of, *versus* current density, 266
Ring expansion, 207
Ring opening, 206, 226
Rosenberg, and polarization in the body, 246

S

SCF-LCAO, 28
Scrubbing, electrochemical, 319
Shashua and his membranes, 65

Shockley states, 14
Silver, 286
Simon Carves process, 318
Simulation of oscillatory behavior, 106
Smog, photochemical, 274
Solutions,
for wave functions at surface, 16
periodic, 66
Species, adsorbed, 20
Spivey, and hydration in biological reactions, 244
Stability of steady-state points, 93
Stainless steels, 295
States, localized, 21
Stereochemistry,
of additions of carbon dioxide, 219
of reductive dimerization, 220
Stereospecificity, 208
Stone Webster process, 316
Substrate, and hydration, 244
Substitution,
aromatic, 173
reactions, 187
side chain, 174
Sulfur dioxide, electrochemical scrubbing, 316
Surface, and electron distribution, 19
Surface compounds, approximation for, 29
Surface oxides, and oscillations, 59
Surface states, in crystals, 13
Syntheses, commercial, 227
Synthesis,
and ATP in enzymes, 263
and difficulties, 125
Systems, autonomous, 66
Szent-Györgyi, and donor–acceptor complexes, 268

T

Tafel equation, and biology, 252
Thyroid function, and electrochemical mechanism, 269
Tin, 299
anodized, corrosion, pitting, 55
Tissues, secretory, and electron transfer rates, 266
Trajectory, of representative points, 92
Transfer coefficients, value of, calculated for secretory tissues in nerve cells, 267
Transition probability, 6

Transport, active, 264
Transport control, 333

V

Valency bond approach, 32
Versatility of electrolytic processes, 277
Voltam linear sweep, 150
Voltammetry, 145
Volterra–Lotka system, 85

W

Wannier functions, 11
Warburg impedance, and oscillations, 107, 108
Wave function, at surface, 9, 10

Z

Ziegler condensation, 223